AF836025

Lactobacillus Genomics and Metabolic Engineering

https://doi.org/10.21775/9781910190890

Edited by

Sandra M. Ruzal[1,2]

[1]Universidad de Buenos Aires
Facultad de Ciencias Exactas y Naturales
Departamento de Química Biológica
Buenos Aires
Argentina

[2]CONICET – Universidad de Buenos Aires
Instituto de Química Biológica de la Facultad de Ciencias Exactas y Naturales (IQUIBICEN)
Buenos Aires
Argentina

Caister Academic Press

Caister Academic Press
Norfolk, UK

www.caister.com

British Library Cataloguing-in-Publication Data
A catalogue record for this book is available from the British Library

ISBN: 978-1-910190-89-0 (paperback)
ISBN: 978-1-910190-90-6 (ebook)

Cover design adapted from images provided by Sandra M. Ruza: scanning electron microscopy images of *Lactobacillus acidophilus* grown with and without high salt.

Contents

Current Books of Interest

Plant-Microbe Interactions in the Rhizosphere	2019
Porcine Viruses: From Pathogenesis to Strategies for Control	2019
Lactobacillus Genomics and Metabolic Engineering	2019
Cyanobacteria: Signaling and Regulation Systems	2018
Viruses of Microorganisms	2018
Genes, Genetics and Transgenics for Virus Resistance in Plants	2018
DNA Tumour Viruses: Virology, Pathogenesis and Vaccines	2018
Pathogenic *Escherichia coli*: Evolution, Omics, Detection and Control	2018
Postgraduate Handbook: A Comprehensive Guide for PhD and Master's Students and their Supervisors	2018
Enteroviruses: Omics, Molecular Biology, and Control	2018
Molecular Biology of Kinetoplastid Parasites	2018
Bacterial Evasion of the Host Immune System	2017
Illustrated Dictionary of Parasitology in the Post-genomic Era	2017
Next-generation Sequencing and Bioinformatics for Plant Science	2017
The CRISPR/Cas System: Emerging Technology and Application	2017
Brewing Microbiology: Current Research, Omics and Microbial Ecology	2017
Metagenomics: Current Advances and Emerging Concepts	2017
Bacillus: Cellular and Molecular Biology (Third Edition)	2017
Cyanobacteria: Omics and Manipulation	2017
Foot-and-Mouth Disease Virus: Current Research and Emerging Trends	2017
Brain-eating Amoebae: Biology and Pathogenesis of *Naegleria fowleri*	2016
Staphylococcus: Genetics and Physiology	2016
Chloroplasts: Current Research and Future Trends	2016
Microbial Biodegradation: From Omics to Function and Application	2016
Influenza: Current Research	2016
MALDI-TOF Mass Spectrometry in Microbiology	2016
Aspergillus and *Penicillium* in the Post-genomic Era	2016
The Bacteriocins: Current Knowledge and Future Prospects	2016
Omics in Plant Disease Resistance	2016
Acidophiles: Life in Extremely Acidic Environments	2016
Climate Change and Microbial Ecology: Current Research and Future Trends	2016
Biofilms in Bioremediation: Current Research and Emerging Technologies	2016
Microalgae: Current Research and Applications	2016
Gas Plasma Sterilization in Microbiology: Theory, Applications, Pitfalls and New Perspectives	2016

Full details at www.caister.com

Preface

The *Lactobacillus* genus comprises more than 200 formally recognized species characterized by their phylogenetic and metabolic diversity. Lactobacilli species are found in a variety of ecological niches such as decomposing plant materials, wine, meat and raw milk and are often commensals to plants and animals including humans. They are food-grade microorganisms widely applied in the fermented food industry due to their technological and health-promoting properties; these bacteria have been extensively used as starter cultures and as probiotics. This ten-chapter book aims to survey the most relevant aspects of the genus. Due to the available genomic information for the *Lactobacillus* genus, comparative genomic approaches have been taken to evaluate strains or species found in different niches, to give an insight into niche adaptation within the genus (see Chapters 1 and 2). A detailed description of the catabolic pathways of complex carbohydrates metabolism (see Chapter 3) and their relation to their main fermentation product, lactic acid (see Chapter 4) are depicted; the ability of *Lactobacillus* to respond to environmental conditions, focusing on osmotic stress, by altering the nature of their cell wall for adaptation are explored (see Chapter 5). In particular, focus is made on S-layer proteins, with relevant and updated concepts regarding genetics, structural features, cell wall and self-assembly, functionality and biotechnological applications (see Chapter 6). Also, an updated revision is presented of phages infecting strains of *Lactobacillus* spp. with particular emphasis on structural studies on phage–host interactions (see Chapter 7). Overview of methods for the introduction of DNA into *Lactobacillus* species are described (see Chapter 8) and also tools and applications in different areas for recombinant gene expression (see Chapter 9). Finally, since commensal and environmental bacteria appear as a reservoir of antibiotic resistance genes, a genomic overview of these resistance genes in *Lactobacillus* are described (see Chapter 10).

I would like to express my gratitude to all the authors for their hard work and effort in contributing to this book. I am also very grateful to Annette Griffin at Caister Academic Press, who invited me to edit this book.

Sandra M. Ruzal

A Genomic Perspective on Niche Adaptability in *Lactobacillus*

1

Ewelina Stefanovic and Olivia McAuliffe*

Teagasc Food Research Centre, Fermoy, Ireland.

*Correspondence: olivia.mcauliffe@teagasc.ie

https://doi.org/10.21775/9781910190890.01

Abstract

The *Lactobacillus* genus comprises more than 200 formally recognized species characterized by a phylogenetic and metabolic diversity that exceeds that of a typical bacterial family. The widespread dissemination of members of the lactobacilli in different ecological niches testifies to their extraordinary niche adaptability. Advances in sequencing technologies have facilitated a comprehensive examination of the characteristics of the *Lactobacillus* genus through large-scale comparative genomics, and aided an understanding of the genomic background underpinning the presence of lactobacilli in such a broad range of habitats. Comparative genomic analysis has revealed that adaptation to such highly variable environments is a result of genome evolution. Gene loss and acquisition, mainly driven by horizontal gene transfer, underlies the remarkable genomic diversity observed, resulting in species which may be considered either niche generalists or niche specialists. Larger genome sizes are associated with ecologically flexible species such as *Lactobacillus casei* and *Lactobacillus plantarum.* These niche generalists have typically acquired or retained the capacity to migrate between different habitats and have recently been described as nomadic. Niche specialists, or host-adapted species such as *Lactobacillus sanfranciscensis,* possess much smaller genomes, reflecting ecological specialization. For many species, sufficient information to infer their real niche preferences remains elusive. In this chapter, we review the available genomic information for the *Lactobacillus* genus and the comparative genomic approaches that have been taken to evaluate strains or species found in different niches, which give an insight into niche adaptation within the genus.

Introduction

An ecological niche can be described as a multidimensional space of resources and environmental conditions that together define where an organism can survive and grow. Bacteria generally share niche and resources with other microbial species as well as more complex organisms, and thus they have evolved mechanisms to communicate, to compete and to adapt to environments as diverse as humans, animals or plants (Aussel *et al.*, 2016). The potential of a bacterial species to exploit a particular niche is dictated by factors such as environmental conditions, nutrient availability and the presence of competitors, predators or bacteriophages (Pereira and Berry, 2017). Evolutionary adaptation to diverse nutritional niches differs fundamentally in bacteria when compared to higher organisms. Bacteria adapt by remodelling their genomes. Expansion and contraction in gene content is driven by horizontal modes of gene transfer, mechanisms that play such a frequent and profound role in the evolution of bacterial genomes that they often mask vertical patterns of descent. Indeed, genome comparisons of many different bacterial species from a range of ecological niches has revealed a prominent role for gene gain and loss in the processes of niche adaptation, ecological specialization, host-switching, and other lifestyle changes (Didelot *et al.*, 2009). However, microbial species can often also grow in a range of conditions much broader than its actual niche. This generalism is widely observed in nature and it is generally accepted that heterogeneity of the natural environments in which organisms grow promotes phenotypic variations giving rise to certain ecological advantages (Woodcock *et al.*, 2017).

The lactic acid bacteria (LAB) are a diverse group of Gram-positive obligately fermentative microorganisms that produce lactic acid as the main product of sugar degradation (for review, see (McAuliffe, 2017)). This group of organisms is currently classified in the order Lactobacillales and comprises species of the genera *Lactococcus, Lactobacillus, Streptococcus, Leuconostoc, Oenococcus* and *Pediococcus* among others. While members of the LAB group are closely related phylogenetically, they occupy a diverse set of ecological niches from plants and fermenting plant material to milk to the gastrointestinal tract and genitourinary tract of animal and humans. Such niche diversity among closely related species suggests considerable genetic adaptation during their evolution. Indeed, comparative and functional genome analysis of multiple LAB species and strains has revealed a central trend in LAB evolution: the loss of ancestral genes and metabolic simplification towards adaptation to nutritionally rich environments (Makarova and Koonin, 2007).

The *Lactobacillus* genus is a broadly defined group within the LAB that are closely associated with terrestrial and marine animals, their environment (plants, materials of plant origin, manure) and their food (cheese, yoghurt) (Sun *et al.*, 2015). It represents the largest and most diverse of the LAB genera, comprising some 200 formally recognized species with applications in industrial, biotechnological and medical fields. This heterogeneous group occupies niches that are largely variable in conditions such as nutrient availability, temperature and the presence of other (competing) bacteria. Nonetheless, lactobacilli are able to adapt to these specific conditions and successfully colonize these varying habitats. This adaptability originates from the remarkable genetic diversity of this genus that is much more extensive than that normally observed at the family level (Salvetti and O'Toole, 2017). Application of high-throughput whole genome sequencing and the associated 'omics' technologies have completely revolutionized the study of this important group of organisms, from providing a comprehensive view of the remarkable diversity within the group through to the development of improved or novel strains and their associated processes in industrial application. In this chapter, we provide an overview of the developments in *Lactobacillus* research during this genomic era, with particular emphasis on the genetic events responsible for the adaptation of these organisms to their specific niches.

Genomics of *Lactobacillus* and insights into niche adaptation

The availability of genome sequences has facilitated a deeper examination of the characteristics of the *Lactobacillus* genus, and enabled an understanding of the genomic background for presence of lactobacilli in such a broad range of habitats. Firstly, the genomes of lactobacilli vary in size between 1.23 Mbp (*L. sanfranciscensis*) up to 4.91 Mbp (*L. parakefiri*) (Sun *et al.*, 2015). Based on the analysis of 213 genomes of lactobacilli and associated genera of LAB, the *Lactobacillus* pangenome, consisting of all genes present in a given set of genomes, was estimated to contain 45,000 gene families, while 73 genes mainly responsible for cell growth and replication make up the core genome, shared by all of the analysed strains (Sun *et al.*, 2015). The comparative genome analysis of lactobacilli confirmed the overall trend observed in other LAB, which is minimization of genomes through loss of ancestral genes, but also acquisition of genes by horizontal gene transfer, as a response to adaptation to a specific habitat of these bacteria (Makarova *et al.*, 2006). Furthermore, an increasing number of *Lactobacillus* sequences revealed the potential link between genomic characteristics and specific environmental niche inhabited. To better understand the connection between the habitat and the genomic content, the focus of genomic studies has moved from the comparison of genomes of a single species or groups of related species to metagenomic analysis of isolates present in a niche. These studies have shown that larger genome sizes characterize species able to survive in niches with varying environmental conditions, while species with smaller genomes are adapted for highly specific environments and less variable conditions, thus leading to simplification of metabolic repertoire, a higher occurrence of pseudogenes and lower GC content (Papizadeh *et al.*, 2017). Both of these trends were confirmed in the genomes of lactobacilli, where larger genomes of lactobacilli corresponded to the free-living and nomadic species, while in host-specialized species, both genome size and GC content were significantly smaller (Fig. 1.1) (Duar *et al.*, 2017b). Higher numbers of genes in free-living organisms are necessary for their survival in habitats differing in nutrients, temperature and the presence of other competing bacteria. On the other hand, a transition to nutritionally rich environments [dairy products and other fermented foods or the gastrointestinal tract (GIT)] has resulted in metabolic simplification and the loss of the ability for biosynthesis of many cofactors, vitamins and amino acids (Papizadeh *et al.*, 2017). In the following sections, characteristics defining adaptation to various niches from which lactobacilli are isolated are discussed.

The dairy niche

In an evolutionary sense, the dairy niche represents a recently occurring man-made niche (Douillard and de Vos, 2014). As such, it does not strictly represent a 'natural' niche for lactobacilli (Duar *et al.*, 2017b) as lactobacilli have been essentially 'domesticated' for use in food and feed production.

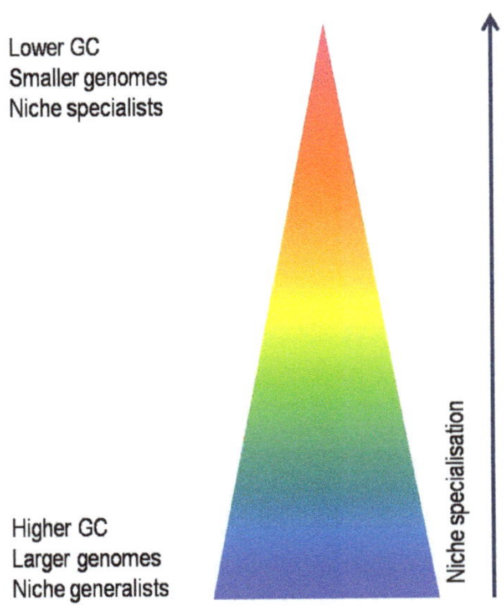

Figure 1.1 Correlation of genome composition to ecological specialization in the *Lactobacillus* genus.

However, genomic studies of species regularly isolated from dairy products have revealed specific characteristics of dairy isolates which cannot be overlooked, as they demonstrate some proof of niche specialization events. Dairy specialization is a common characteristic of several species that are not exclusively isolated from dairy products, and it unavoidably occurred through continuous propagation of these strains in milk fermentations. Numerous comparative genomic studies of species isolated from the dairy niche, among others, reveal several important genomic characteristics of dairy isolates. In these strains, a very limited spectrum of carbohydrate utilization genes is present, due to the fact that in the milk environment, lactose is the dominant sugar. A comparative study of *Lactobacillus casei* strains showed distinct differences in the metabolic characteristics of nine strains isolated from various sources, where dairy isolates were able to utilize the most limited carbohydrate profile, compared to plant or mammalian isolates (Broadbent *et al.*, 2012). Among *Lactobacillus paracasei* genomes, it was shown that strains of dairy origin possess lower numbers of phosphotransferase system (PTS) cassettes which are involved in carbohydrate utilization (Smokvina *et al.*, 2013). Similar findings were confirmed for the dairy adapted species *Lactobacillus delbrueckii* subsp. *bulgaricus*, commonly used in yoghurt production. It was shown for this species that, in addition to a limited number of genes for sugar utilization, a high number of pseudogenes were present (van de Guchte *et al.*, 2006). Genome analysis of the dairy isolate *Lactobacillus helveticus* DPC4571 revealed that this strain lacked genes of the PTS system, cell wall anchoring proteins and mucus-binding proteins, all of which would be redundant in the dairy niche (Callanan *et al.*, 2008). In a comparative study that analysed 100 *Lactobacillus rhamnosus* strains, most of the dairy isolates grouped together, and were characterized by the presence of genes encoding lactose, maltose and rhamnose metabolism and the absence of pili, antimicrobial resistance genes, stress resistance genes and other functions that provide the adaptability to various range of habitats (Douillard *et al.*, 2013). All of these studies suggested that adaptation to the nutritionally stable and rich dairy niche facilitates loss of superfluous genes, as there is no requirement to maintain uptake systems for a variety of substrates.

Another phenomenon that contributes to nutrient-rich habitat adaptation of lactobacilli is the acquisition of new genes and an increase in number of paralogous genes. These can emerge in specific gene duplication events or through horizontal gene transfer (HGT) (Makarova and Koonin, 2007). For instance, a genomic island with different GC content compared to the rest of the genome was detected in genome of *L. helveticus* DPC4571, and included genes encoding enzymes of fatty-acid and amino acid metabolism (Callanan *et al.*, 2008). Similarly, the multiple peptidases or proteases encoded in the genomes of dairy *L. helveticus*, which are thought to be the result of gene acquisition events, along with the decay of the genes associated with survival in the GIT, explain the high affinity of these strains to the cheese environment as a protein-rich niche. The presence of a higher number of proteinases with different but complimentary substrate and cleavage specificities could explain the efficiency of the *L. helveticus* proteolytic system which provides an adaptive advantage regarding milk protein hydrolysis (Genay *et al.*, 2009). The variability in proteinase gene content could be seen as an important industrial feature, contributing to the desired flavour development (Broadbent *et al.*, 2011).

Other food-related niches

The meat environment

Certain *Lactobacillus* species are part of the indigenous microbiota of meat and may also be used in the fermentation and preservation of meat products. Meat is a rich source of peptides and amino acids, but a relatively poor source of sugars. An example of a species which has demonstrated specific metabolic adaptations to life in the meat environment is *Lactobacillus sakei* (Nyquist *et al.*, 2011). In response to the presence of meat proteins, *L. sakei* has been shown to up-regulate genes encoding oligopeptide transporters and intracellular peptidases to take advantage of these substrates (Fadda *et al.*, 2010; Xu *et al.*, 2015). The capacity to metabolize certain amino acids that are abundant in meat is also a feature of this species. *L. sakei* use arginine, via the arginine deiminase pathway (ADI), to generate ornithine, ammonia, CO_2 and ATP (Rimaux *et al.*, 2012) (Fig. 1.2). The capacity to metabolize arginine is thought to enhance the survival of *L. sakei* in the meat environment as a result of the generation of ATP. This pathway is induced by the presence of arginine and by anaerobiosis, conditions common during meat storage. *L. sakei* is one of the best equipped of the *Lactobacillus* species to cope with growth under oxidative stress

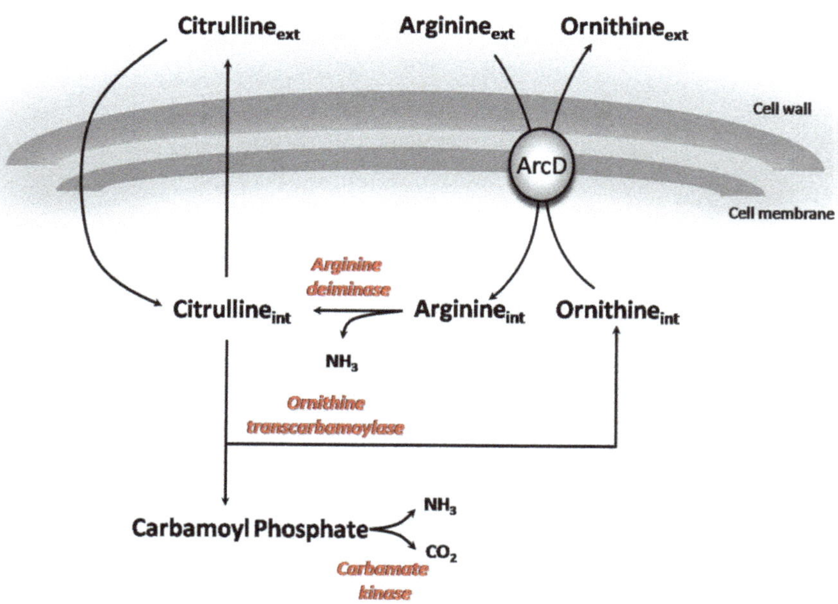

Figure 1.2 Catabolic pathway for arginine in *Lactobacillus sakei* through the arginine deiminase (ADI) pathway. Enzymes are indicated in red. ArcD represents arginine/ornithine antiporter pathway.

as occurs with anaerobiosis, despite the fact that it is a facultative anaerobe and its main metabolism is fermentation. It has been proposed that the presence of a mutated cytochrome P450 gene in some *L. sakei* strains may enable them capable of some form of respiration (Nyquist *et al.*, 2011).

As previously mentioned, meat is a poor source of sugars as carbon sources. One of the sugars present in meat is ribose, and *L. sakei* has been shown to be capable of using ribose through an ATP-dependent system. However, unlike other species, ribose uptake in *L. sakei* is not via ATP-binding cassette (ABC) transporters but via a ribose transporter suspected to function as a facilitator encoded by the *rbsU* gene (McLeod *et al.*, 2011). Nucleosides are also an important carbon source for this species, as evidenced by the redundancy of genes involved in exogenous nucleoside scavenging in the *L. sakei* 23K genome (Nyquist *et al.*, 2011). The ability of *L. sakei* 23K to use N-acetylneuraminic acid (NANA) as a carbon source has also been documented (Anba-Mondoloni *et al.*, 2013). A pathway (*nanTEAR–nanK*) encoding transport and the early steps of catabolism of this amino sugar, which is present in meat, has been identified in *L. sakei* 23K but absent from the genomes of other *L. sakei* strains that were shown to be unable to grow on NANA (Anba-Mondoloni *et al.*, 2013). Other genes thought to play a role in adaptation to the meat niche include bacteriocin-associated genes, which allow these strains to compete on the meat surface, and transporters involved in iron or haem uptake (Chaillou *et al.*, 2009). Red meat is particularly rich in iron and *L. sakei* has been shown to transport haem or haem-carrier molecules (Zagorec and Champomier-Vergès, 2017).

The sourdough environment

Lactobacillus species are commonly associated with natural sourdough fermentation and are also used as biopreservative agents for extending the quality and shelf-life of sourdough breads. More than 60 different species of lactobacilli have been identified from sourdoughs (Gobbetti *et al.*, 2016), including *Lactobacillus sanfransicences, Lactobacillus reuteri, Lactobacillus brevis* and *Lactobacillus zymae* among others. *L. sanfransicences* is solely associated with the man-made environment of traditional sourdough fermentations. The species has not been isolated from other environmental niches and is therefore, quite specialized in its adaptation (Vogel *et al.*, 2011). Genome sequence analysis of *L. sanfranciscensis* TMW 1.1304 isolated from an industrial sourdough fermentation revealed several features which suggest adaptation to the sourdough environment (Vogel *et al.*, 2011). The organism contained one of the smallest genomes within the *Lactobacillus genus*, at 1.3 Mb, and the highest density of rRNA operons per Mb genome among all known bacteria capable of autonomous growth (Vogel *et al.*, 2011). It has been suggested that these characteristics are important for the rapid growth rate of the species in sourdough and the ability to respond to favourable growth conditions in the environment. In terms of metabolic capacity, the sequenced strain had the genetic potential to synthesize *de novo* several amino acids which are scarce in the wheat-based habitat, including aspartate, asparagine and glutamate. The proteolytic system of this species comprises a large number of peptidases, proteases and transport systems but lacks an extracellular protease as would be found in many dairy lactobacilli, reflecting the high dependency of *L. sanfranciscensis* on the protein-rich environment of sourdough.

In terms of niche adaptation, there is significant overlap between strains from sourdough and human intestine ecosystems. *L. reuteri* is an interesting case as this species, while present in sourdough, is also a dominant member of the intestinal microbiota of vertebrates (Duar *et al.*, 2017a). The species is phylogenetically differentiated into host-adapted lineages comprising rodent isolates (lineage I and III), human isolates (lineage II and IV), pig isolates (lineage IV and V) and poultry isolates (lineage VI). Sourdough isolates cluster with one or other of these host-adapted lineages, suggesting that adaptation to the sourdough environment is a recent evolutionary event. In terms of gene content, however, sourdough isolates cluster separately. A 2015 comparative genomics study of 16 strains of *L. reuteri* attempted to understand how the intestinal strains adapted to the sourdough environment and to identify genes which were unique to the sourdough isolates (Zheng *et al.*, 2015). Overall, it was demonstrated that gene loss and horizontal gene transfer played a major role in the adaptation of the species to the sourdough environment. The enrichment for genes coding for energy conversion and carbohydrate metabolism in the genomes of sourdough isolates indicated positive selection in this environment. These genes included maltose phosphorylase, alcohol dehydrogenase and lactate dehydrogenase, genes known to contribute to competitiveness in the cereal environment. Genes common to all sourdough isolates but absent from other strains included those responsible for reutericyclin production. Reutericyclin is an antimicrobial compound that is structurally but not functionally related to naturally occurring tetramic acids and it was proposed that it may contribute to the competitiveness of the organism in sourdough. However, this has not been demonstrated in studies where a reutericyclin-negative derivative of the sourdough isolate TMW 1.656 was found to be as competitive as its wild-type parental strain (Lin *et al.*, 2015).

The plant niche

The microbes that are found living on plants are known as endophytes. These organisms generally live in a symbiotic relationship with the plant, where the microbes contribute to plant fitness and development and the plant provides the microbes with a reliable and constant source of nutrients and protection from stress (Mercado-Blanco, 2015). Plant matter is associated with native LAB including *Lactobacillus* species, few of which are known to be endophytic, while most are contaminants or are deliberately introduced, in particular for the preservation of vegetables for human consumption, and grass and maize for animal feed. Whichever the route to their presence in this environment, plant-associated species display particular traits that contribute to improving the quality, shelf-life, nutritional, and sensory attributes of plant-based food products for human and animal consumption.

Preservation of plant-based substrates through fermentation includes the process of ensiling, an effective method for maintaining the integrity and quality of forage crops used as livestock feed. During the process, epiphytic microbes ferment the plant material by acidification under anaerobic conditions, inhibiting the growth of undesirable species. *Lactobacillus buchneri* is a common member of the silage microbiome. Genome sequence analysis of this heterofermentative species has shown the absence of key enzymes necessary for glycolysis but the presence of a complete phosphoketolase pathway, thus confirming the species as an obligate heterofermenter (Heinl and Grabherr, 2017). As such, the organism, along with other lactobacilli, has been applied as a supplement during silage fermentation to improve the fermentation process and the stability of the silage. Heterofermentative species are more suitable than homofermentative species for this purpose as the additional by-products of fermentation provide additional protection against undesirable microbes, e.g. the anti-fungal properties of acetic acid. Niche adaptation in *L. buchneri* can also be demonstrated by its resistance to stresses encountered in the silage environment such as ethanol and oxygen. Transcriptomic studies revealed the up-regulation of oxidases, peroxidases and chaperone proteins in response to stresses of this nature. The presence of an S-layer is also thought to contribute to the physical robustness of the organism in this challenging environment (Heinl and Grabherr, 2017). Other species associated with silage fermentation include *Lactobacillus plantarum*, *L. brevis* and *Lactobacillus hilgardii*. A recently discovered species, *Lactobacillus hokkaidonensis*, a member of the *Lactobacillus vaccinostercus* group isolated from silage fermentation in a subarctic region of Japan (Tanizawa *et al.*, 2015), demonstrates remarkable psychrotolerance. In cold weather conditions, the impaired ability of LAB to produce sufficient acid to lower the pH can result in lower quality silage. While protection from low temperatures involves a wide range of biological systems which are difficult to identify through genomics alone, several factors were identified on the genome of *L. hokkaidonensis*, including the potential for accumulation of compatible solutes and the synthesis of glutathione, which could contribute to its metabolic activity at cold temperatures. Four transporters most likely responsible for the uptake of osmolytes were identified, similar to transporters of *L. sakei*, an organism that is also considered psychrotrophic. A bifunctional glutathione synthase encoded in the integrated and conjugative elements (ICE) region was also discovered. The presence of this gene system in *L. hokkaidonensis* and other psychrotolerant species such as *L. coryniformis* suggests that glutathione may facilitate psychrotolerance in both species (Tanizawa *et al.*, 2015).

Lactobacillus plantarum is a versatile species found in a variety of ecological niches, ranging from plants, to the gastro-intestinal tracts of human and animals, as well as food materials, such as meat, fish, vegetables and raw or fermented dairy products. Its genome size of approximately 3.3 Mb is among the largest known for an LAB (Makarova *et al.*, 2006; Makarova and Koonin, 2007) and as previously mentioned, it is proposed that a larger genome is related to the diversity of environmental niches in which *L. plantarum* is encountered. Nonetheless, this bacterium is most frequently found in the fermentation of plant-derived raw material. It is the species most often found associated with the fermentation of plant material such as cocoa, olives, grape musts and sorghum, due to its ability to degrade tannins (Jiménez *et al.*, 2013). The species possesses the enzyme tannase (tannin acyl hydrolase, EC 3.1.1.20), which catalyses the hydrolysis of ester linkages in tannin to produce gallic acid and glucose. The gallic acid is then further decarboxylated to the less-toxic pyrogallol. Genes encoding an intracellular tannase and a gallate decarboxylase have been identified on the chromosome of *L. plantarum* WCFS1. While these genes are physically separated on the genome and transcribed independently, they are subject to common regulation (Jiménez *et al.*, 2013, 2014). A study on the adaptation of *L. plantarum* C2 to the plant-like conditions of carrot juice and pineapple juice demonstrated that the strain senses the plant stimulus and adjusts its carbohydrate metabolism to fit, towards pathways involving the metabolism and catabolism of amino acids (Filannino *et al.*, 2016). This could increase the strain's capacity to compete in diverse environmental conditions. Comparative genomic analysis of 54 strains of *L. plantarum* isolated from a variety of sources attempted to understand the link between evolution and the ecological versatility of this organism (Martino *et al.*, 2016). The study revealed an absence of specific signatures in the genome associated with a particular ecological niche, reflecting the nomadic lifestyle of this particular species. This genomic flexibility allows the organism to effectively grow in a wide range of habitats, highlighting the generalist nature of *L. plantarum* (Martino *et al.*, 2016).

The human gastrointestinal tract as an environmental niche

Lactobacilli are essential members of a healthy human microbiota and are found at various niches in the human body (Fig. 1.3). Undoubtedly, the adaptation of lactobacilli to the human gut is the most analysed of all environmental niches, with many studies performed on probiotic strains, which are mostly used for treatment of gut disorders. Although recently it was proposed that colonization of neonates occurs *in utero* (Collado *et al.*, 2016), currently it is still widely accepted that the colonization of the GIT starts after birth and its development depends on the infant's diet, hygiene level and other factors (Azcarate-Peril *et al.*, 2008). The human GIT presents a challenging environment and strains successfully colonizing this niche require the ability to adhere to the mucosal surfaces of the gut, to compete with other bacteria for the available nutrients and to adapt to the presence of host-derived molecules such as bile salts (Duar *et al.*, 2017b). In terms of their original habitat, strains present in the GIT can roughly be divided in two groups. The majority of GIT isolates detectable in faeces, especially of humans, comprise transient, allochthonous microbiota originating from different sources, mostly food or saliva. In contrast, a smaller group of GIT isolates comprise actual gut inhabitants, often designated as autochthonous species and rarely isolated from other sources not connected with the GIT (Reuter, 2001). For example, 17 *Lactobacillus* species are putative inhabitants of the human gut, but only a limited number (*Lactobacillus gasseri, Lactobacillus salivarius, Lactobacillus ruminis* and *L. reuteri*) are actual inhabitants of the GIT, while most others are transient microorganisms originating from fermented food products (Walter, 2008).

The oral cavity is part of the GIT in a broader sense, and the mouth represents the entry port of food to the lower parts of the digestive tract. Numerous bacterial species including lactobacilli colonize the teeth, gums, saliva and tongue (Douillard and de Vos, 2014). At the birth of vaginally derived neonates, the sterile mouth of the baby is colonized by lactobacilli present in mother's vagina. However, these lactobacilli are transient and not sustained in the baby's mouth after one month. Lactobacilli are also found in the mouths of breastfed infants, but after weaning and prior to tooth emergence, lactobacilli are rarely found in the oral cavity of infants. Later in life, the most obvious sources of lactobacilli in the oral niche are food or other infected humans (Caufield, *et al.*, 2015).

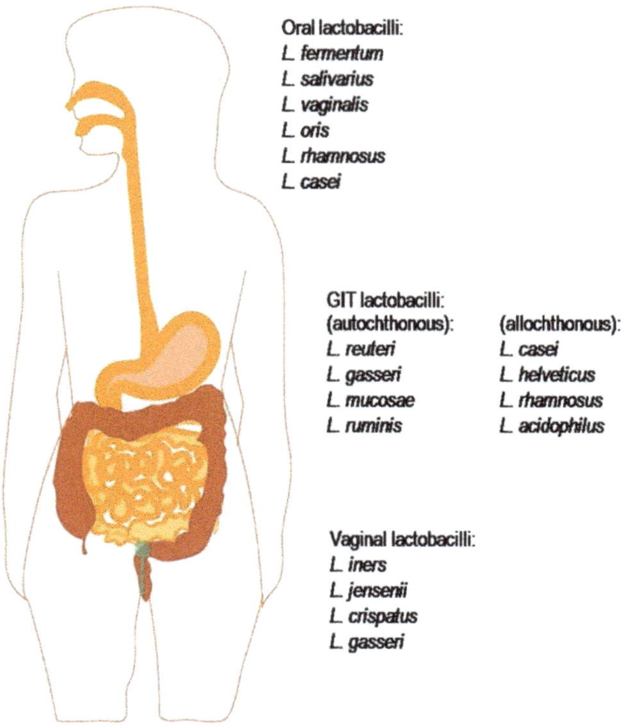

Figure 1.3 Locations of *Lactobacillus* species in various niches of the human body.

Upon analysis of the oral microbiota of older children and adults, two major populations of oral lactobacilli are generally distinguishable. The first of them consists of transient lactobacilli connected with food intake, and these are often a source of lactobacilli isolated from faeces (Dal Bello and Hertel, 2006). However, a key ecological determinant for the sustained colonization of lactobacilli in the oral cavity is the presence of dental caries (Caufield, *et al.*, 2015). Species usually connected with caries lesions belong to *L. gasseri*, *L. rhamnosus*, *Lactobacillus vaginalis*, *Lactobacillus crispatus*, *L. fermentum*, *Lactobacillus salivarius*, *Lactobacillus oris* and *L. casei* (Caufield, *et al.*, 2015, Yang, *et al.*, 2010). The phylogenetic relationships between these species show that lactobacilli associated with dental caries belong to different phylogenetic groups, and adaptation to the caries niche appears to be independent in different lineages (Caufield, *et al.*, 2015). Dental caries are caused by acidogenic bacteria that produce lactic acid as a result of anaerobic fermentation of carbohydrates. However, lactobacilli are not detected as a major cause of initial caries lesions development, but rather considered as secondary invaders of lesions already formed by *Streptococcus mutans*. The initial colonization of the teeth surface by *S. mutans* creates the necessary niche capable of mechanistically retaining both lactobacilli ('retentative' niche) and food, a source of carbohydrates. In light of caries developments, several characteristics enable survival and persistence of oral lactobacilli in caries lesions. Apart from the obvious tolerance to low pH, some isolates connected with dental caries possess collagen-binding proteins, which could help them to endure in caries lesions. Another feature contributing to caries adaptation of lactobacilli is their relative tolerance to fluorides, and the ability to metabolize xylitol, mannitol and sorbitol, often used as sweeteners and anticaries agents, thus actually promoting caries progression in lesions dominated by lactobacilli (Caufield, *et al.*, 2015). When genomes of *L. rhamnosus* strains isolated from dental pulp were analysed, it was shown they possessed up to 250 unique genes, coding for transcriptional regulators, ferric ion ABC transporters and exopolysaccharides (EPS), but no genes encoding pili components were detected. This suggested that EPS could be important for biofilm formation and promotion of caries, while pili seem not to be essential for the persistence in the oral niche (Douillard and de Vos, 2014). From the available data, it could be concluded that caries lesions can promote the development of lactobacilli populations with specific phenotypes supporting their persistence in the oral cavity and constantly supplying lower parts of GIT with lactobacilli. However, oral niche adaptability has been the focus of a small number of studies and further analysis of the specific genomic content of common caries-related lactobacilli is necessary to describe specific mechanisms of accommodation to the oral niche.

Some of the common features of gut isolates were identified through a series of genomic comparison studies of *Lactobacillus* species isolated from various niches. These studies aimed to distinguish between isolates of different origin and are beneficial in pointing out some features that are important for gut survival. They do not, however, provide evidence of typical niche specialization, since as previously mentioned, the GIT represents a transient environmental niche rather than the true niche for these species. In species that are considered as free-living or nomadic (*L. plantarum*, *L. casei* group), or frequently used in dairy processing and as probiotics (*L. helveticus*, *L. acidophilus*), several features were distinguished among gut isolates or strains used as probiotics. A genome-scale study of *L. helveticus* strains MTCC 5463 (a probiotic strain) and DPC4571 (a dairy isolate) showed that the MTCC 5463 genome possessed multiple genes for bile salt hydrolase (bsh), important for survival in the bile-rich environment of the GIT. In order to survive in conditions of constant competition for nutrients, the probiotic strain carries a higher number of starvation-induced genes compared to the dairy isolate (Senan *et al.*, 2014). Another study performed on *L. helveticus* strains confirmed that, in probiotic strain R0052, mucus-binding proteins were present, since they would be essential for survival and residence of the strain in the gut (Cremonesi *et al.*, 2012). In the previously mentioned study that analysed 100 *L. rhamnosus* strains, intestinal and probiotic strains shared similarities with other human isolates. These strains were bile resistant, pili-possessing and L-fucose-utilizing, all characteristics important for intestinal tract survival (Douillard *et al.*, 2013). In the study of 11 fully sequenced LAB (10 of which were lactobacilli) originating from different sources (three dairy, five gut and three multiniche LAB), nine genes grouped in four classes (sugar metabolism, the proteolytic system, restriction–modification systems and bile salt hydrolysis) were identified as niche-determinative as they insured survival in the gut or dairy environments (O'Sullivan *et al.*, 2009). All of these studies pointed to characteristics that enable survival of lactobacilli associated with the GIT niche: a wide spectrum of genes encoding transporters to make the best out of available nutrients; genes encoding for acid and bile resistance and finally genes conferring interaction and signalling with the host such as pili containing mucus-binding proteins, but these are only providing evidence of strain (rather than species) adaptation. Pili and mucus-binding proteins and other mediators of bacterial adhesion to the gut mucosa have been confirmed in strains isolated from the GIT, and are often seen as one of the prerequisites for the persistence of probiotic strains. The stratified squamous epithelia is not present in the human stomach, and the colonization of mucosal surfaces by GIT microbiota is very limited in humans since significant epithelial associations of gut bacteria or biofilms have not been described. Instead, commensal bacteria appear to live in suspension with limited contact with epithelial cells. Rapid generation times are therefore vital for the bacteria to avoid washout. How lactobacilli facilitate rapid growth in the human intestinal tract remains unclear. This is why the majority of the traditional probiotic strains (*L. acidophilus*, *L. casei* group, *L. delbrueckii*, *L. brevis*, *L. plantarum*, *L. johnsonii*, *L. fermentum*) are probably allochthonous to the human intestinal tract and although they are present in faecal samples, they are not able to form stable populations in the human gut. However, since probiotic effects do not necessarily depend on the persistence, their probiotic function remains unimpaired if constant supply of probiotics is maintained (Walter, 2008). Interestingly, even when autochthonous strains are used as probiotics, their persistence is not significantly better than other allochthonous probiotics. When the persistence of the probiotic species *L. reuteri*, *L. mucosae* (both of which are recognized as autochthonous) and *L. acidophilus* (allochthonous) were analysed, the cell numbers of two autochthonous strains were higher than for the allochthonous strain, suggesting that autochthonous probiotics were able to reach a higher 'effective' dose in the gut. This is important in cases where the probiotic effect depends on the number of present bacteria. The restricted ability of *L. acidophilus* to form stable populations in the human gut is likely due to the absence of specific adaptive features, which are present in *L. reuteri* and *L. mucosae*. Nevertheless, autochthonous strains were not more persistent over time. This implies that even if a probiotic strains belongs to the autochthonous microbiota of an individual, it might be impossible to establish a stable population in the gut (Frese *et al.*, 2012).

One of the human autochthonous species is *L. gasseri*, and the molecular basis for its autochthony has been reported in analysis of the genome sequence of human isolate *L. gasseri* ATCC33323. With regard to sugar metabolism, its diverse range of PTS transporters and sugar hydrolases for mono- and di-saccharides imply its adaptation to the upper GIT, where these sugars are present. However, complex sugars are not metabolised by this strain. In addition, two *bsh* genes detected increase the competitiveness of the strains and support its survival in the gut. Up to 14 putative mucus-binding proteins were detected, facilitating the adhesion of the strain, but not all of them possessed a signal peptide, meaning they might be either non-functional or secreted by other mechanism. In addition, an EPS cluster of 16 genes was detected, and EPS polymers have been shown to contribute to successful adherence. Additional gut-specialization features included the ability to degrade oxalate, commonly present in dietary sources and also produced by intestinal microbiota from precursors. Besides that, a broad range of molecular chaperones and chaperonins that protect proteins from irreversible aggregation during stress conditions were detected. Interestingly, although bacteriocins are helpful in maintaining stable populations and in bacterial competition, *L. gasseri* does not encode for bacteriocin peptides (Azcarate-Peril *et al.*, 2008). Similarly, a lack of bacteriocin production or transport was observed in another human autochthonous strain, *L. ruminis* ATCC 25644 (Forde *et al.*, 2011), suggesting that this trait is not an essential feature of genuine gut lactobacilli.

Another species considered to be autochthonous for the human (and animal) GIT is *L. reuteri* (Reuter, 2001) but ecological strategies of this species are different in humans and animals. For example, while in animals *L. reuteri* forms large populations in the proximal regions of GIT, in humans they use mainly mucus-binding adhesion, which results in relatively low prevalence in the human population (Wegmann *et al.*, 2015). According to multi-locus sequence analysis (MLSA) of *L. reuteri* from different hosts (Oh *et al.*, 2010), human isolates of *L. reuteri* belong to two distinct clades. Group II is solely comprised of human isolates, while group VI includes both human and poultry isolates. In group II, one of the most important observations was the absence of many genes encoding for biofilm formation and adhesion, which could mean that in humans, *L. reuteri* develops a 'planktonic' lifestyle in more distal regions of the human gut and limited interactions with gut epithelium. Since no large surface proteins were detected, group II isolates probably persists in the gut by fast multiplication. Interestingly, the *pdu–cbi–cob–hem* cluster was conserved within the human strains. This cluster encodes for cobalamin synthesis, 1,2-propanediol fermentation and reuterin production. Because nutrients are scarcely available in the colon, the ability of *L. reuteri* to use 1,2-propanediol might be an important factor of survival in the colon. The production of reuterin might contribute to the fitness of *L. reuteri* in the human gut through inhibition of competitors. Strains belonging to human II group have highly similar genetic content, are highly conserved and are clonally related. In addition, they show signs of reductive evolution, since pseudogene formation and gene deletions have been detected (Frese *et al.*, 2011). In a subsequent study, that aimed to confirm host specialization, the administration of human II isolates facilitated the presence of the isolates in human faecal samples, while after administration of human VI isolates, they were undetectable in human faeces. The genomic analysis of two subgroups of group VI (chicken and human) showed that while poultry VI strains possessed an open genome with a large number of strain-specific genes, human VI isolates possessed very few unique genes and showed essentially a closed pangenome and showed very little genomic variation, similarly to group II. It was proposed that strains from lineage VI can become transiently associated with humans. These findings suggest that a single clone acquired specific traits that allowed a temporary migration to humans, such as genes encoding antibiotic resistance and capsular polysaccharide synthesis (Duar *et al.*, 2017a).

The animal gastrointestinal tract as an environmental niche

The vertebrate GIT niche

Although most often studied in relation to the GIT of humans, lactobacilli are also regularly isolated from the digestive tract of a variety of insects and vertebrates, such as birds, rodents, and farm animals, but the host range is probably broader (Duar, *et al.*, 2017a). Unlike in humans, where limited gut niche specialization has been observed, most lactobacilli present in the gut of mice, rats, pigs and chickens are autochthonous, since they form stable populations throughout the life of the host. The stomach of pigs, mice and rats and the crops of birds are lined, at least partially, with squamous stratified epithelium and form a layer of bacterial cells, often embedded in a matrix of extracellular polymeric substances. The features of the autochthonous species that enable specialization to animal hosts include adherence due to large surface proteins and EPS production that contributes to cell aggregation. In addition, some features, such as IgA protease are important for survival of innate or adaptive immune response of the host (Walter, 2008).

In the above mentioned study (Oh *et al.*, 2010), the phylogeny of *L. reuteri* strains isolated from six different host species (human, mouse, rat, pig, chicken and turkey) was assessed by MLSA. A significant level of heterogeneity that corresponded to the host origin was observed. Apart from two previously mentioned groups (group II of human isolates, and group VI of human and poultry isolates), groups I and III included rodent isolates, groups IV and V included pig isolates. In addition, this grouping was later confirmed in genomic analysis (Duar *et al.*, 2017a). This result suggested that different conditions present within GIT of different hosts facilitated diversification of strains in host specific lineages and the development of a stable relationship of *L. reuteri* with vertebrate hosts during a long period of close associations (Frese *et al.*, 2011). For example, the urease cluster was highly conserved among rodent isolates, and urease is important for survival in acidic conditions in forestomach. In addition, 11 large surface proteins were detected almost exclusively in rodent strains. Most of them are potentially involved in epithelial adhesions and biofilm formations. The xylose operon was highly conserved in rodents and porcine strains. Xylose could be an important substrate for gut bacteria as a plant-derived sugar present in straw and bran. The SecA2 cluster was detected in most strains from rodents and pigs, while it is rare in human and poultry isolates. Since mobile elements were present within the cluster, it is assumed that it was acquired in HGT event. This secretion system could be involved in secretion of several adjacent proteins, one of which is serine rich, and probably possesses adhesion function. An important note is that rodent-specific genes were not conserved among all rodent strains, and adaptation to murine GIT is defined by a rodent-specific accessory genome, reflecting a plasticity of rodent *L. reuteri* genomes and higher diversity among the host population, in contrast to previously described very limited variations among human isolates (Frese *et al.*, 2011).

A pangenome study of *L. reuteri* based on five genomes of pig GIT isolates (Wegmann *et al.*, 2015) confirmed that two distinct clades are observed (IV and V), as previously suggested (Oh *et al.*, 2010). No genes conserved in all pig isolates were found but genes specific for strains in one group were detected. They included genes encoding cell wall -associated proteins, EPS biosynthetic enzymes, phage-related functions, mobile elements and DNA metabolic enzymes. This means that the two populations of pig-associated *L. reuteri* evolved separately, through a process driven by differences in host genotypes or dietary components (Wegmann *et al.*, 2015). In another study of *L. reuteri* strains, the total pangenome of 42 strains of different isolation source (26 of them of various origin previously sequenced and 16 pig isolates from the current study) was assessed in order to find specific genomic characteristics associated with porcine adaptation. All porcine isolates belonged to group IV according to the MLSA clustering. The phylogenetic clustering based on RAST subsystem comparison showed that all porcine strains grouped together, while strains of other origins were dispersed and not categorized for each host. Six genes (three encoding hypothetical and three encoding cell surface proteins) were identified only in the porcine group, suggesting that surface proteins contribute to host adaptation. Also, these host-specific genes were probably obtained by horizontal gene transfer, since transposase genes were co-localized. Compared with strains originating from other niches, pig isolates were missing six genes, four of which were associated with several pathways for energy production, but they were not found to be essential for growth and survival. Interestingly, differentiation between pig isolates with high (HS) and low (LS) anti-pathogenic effect (determined based on the activity against two frequent swine pathogens) showed that the HS group had the full cluster for production of reuterin, while in some strains in the LS group, a subsystem for citrate metabolism transport and regulation was determined. Both of these traits could represent specific adaptation leading to survival in the pig as a host (Lee *et al.*, 2017a).

Another species of *Lactobacillus* that is commonly isolated from the GIT of vertebrates is *L. salivarius.* In a comparative study that analysed 35 *L. salivarius* strains from pigs, humans and chicken, strains clustered based on the host from they were isolated. Two main strategies to adapt to a specific host were increasing adhesion and colonization abilities and producing more energy through changes in carbohydrate utilization. Several molecules contributed to this diversification of isolates. In each of the groups of isolates, specific genes for cell surface proteins that facilitate adhesion to a host habitat were detected. These include mucus-binding proteins, or less precisely annotated molecules, such as cell surface proteins or hypothetical secreted protein, for which adhesion function has been assigned according to functional domains detected in proteins. Additionally, extracellular components are considered responsible for niche adaptation through interactions with hosts. The clustering based on orthologues encoding extracellular proteins showed that bacteria grouped by the hosts from which they were isolated. Similarly, the divergence of strains associated with the host was observed based on the comparison of 24 conserved EPS genes. Additionally, differences in a host's dietary habits facilitated differences in energy consumption in specific lineages. For instance, in pig isolates, genes involved in carbohydrate metabolism via the citric-acid cycle were detected, indicating that these genes contribute to the energy production for the growth of pig isolates. In the genomes of chicken isolates, genes related to acetate and lactate metabolism were specific. Interestingly, *L. salivarius* strains isolated from humans have no host specific genes for adhesion to the habitat and for energy production. Because humans intake various foods, genes required only in certain circumstances are not essential. In contrast, the diet of pigs and chickens is based on uniform formulated feed, so the genomic content for energy metabolism of commensal bacteria is much less variable. In terms of bile salt hydrolases, while human and chicken isolates possess one or two *bsh* genes, most of pig isolates have three genes (Lee *et al.*, 2017b).

The invertebrate GIT niche

In its prominent role as a pollinator, the honey bee is the key species for agricultural production. The recent decline of populations of honey bees and bumble bees triggered the need for a better understanding of the bees' microbiota and its potential to benefit the host health (Engel *et al.*, 2012). The honey-bee gut microbiota consists of eight distinct phylotypes, five Gram negative (which will not be discussed here), and three Gram positive, two of which refer to lactobacilli, designated as Firm 4 and Firm 5 (Firmicutes) and one referring to bifidobacteria (Engel *et al.*, 2012). Lactobacilli in the Firm 4 group include *L. mellifer* and *L. mellis,* while Firm 5 includes *L. helsinborgensis, L. melliventris, L. kimbladii, L. kullabergenis* and *L. apis* (Ellegaard *et al.*, 2015; Moran, 2015). In the gut of the honey bee, lactobacilli are isolated from the crop (*L. kunkeei,* and both Firm 4 and Firm 5 groups) (Ellegaard *et al.*, 2015), but they mostly dominate in distal parts of the insect's gut (hindgut) (Kwong and Moran, 2016). The genomes of lactobacilli commonly associated with honey bees are characterized by small sizes (1.54 Mbp in *L. apis* to 2.13 Mbp in *L. kimbladii* (Ellegaard *et al.*, 2015)). The genomic analysis of lactobacilli and bifidobacteria revealed several features important for adaptation to this host. Interestingly, in all three groups, a significant level of synteny was observed. Each group was characterized by novel outer surface protein families probably involved in the interactions with the host or other bacteria.

With regard to differences among the two groups of lactobacilli, one of the gene groups that was present in Firm 4 (and Bifido) group, but not in Firm 5, was *cydABCD,* which is involved in aerobic respiration and thus is important for colonization of the gut and possibly reflects the adaptation to different microhabitats. Firm 4 group also possessed

11 conserved group-specific protein families, which are, based on their domains, involved in co-aggregation with other bacteria. Additionally, some of these domains have been previously identified in *L. plantarum* and were connected with utilization of complex plant polysaccharides. However, the most important feature contributing to niche specialisation of honey-bee lactobacilli refers to their sugar metabolism capacities. In Firm 5 group, the accessory genes were predominantly PTS transporters, and large majority of them coded for mannose–sorbose–fructose family transporters. Some of the species (*L. kullabergensis* and *L. kimbladii*) possessed up to 70 genes for this transporter family. Interestingly, PTS transporters were often found in genomic islands (GI) with a lack of sequence similarity between the genomes, and were co-localized with other sugar-metabolism-related genes (glucosidases, hydrolases, isomerases etc) suggesting that these GI code mainly for carbohydrate-related functions. Deeper analysis of the evolutionary relationships of the PTS genes showed that they have undergone an extreme expansion which preceded the diversification of Firm 5 strains and was followed by loss, recombination, diversification and transfer between groups 4 and 5 (Ellegaard *et al.*, 2015). Similar findings were observed in the metagenomic study of gut microbiota of honey bee, which found that cluster of orthologous genes (COG) for carbohydrate metabolism and transport was significantly enriched, referring to the adaptation to the diet and gut environment. The large number of different PTS were present in Firm groups, suggesting that they metabolize a variety of sugars, and mannose transporters are the most abundant and characterized by wide substrate specificity. Various nectar and pollen sugars cannot be metabolized or are toxic to the honey bees, and metabolism of these sugars by the gut microbiota could be critical for detoxification of food components. In regard to EPS production no *eps* clusters were identified within Firm 5 group, but Firm 4 contained genes for dTDP-rhamnose synthesis, and multiple glycosyl transferases, both of which are involved in EPS production (Engel *et al.*, 2012).

As previously said, *Lactobacillus kunkeei* is usually present in the crop of the honey bee, and also in the larval gut, nectar, honey and hive, however it is absent from adult hindgut (Moran, 2015). It is classified as a sole obligate fructophilic LAB (FLAB), meaning that it grows well on D-fructose, but poorly on D-glucose. Analysis of the 16 sequenced genomes of the species showed that the genome sizes were approximately 1.54 Mbp, significantly smaller compared to the majority of other lactobacilli and among the smallest genomes in bee isolates. In correlation with poor carbohydrate metabolic ability of the species, *L. kunkeei* genomes also had fewer genes for carbohydrate metabolism and transport compared to other small *Lactobacillus* genomes, lacking most of the genes used for TCA cycle and complete missing of PTS system. Additionally, over-representation of genes for catabolism and anabolism of fatty acids and amino acid biosynthesis was observed. An important observation was made in regard to ADH/ALDH protein, a bifunctional alcohol/acetaldehyde-dehydrogenase, one of the key enzymes for D-glucose metabolism and ethanol production in the phosphoketolase pathway of heterofermentative LAB. *Fructobacillus* spp. have been reported as the only heterofermentative LAB that lack *adhE* gene and ADH/ALDH activities, which is in connection with their poor glucose metabolism and low ethanol production. Although *L. kunkeei* has similar biochemical characteristics, shorter *adhE* genes encoding only ALDH domain, but not ADH domain were identified. This means that in glucose metabolism of *L. kunkeei*, the lack of ADH activity is followed by acetate production (Maeno *et al.*, 2016). More recently, another *Lactobacillus* (*L. apinorum*) has been included in the group of fructophilic LAB (Maeno *et al.*, 2017), contributing to observation of even higher level of specialization of some insect-related lactobacilli.

The vaginal niche

Lactobacilli represent a common member of the complex vaginal microbiota in women from different ethnic groups and living in different geographical locations (Douillard and de Vos, 2014). The main *Lactobacillus* species that constitute the vaginal flora are *L. crispatus*, *L. jensenii*, *L. gasseri* and *L. iners*, and interestingly, these species are rarely isolated from other habitats (Martín *et al.*, 2008; Mendes-Soares *et al.*, 2014). Unlike in the GIT where different lactobacilli cohabit due to the nutritional adaptations and resource partitioning (Tannock *et al.*, 2012), in the vaginal cavity one species usually dominates the community and stable co-existence of vaginal lactobacilli is not often observed. The four most commonly found vaginal *Lactobacillus* species interact with the environment in different ways which enable them to establish a stable population of single species dominance (Mendes-Soares *et al.*, 2014). On the other hand, some authors argue that species can also share niche space temporarily through conditional differentiation, taking advantage of the ability to accommodate the range of environmental conditions, due to which the abundance of the species is determined by factors that influence their competitive interactions (France *et al.*, 2016).

The vaginal microbiome in healthy women consists of five cluster groups, four of which are dominated by lactobacilli, while the fifth cluster has low numbers of LAB and high numbers of strict anaerobes. The key ecological stamp in all of the identified clusters is the production of lactic acid (Ravel *et al.*, 2011). In addition, no single core vaginal microbiome was observed, but more precisely, core genomes were defined in the community groups. The core functions are conserved among these communities, and that functional redundancy would be associated with increased community reliability to adapt to the environmental changes (Ravel *et al.*, 2011).

Analysis of the genomic features of vaginal-related species showed that their genomes are significantly smaller and with lower GC content than species characteristic for other ecological niches, suggesting a high level of host adaptation and the development of a symbiotic lifestyle. Analysis of the protein families revealed that the species possess different mechanisms to interact with their environment. The number of proteins present exclusively in each of the vaginal species was the highest in *L. crispatus* and the lowest in *L. iners*, which also have the largest and the smallest genomes of vaginal lactobacilli, respectively. It was found, for example, that *L. iners* possesses the gene for thiol-activated cytolysin (inerolysin), which is identified as up-regulated during bacterial vaginosis (BV) (Mendes-Soares *et al.*, 2014). This raises the possibility that the *L. iners*-encoded CDS may play an unappreciated role in BV and might contribute to the pathogenesis of the condition (Macklaim *et al.*, 2013), although another study observed this trait as an adaptation to cases of scare nutrient availability in the vaginal niche (France *et al.*, 2016). Interestingly, all analysed vaginal *L. gasseri* strains encode for the bacteriocin pediocin (Mendes-Soares *et al.*, 2014). The *L. crispatus* strains were found to have a unique DNA polymerase, bacteriocin and toxin–antitoxin systems and genes encoding mobile genetic elements, especially transposases that contribute to its large genome size. Additionally, *L. crispatus* had 26 protein families that are not found in other vaginal lactobacilli, two of which possess cell wall anchoring domains (Mendes-Soares *et al.*, 2014). Additionally, lysogeny of phage particles was observed recently in most of the vaginal isolates of *L. crispatus*, which could explain the large number of genes encoding mobile genetic elements (Derrien and van Hylckama Vlieg, 2015).

In a comparative-genome study of 10 *L. crispatus* strains, several features important for successful colonization of the vaginal cavity were reported. In regard to vaginal health, the most interesting features include genes for EPS biosynthesis, which could be important for adhesion, biofilm formation and competitive exclusion of pathogens. Eight out of ten analysed strains produced EPS, but strains produced EPS with different sugar monomers and glycosidic linkages. In addition, *L. crispatus* were shown to possess genes encoding bacteriocins similar to enterolysin A, helveticin J and pediocin J, thus contributing to protective function against pathogens. The metabolic route for generation of hydrogen peroxide from pyruvate was ubiquitously present, further enhancing the protective role of *L. crispatus* against pathogens. The conserved pathways were annotated for the metabolism glucose and mannose. Although no complete routes for the metabolism of glycogen were identified, seven vaginal strains were discovered to carry a type I pullulanase gene, which could contribute to the degradation of glycogen. Although strains were able to synthesize seven amino acids *de novo*, overall the dependency on external supplies of amino acids was shown. In terms of adhesion to the surface of the vagina, 103 proteins were identified with the potential to enable colonization of the host. Approximately 30 putative S-layer protein-encoding genes that could potentially contribute to the bacterial adhesion were reported (Ojala *et al.*, 2014). Importantly, several molecular factors in *L. crispatus* were shown to be antagonistic to the vaginal pathogen *Gardenella vaginalis*, and prevent its adhesion, most important of which is *Lactobacillus* epithelium adhesin (LEA), which is shown to be universally present in *L. crispatus* strains (Ojala *et al.*, 2014).

The comparison of genome characteristics of *L. crispatus* and *L. iners* revealed ecological factors that drive niche partitioning by these species. Both of the species rely on fermentation of carbohydrates as sources of energy. Genes encoding glucose, trehalose, maltose and mannose were identified in both species, while *L. crispatus* additionally can ferment lactose, galactose, fructose and sucrose. *L. crispatus* and *L. iners* have similar numbers of genes related to metabolism of some amino acids, but *L. crispatus* also has the genetic capability to transport and breakdown putrescine, a product of ornithine catabolism, suggesting that *L. iners* is more dependent on exogenous sources of amino acids. However, *L. iners* is able to produce inerlysin, a pore-forming cytolysin, which enables *L. iners* to liberate resources from the host cell, giving competitive advantage to this species in the vaginal environment when nutrients are scarce. The reported differences in the genetics of metabolism could be the determining factor competitive interactions between these two species (France *et al.*, 2016).

Concluding remarks

The extraordinary genomic diversity that defines the *Lactobacillus* genus has resulted from interactions with different environments and different genomes within those environments as these organisms have evolved to adapt to the various niches in which they are found. While no specific signature sequences have been identified which associate a particular species with a particular niche (Sun *et al.*, 2015), there is overwhelming evidence that remodelling of genomes through gene loss and acquisition has played a major role in the adaptation of this genus to nutritionally rich environments. Phylogenomic analysis has revealed that lifestyle categories, whether free-living, host adapted or 'nomadic', correlate with phylogenetic groupings (Duar *et al.*, 2017b). Many of the present-day species of *Lactobacillus* exhibit variation in their reliance on specific niches, having evolved from free-living ancestors. In general, the metabolic capabilities of species reflect their lifestyle adaptations. These capabilities have the potential to be exploited in a wide range of applications, as our growing knowledge on the capacities of the *Lactobacillus* genus reveal a largely, as yet, untapped resource. Exploitation of the diversity of lactobacilli could have significant implications for new product development. There are exciting new opportunities in the field of biotransformation, and additional opportunities to improve the application of this important genus in industrial and therapeutic applications.

References

Anba-Mondoloni, J., Chaillou, S., Zagorec, M., and Champomier-Vergès, M.C. (2013). Catabolism of N-acetylneuraminic acid, a fitness function of the food-borne lactic acid bacterium *Lactobacillus sakei*, involves two newly characterized proteins. Appl. Environ. Microbiol. *79*, 2012–2018. https://doi.org/10.1128/AEM.03301-12

Aussel, L., Beuzón, C.R., and Cascales, E. (2016). Meeting report: Adaptation and communication of bacterial pathogens. Virulence 7, 481–490. https://doi.org/10.1080/21505594.2016.1152441

Azcarate-Peril, M.A., Altermann, E., Goh, Y.J., Tallon, R., Sanozky-Dawes, R.B., Pfeiler, E.A., O'Flaherty, S., Buck, B.L., Dobson, A., Duong, T., *et al.* (2008). Analysis of the genome sequence of *Lactobacillus gasseri* ATCC 33323 reveals the molecular basis of an autochthonous intestinal organism. Appl. Environ. Microbiol. *74*, 4610–4625. https://doi.org/10.1128/AEM.00054-08

Broadbent, J.R., Cai, H., Larsen, R.L., Hughes, J.E., Welker, D.L., De Carvalho, V.G., Tompkins, T.A., Ardö, Y., Vogensen, F., De Lorentiis, A., *et al.* (2011). Genetic diversity in proteolytic enzymes and amino acid metabolism among *Lactobacillus helveticus* strains. J. Dairy Sci. *94*, 4313–4328. https://doi.org/10.3168/jds.2010-4068

Broadbent, J.R., Neeno-Eckwall, E.C., Stahl, B., Tandee, K., Cai, H., Morovic, W., Horvath, P., Heidenreich, J., Perna, N.T., Barrangou, R., *et al.* (2012). Analysis of the *Lactobacillus casei* supragenome and its influence in species evolution and lifestyle adaptation. BMC Genomics *13*, 533. https://doi.org/10.1186/1471-2164-13-533

Callanan, M., Kaleta, P., O'Callaghan, J., O'Sullivan, O., Jordan, K., McAuliffe, O., Sangrador-Vegas, A., Slattery, L., Fitzgerald, G.F., Beresford, T., *et al.* (2008). Genome sequence of *Lactobacillus helveticus*, an organism distinguished by selective gene loss and insertion sequence element expansion. J. Bacteriol. *190*, 727–735. https://doi.org/10.1128/JB.01295-07

Chaillou, S., Daty, M., Baraige, F., Dudez, A.M., Anglade, P., Jones, R., Alpert, C.A., Champomier-Vergès, M.C., and Zagorec, M. (2009). Intraspecies genomic diversity and natural population structure of the meat-borne lactic acid bacterium *Lactobacillus* sakei. Appl. Environ. Microbiol. *75*, 970–980. https://doi.org/10.1128/AEM.01721-08

Collado, M.C., Rautava, S., Aakko, J., Isolauri, E., and Salminen, S. (2016). Human gut colonisation may be initiated *in utero* by distinct microbial communities in the placenta and amniotic fluid. Sci. Rep. *6*, 23129. https://doi.org/10.1038/srep23129

Cremonesi, P., Chessa, S., and Castiglioni, B. (2012). Genome sequence and analysis of *Lactobacillus helveticus*. Front. Microbiol. *3*, 435. https://doi.org/10.3389/fmicb.2012.00435

Derrien, M., and van Hylckama Vlieg, J.E. (2015). Fate, activity, and impact of ingested bacteria within the human gut microbiota. Trends Microbiol. *23*, 354–366. https://doi.org/10.1016/j.tim.2015.03.002

Didelot, X., Darling, A., and Falush, D. (2009). Inferring genomic flux in bacteria. Genome Res. *19*, 306–317. https://doi.org/10.1101/gr.082263.108

Douillard, F.P., and de Vos, W.M. (2014). Functional genomics of lactic acid bacteria: from food to health. Microb. Cell Fact. 13 (Suppl. 1), S8. https://doi.org/10.1186/1475-2859-13-S1-S8

Douillard, F.P., Ribbera, A., Kant, R., Pietilä, T.E., Järvinen, H.M., Messing, M., Randazzo, C.L., Paulin, L., Laine, P., Ritari, J., *et al.* (2013). Comparative genomic and functional analysis of 100 *Lactobacillus rhamnosus* strains and their comparison with strain GG. PLOS Genet. *9*, e1003683. https://doi.org/10.1371/journal.pgen.1003683

Duar, R.M., Frese, S.A., Lin, X.B., Fernando, S.C., Burkey, T.E., Tasseva, G., Peterson, D.A., Blom, J., Wenzel, C.Q., Szymanski, C.M., *et al.* (2017a). Experimental evaluation of host adaptation of *Lactobacillus reuteri* to different vertebrate species. Appl. Environ. Microbiol. *83*. https://doi.org/10.1128/AEM.00132-17

Duar, R.M., Lin, X.B., Zheng, J., Martino, M.E., Grenier, T., Pérez-Muñoz, M.E., Leulier, F., Gänzle, M., and Walter, J. (2017b). Lifestyles in transition: evolution and natural history of the genus *Lactobacillus*. FEMS Microbiol. Rev. *41*, S27–S48. https://doi.org/10.1093/femsre/fux030

Ellegaard, K.M., Tamarit, D., Javelind, E., Olofsson, T.C., Andersson, S.G., and Vásquez, A. (2015). Extensive intra-phylotype diversity in lactobacilli and bifidobacteria from the honeybee gut. BMC Genomics *16*, 284. https://doi.org/10.1186/s12864-015-1476-6

Engel, P., Martinson, V.G., and Moran, N.A. (2012). Functional diversity within the simple gut microbiota of the honey bee. Proc. Natl. Acad. Sci. U.S.A. *109*, 11002–11007. https://doi.org/10.1073/pnas.1202970109

Fadda, S., Anglade, P., Baraige, F., Zagorec, M., Talon, R., Vignolo, G., and Champomier-Vergès, M.C. (2010). Adaptive response of *Lactobacillus sakei* 23K during growth in the presence of meat extracts: a proteomic approach. Int. J. Food Microbiol. *142*, 36–43. https://doi.org/10.1016/j.ijfoodmicro.2010.05.014

Filannino, P., Di Cagno, R., Crecchio, C., De Virgilio, C., De Angelis, M., and Gobbetti, M. (2016). Transcriptional reprogramming and phenotypic switching associated with the adaptation of *Lactobacillus plantarum* C2 to plant niches. Sci. Rep. *6*, 27392. https://doi.org/10.1038/srep27392

Forde, B.M., Neville, B.A., O'Donnell, M.M., Riboulet-Bisson, E., Claesson, M.J., Coghlan, A., Ross, R.P., and O'Toole, P.W. (2011). Genome sequences and comparative genomics of two *Lactobacillus ruminis* strains from the bovine and human intestinal tracts. Microb. Cell Fact. 10 (Suppl. 1), S13. https://doi.org/10.1186/1475-2859-10-S1-S13

France, M.T., Mendes-Soares, H., and Forney, L.J. (2016). Genomic Comparisons of *Lactobacillus crispatus* and *Lactobacillus iners* reveal potential ecological drivers of community composition in the vagina. Appl. Environ. Microbiol. *82*, 7063–7073. https://doi.org/10.1128/AEM.02385-16

Frese, S.A., Benson, A.K., Tannock, G.W., Loach, D.M., Kim, J., Zhang, M., Oh, P.L., Heng, N.C., Patil, P.B., Juge, N., *et al.* (2011). The evolution of host specialization in the vertebrate gut symbiont *Lactobacillus reuteri*. PLOS Genet. *7*, e1001314. https://doi.org/10.1371/journal.pgen.1001314

Frese, S.A., Hutkins, R.W., and Walter, J. (2012). Comparison of the colonization ability of autochthonous and allochthonous strains of lactobacilli in the human gastrointestinal tract. Adv. Microbiol. *2*, 399–409. https://doi.org/10.4236/aim.2012.23051

Genay, M., Sadat, L., Gagnaire, V., and Lortal, S. (2009). *prtH2*, not *prtH*, is the ubiquitous cell wall proteinase gene in *Lactobacillus helveticus*. Appl. Environ. Microbiol. *75*, 3238–3249. https://doi.org/10.1128/AEM.02395-08

Gobbetti, M., Minervini, F., Pontonio, E., Di Cagno, R., and De Angelis, M. (2016). Drivers for the establishment and composition of the sourdough lactic acid bacteria biota. Int. J. Food Microbiol. *239*, 3–18. https://doi.org/10.1016/j.ijfoodmicro.2016.05.022

Heinl, S., and Grabherr, R. (2017). Systems biology of robustness and flexibility: *Lactobacillus buchneri*-A show case. J. Biotechnol. *257*, 61–69. https://doi.org/10.1016/j.jbiotec.2017.01.007

Jiménez, N., Curiel, J.A., Reverón, I., de Las Rivas, B., and Muñoz, R. (2013). Uncovering the *Lactobacillus plantarum* WCFS1 gallate decarboxylase involved in tannin degradation. Appl. Environ. Microbiol. *79*, 4253–4263. https://doi.org/10.1128/AEM.00840-13

Jiménez, N., Esteban-Torres, M., Mancheño, J.M., de Las Rivas, B., and Muñoz, R. (2014). Tannin degradation by a novel tannase enzyme present in some *Lactobacillus plantarum* strains. Appl. Environ. Microbiol. *80*, 2991–2997. https://doi.org/10.1128/AEM.00324-14

Kwong, W.K., and Moran, N.A. (2016). Gut microbial communities of social bees. Nat. Rev. Microbiol. *14*, 374–384. https://doi.org/10.1038/nrmicro.2016.43

Lee, J.Y., Han, G.G., Choi, J., Jin, G.D., Kang, S.K., Chae, B.J., Kim, E.B., and Choi, Y.J. (2017a). Pan-genomic approaches in *Lactobacillus reuteri* as a porcine probiotic: investigation of host adaptation and antipathogenic activity. Microb. Ecol. *74*, 709–721. https://doi.org/10.1007/s00248-017-0977-z

Lee, J.Y., Han, G.G., Kim, E.B., and Choi, Y.J. (2017b). Comparative genomics of *Lactobacillus salivarius* strains focusing on their host adaptation. Microbiol. Res. *205*, 48–58. https://doi.org/10.1016/j.micres.2017.08.008

Lin, X.B., Lohans, C.T., Duar, R., Zheng, J., Vederas, J.C., Walter, J., and Gänzle, M. (2015). Genetic determinants of reutericyclin biosynthesis in *Lactobacillus reuteri*. Appl. Environ. Microbiol. *81*, 2032–2041. https://doi.org/10.1128/AEM.03691-14

Macklaim, J.M., Fernandes, A.D., Di Bella, J.M., Hammond, J.A., Reid, G., and Gloor, G.B. (2013). Comparative meta-RNA-seq of the vaginal microbiota and differential expression by *Lactobacillus iners* in health and dysbiosis. Microbiome *1*, 12. https://doi.org/10.1186/2049-2618-1-12

Maeno, S., Dicks, L., Nakagawa, J., and Endo, A. (2017). *Lactobacillus apinorum* belongs to the fructophilic lactic acid bacteria. Biosci. Microbiota Food Health *36*, 147–149. https://doi.org/10.12938/bmfh.17-008

Maeno, S., Tanizawa, Y., Kanesaki, Y., Kubota, E., Kumar, H., Dicks, L., Salminen, S., Nakagawa, J., Arita, M., and Endo, A. (2016). Genomic characterization of a fructophilic bee symbiont *Lactobacillus kunkeei* reveals its niche-specific adaptation. Syst. Appl. Microbiol. *39*, 516–526. https://doi.org/10.1016/j.syapm.2016.09.006

Makarova, K., Slesarev, A., Wolf, Y., Sorokin, A., Mirkin, B., Koonin, E., Pavlov, A., Pavlova, N., Karamychev, V., Polouchine, N., *et al.* (2006). Comparative genomics of the lactic acid bacteria. Proc. Natl. Acad. Sci. U.S.A. *103*, 15611–15616. https://doi.org/10.1073/pnas.0607117103

Makarova, K.S., and Koonin, E.V. (2007). Evolutionary genomics of lactic acid bacteria. J. Bacteriol. *189*, 1199–1208. https://doi.org/10.1128/JB.01351-06

Martín, R., Soberón, N., Vaneechoutte, M., Flórez, A.B., Vázquez, F., and Suárez, J.E. (2008). Characterization of indigenous vaginal lactobacilli from healthy women as probiotic candidates. Int. Microbiol. *11*, 261–266. https://doi.org/10.2436/20.1501.01.70

Martino, M.E., Bayjanov, J.R., Caffrey, B.E., Wels, M., Joncour, P., Hughes, S., Gillet, B., Kleerebezem, M., van Hijum, S.A., and Leulier, F. (2016). Nomadic lifestyle of *Lactobacillus plantarum* revealed by comparative genomics of 54 strains isolated from different habitats. Environ. Microbiol. *18*, 4974–4989. https://doi.org/10.1111/1462-2920.13455

McAuliffe, O. (2017). Genetics of lactic acid bacteria. In Cheese: Chemistry, Physics and Microbiology, 4th edn, McSweeney, P.L., Fox, P.F., Cotter, P.D., and Everett, D.W., eds. (Academic Press, London), pp. 227–243. https://doi.org/10.1016/B978-0-12-417012-4.00047-8

McLeod, A., Snipen, L., Naterstad, K., and Axelsson, L. (2011). Global transcriptome response in *Lactobacillus sakei* during growth on ribose. BMC Microbiol. *11*, 145. https://doi.org/10.1186/1471-2180-11-145

Mendes-Soares, H., Suzuki, H., Hickey, R.J., and Forney, L.J. (2014). Comparative functional genomics of *Lactobacillus* spp. reveals possible mechanisms for specialization of vaginal lactobacilli to their environment. J. Bacteriol. *196*, 1458–1470. https://doi.org/10.1128/JB.01439-13

Mercado-Blanco, J. (2015). Life of microbes inside the plant. In Principles of Plant–Microbe Interactions, Lugtenberg, B., ed. (Springer, Switzerland), pp. 25–32. https://doi.org/10.1007/978-3-319-08575-3_5

Moran, N.A. (2015). Genomics of the honey bee microbiome. Curr. Opin. Insect Sci. *10*, 22–28. https://doi.org/10.1016/j.cois.2015.04.003

Nyquist, O.L., McLeod, A., Brede, D.A., Snipen, L., Aakra, Å., and Nes, I.F. (2011). Comparative genomics of *Lactobacillus sakei* with emphasis on strains from meat. Mol. Genet. Genomics *285*, 297–311. https://doi.org/10.1007/s00438-011-0608-1

O'Sullivan, O., O'Callaghan, J., Sangrador-Vegas, A., McAuliffe, O., Slattery, L., Kaleta, P., Callanan, M., Fitzgerald, G.F., Ross, R.P., and Beresford, T. (2009). Comparative genomics of lactic acid bacteria reveals a niche-specific gene set. BMC Microbiol. *9*, 50. https://doi.org/10.1186/1471-2180-9-50

Oh, P.L., Benson, A.K., Peterson, D.A., Patil, P.B., Moriyama, E.N., Roos, S., and Walter, J. (2010). Diversification of the gut symbiont *Lactobacillus reuteri* as a result of host-driven evolution. ISME J. *4*, 377–387. https://doi.org/10.1038/ismej.2009.123

Ojala, T., Kankainen, M., Castro, J., Cerca, N., Edelman, S., Westerlund-Wikström, B., Paulin, L., Holm, L., and Auvinen, P. (2014). Comparative genomics of *Lactobacillus crispatus* suggests novel mechanisms for the competitive exclusion of *Gardnerella vaginalis*. BMC Genomics *15*, 1070. https://doi.org/10.1186/1471-2164-15-1070

Papizadeh, M., Rohani, M., Nahrevanian, H., Javadi, A., and Pourshafie, M.R. (2017). Probiotic characters of *Bifidobacterium* and *Lactobacillus* are a result of the ongoing gene acquisition and genome minimization evolutionary trends. Microb. Pathog. *111*, 118–131. https://doi.org/10.1016/j.micpath.2017.08.021

Pereira, F.C., and Berry, D. (2017). Microbial nutrient niches in the gut. Environ. Microbiol. *19*, 1366–1378. https://doi.org/10.1111/1462-2920.13659

Ravel, J., Gajer, P., Abdo, Z., Schneider, G.M., Koenig, S.S., McCulle, S.L., Karlebach, S., Gorle, R., Russell, J., Tacket, C.O., *et al.* (2011). Vaginal microbiome of reproductive-age women. Proc. Natl. Acad. Sci. U.S.A. *108* (Suppl. 1), 4680–4687. https://doi.org/10.1073/pnas.1002611107

Reuter, G. (2001). The *Lactobacillus* and *Bifidobacterium* microflora of the human intestine: composition and succession. Curr. Issues Intest. Microbiol. *2*, 43–53.

Rimaux, T., Rivière, A., Illeghems, K., Weckx, S., De Vuyst, L., and Leroy, F. (2012). Expression of the arginine deiminase pathway genes in *Lactobacillus sakei* is strain dependent and is affected by the environmental pH. Appl. Environ. Microbiol. *78*, 4874–4883. https://doi.org/10.1128/AEM.07724-11

Salvetti, E., and O'Toole, P.W. (2017). When regulation challenges innovation: The case of the genus *Lactobacillus*. Trends Food Sci. Tech. *66*, 187–194. https://doi.org/10.1016/j.tifs.2017.05.009

Senan, S., Prajapati, J.B., and Joshi, C.G. (2014). Comparative genome-scale analysis of niche-based stress-responsive genes in *Lactobacillus helveticus* strains. Genome *57*, 185–192. https://doi.org/10.1139/gen-2014-0020

Smokvina, T., Wels, M., Polka, J., Chervaux, C., Brisse, S., Boekhorst, J., van Hylckama Vlieg, J.E., and Siezen, R.J. (2013). *Lactobacillus paracasei* comparative genomics: towards species pan-genome definition and exploitation of diversity. PLOS ONE *8*, e68731. https://doi.org/10.1371/journal.pone.0068731

Sun, Z., Harris, H.M., McCann, A., Guo, C., Argimón, S., Zhang, W., Yang, X., Jeffery, I.B., Cooney, J.C., Kagawa, T.F., *et al.* (2015). Expanding the biotechnology potential of lactobacilli through comparative genomics of 213 strains and associated genera. Nat. Commun. *6*, 8322. https://doi.org/10.1038/ncomms9322

Tanizawa, Y., Tohno, M., Kaminuma, E., Nakamura, Y., and Arita, M. (2015). Complete genome sequence and analysis of *Lactobacillus hokkaidonensis* LOOC260^T, a psychrotrophic lactic acid bacterium isolated from silage.

BMC Genomics *16*, 240. https://doi.org/10.1186/s12864-015-1435-2

Tannock, G.W., Wilson, C.M., Loach, D., Cook, G.M., Eason, J., O'Toole, P.W., Holtrop, G., and Lawley, B. (2012). Resource partitioning in relation to cohabitation of *Lactobacillus* species in the mouse forestomach. ISME J. *6*, 927–938. https://doi.org/10.1038/ismej.2011.161

van de Guchte, M., Penaud, S., Grimaldi, C., Barbe, V., Bryson, K., Nicolas, P., Robert, C., Oztas, S., Mangenot, S., Couloux, A., *et al.* (2006). The complete genome sequence of *Lactobacillus bulgaricus* reveals extensive and ongoing reductive evolution. Proc. Natl. Acad. Sci. U.S.A. *103*, 9274–9279. https://doi.org/10.1073/pnas.0603024103

Vogel, R.F., Pavlovic, M., Ehrmann, M.A., Wiezer, A., Liesegang, H., Offschanka, S., Voget, S., Angelov, A., Böcker, G., and Liebl, W. (2011). Genomic analysis reveals *Lactobacillus sanfranciscensis* as stable element in traditional sourdoughs. Microb. Cell Fact. 10 (Suppl. 1), S6. https://doi.org/10.1186/1475-2859-10-S1-S6

Walter, J. (2008). Ecological role of lactobacilli in the gastrointestinal tract: implications for fundamental and biomedical research. Appl. Environ. Microbiol. *74*, 4985–4996. https://doi.org/10.1128/AEM.00753-08

Wegmann, U., MacKenzie, D.A., Zheng, J., Goesmann, A., Roos, S., Swarbreck, D., Walter, J., Crossman, L.C., and Juge, N. (2015). The pan-genome of *Lactobacillus reuteri* strains originating from the pig gastrointestinal tract. BMC Genomics *16*, 1023. https://doi.org/10.1186/s12864-015-2216-7

Woodcock, D.J., Krusche, P., Strachan, N.J.C., Forbes, K.J., Cohan, F.M., Méric, G., and Sheppard, S.K. (2017). Genomic plasticity and rapid host switching can promote the evolution of generalism: a case study in the zoonotic pathogen *Campylobacter*. Sci. Rep. 7, 9650. https://doi.org/10.1038/s41598-017-09483-9

Xu, H.Q., Gao, L., Jiang, Y.S., Tian, Y., Peng, J., Xa, Q.Q., and Chen, Y. (2015). Transcriptome response of *Lactobacillus sakei* to meat protein environment. J. Basic Microbiol. *55*, 490–499. https://doi.org/10.1002/jobm.201400540

Zagorec, M., and Champomier-Vergès, M.C. (2017). *Lactobacillus sakei:* A Starter for Sausage Fermentation, a Protective Culture for Meat Products. Microorganisms 5, 56. https://doi.org/10.3390/microorganisms5030056

Zheng, J., Zhao, X., Lin, X.B., and Gänzle, M. (2015). Comparative genomics *Lactobacillus reuteri* from sourdough reveals adaptation of an intestinal symbiont to food fermentations. Sci. Rep. *5*, 18234. https://doi.org/10.1038/srep18234

Genetics and Genomics of *Lactobacillus sakei* and *Lactobacillus curvatus* 2

Lucrecia C. Terán[1], Raúl R. Raya[1]*, Monique Zagorec[2] and Marie-Christine Champomier-Vergès[3]*

[1]Reference Centre for Lactobacilli (CERELA-CONICET), Tucumán, Argentina.
[2]UMR1014 SECALIM, INRA, Oniris, Nantes, France.
[3]MICALIS Institute, INRA, AgroParisTech, University of Paris-Saclay, Jouy-en-Josas, France.

*Correspondence: marie-christine.champomier-verges@inra.fr and rraya@cerela.org.ar

https://doi.org/10.21775/9781910190890.02

Abstract

Lactobacillus sakei and *Lactobacillus curvatus* are two lactic acid bacteria widely used worldwide as starters for meat fermentation. They are phenotypically closely related species and have often been associated as a '*sakei/curvatus*' group in the past. These species have also been more recently described to be phylogenetically closely related to other *Lactobacillus* species such as *L. fuchuensis* and *L. graminis*. Genomic studies have contributed to a better characterization of each species and enlightened their specificities. *L. sakei* exhibits a high genomic diversity and several genomes are now available. Complete and draft genome sequences are also available from *L. curvatus* strains originating from various origins. This chapter will provide information on genomic repertoires of each species and illustrate their main common features.

Introduction: generalities of lactic acid bacteria

Lactic acid bacteria (LAB) have a great importance in the food industry, mainly due to their capacity of fermenting milk, meat and vegetables. Their fermentative metabolism leads to acidification thus affecting the growth of unwanted microorganisms. LAB are recognized as safe by the Food and Drugs Administration (FDA, USA) and by the European Food Safety Authority (EFSA). They are often used as starter cultures or probiotics. LAB rely greatly on exogenous sources of nutrients, and are adapted to grow in rich media. As a consequence their metabolic capacities are reduced, as is revealed by their genomes.

Lactobacillus is a highly diverse and paraphyletic genus that includes over 200 species and subspecies, isolated from many different niches. The genetic diversity of the genus *Lactobacillus* is larger than the one described for a family (a superior level of taxonomy), illustrated by the wide genomic variability observed within the genus (Sun *et al.*, 2015). In a recent taxonomic revision the genera *Lactobacillus, Pediococcus, Weissella, Leuconostoc, Oenococcus* and *Fructobacillus* were grouped by phylogenetic evidence as part of the *Lactobacillus* genus complex (Sun *et al.*, 2015). Taxonomic studies performed through nucleotide-sequence analysis can be assessed using different parameters. One of them is ANI (average nucleotide identity), that is used for the species definition and is defined by the average of the calculated nucleotide identity between two homologous sequences (Goris *et al.*, 2007) (Table 2.1).

Among the *Lactobacillus* genus complex, the small clade of psychrotrophic *Lactobacillus* includes *L. sakei, L. curvatus, L. graminis* and *L. fuchuensis* (Felis and Dellaglio, 2007; Salvetti *et al.*, 2012; Zheng *et al.*, 2015). Phylogenetic analyses have shown that the closest species to this group is *L.*

Table 2.1 ANI values* for type strains in *L. sakei* clade (Sun *et al.*, 2015)

	L. sakei subsp. *sakei*	*L. sakei* subsp. *carnosus*	*L. curvatus*	*L. graminis*	*L. fuchuensis*
L. sakei subsp. *sakei*	100%				
L. sakei subsp. *carnosus*	96.9%	100%			
L. curvatus	78.9%	78.8%	100%		
L. graminis	77.4%	77.2%	83.7%	100%	
L. fuchuensis	76.2%	76.8%	76.3%	75.4%	100%

*The cut-off for species definition is an ANI value >95% (Goris *et al.*, 2007; Richter and Rosselló-Móra, 2009).

selangorensis, a species first isolated from Malaysian food products (Leisner *et al.*, 1999; Sun *et al.*, 2015; Zagorec and Champomier-Vergès, 2017). *L. sakei* and *L. curvatus* are strongly related to meat environments and to other food products, like seafood and kimchi. This chapter describes, from a comparative genetic and genomic perspective, their ability to survive in these environments and their ubiquity. We also describe the genomic diversity of both species which supports their great biotechnological potential as starter and bioprotective cultures in the food industry.

Lactobacillus sakei clade

L. sakei was first isolated from rice wine 'saké' in 1934 (Katagiri *et al.*, 1934), while *L. curvatus* was first isolated from milk in 1965 and named after it curvy morphological shape ('croissant-like'). *L. sakei* and *L. curvatus* are the most studied species among this small clade. Historically, although DNA–DNA hybridization experiments showed *L. sakei* and *L. curvatus* were separated species, it has been an issue distinguishing between these two species by biochemical similarities based on sugar-utilization patterns. Many controversial studies were published on the characterization and differentiation of these two species. Thus, two subgroups in each species were identified by whole protein analysis (Klein *et al.*, 1996), which was confirmed by randomly amplified polymorphic DNA (RAPD). The two proposed subspecies of *L. sakei* (*L. sakei* subsp. *sakei* and *L. sakei* subsp. *carnosus*) could not be differentiated phenotypically, while the proposed subspecies of *L. curvatus* (*L. curvatus* subsp. *curvatus* and *L. curvatus* subsp. *melibiosus*) were distinguished by their capacity to ferment melibiose (Klein *et al.*, 1996; Torriani *et al.*, 1996). It was also revealed that *L. curvatus* subsp. *melibiosus* was closer to the *L. sakei* subgroup than to *L. curvatus* subsp. *curvatus*. Berthier and Ehrlich (1999), based on their RAPD studies, separated *L. curvatus* strains into two subgroups by the haem-dependent catalase activity, a characteristic used at this moment for discriminating the two species, *L. sakei* being catalase positive and *L. curvatus* catalase negative. Again, the *L. sakei* subgroups could not be differentiated by any phenotypic feature, while all studied *L. curvatus* strains could be differentiated due to their melibiose negative phenotype. The two subgroups of *L. sakei* clustered together in the phylogenetic tree as well as the two subgroups of *L. curvatus*. Lyhs *et al.* (2002) showed that these patterns were not applicable for all strains, revealing that hydrolysis of arginine by the arginine–deiminase pathway is likely to be the only phenotypic feature to distinguish between *L. sakei* and *L. curvatus*. New studies of *L. sakei*, regarding its intraspecific diversity, were conducted by pulsed-field gel electrophoresis and by 2D proteomic analysis with the aim of finding associations between the genotypic cluster and both subspecies (Chaillou *et al.*, 2009). In this study only differences in the migration patterns of four isoforms of glyceraldehyde 3P dehydrogenase (GapA) were found.

Chaillou *et al.* (2013), in an exhaustive study using multilocus sequence typing (MLST), showed that the complex population structure within *L. sakei* goes beyond the first two subspecies previously described. The study included 232 strains using three housekeeping genes (*recA*, *tuf*, *rpoB*), three catabolic genes (*pepV*, *glpF*, *ldhL*), one putative anabolic gene (*hemN*), and one stress-response gene (*dnaK*). Three separate lineages were identified for *L. sakei*, each with a different population structure (Chaillou *et al.*, 2013). Lineage 1 includes a panmictic subpopulation. It is highly diverse and has high recombination rates with high occurrences of intra-lineage recombination; Lineage 2 includes a subpopulation where recombination has less importance and contains strains with the smallest genomes among *L. sakei* which usually reflects ecological specialization. Lineage 3 represents the earliest branch in the *L. sakei* genealogy tree and has evolved equally by recombination and mutation.

Recently, Huang *et al.* (2017) designed three pairs of primers, chosen from unique bands of RAPD fingerprints, with a length of 281, 278, and 472 bp, being specific for *L. curvatus*, *L. graminis*, and *L. sakei*, respectively, to differentiate among these species by PCR. This represents an easy and low-cost method for differentiating these three species, to overcome the previously mentioned shortcut of biochemical methods as well as the low resolution of 16S rRNA analysis commonly associated with the *L. sakei* clade (Huang *et al.*, 2017).

Genomic context

L. sakei has largely been studied for its biotechnological potential for biopreservation (Bredholt *et al.*, 2001; Chaillou *et al.*, 2014; Leroi *et al.*, 2015). The first complete genome sequence within this clade was that of the strain 23K of *L. sakei* (Chaillou *et al.*, 2005). *L. sakei* 23K was originally isolated from a French sausage. Nine years after the publication of the complete genome of *L. sakei* 23K, the second *L. sakei* sequence uploaded in the NCBI was that of *L. sakei* subsp. *sakei* DSM20017 (Fig. 2.1).

Until November 2017 there were 28 *L. sakei* genomes uploaded in total to the NCBI database, five of them being complete. Their size range from 1.88 to 2.18 Mb and the percentage of GC content goes from 40.60 to 41.18. The strain Wikim22, isolated from kimchi has the lowest GC content among *L. sakei* while the highest is from strain J54 isolated from a traditional French-type dry sausage.

The first sequenced strain of *L. curvatus* was CRL 705, isolated from an Argentinian artisanal sausage, presented as a draft genome (Hebert *et al.*, 2012). Currently, 13 genome sequences of *L. curvatus* are publicly available, from which five are complete. The first complete genome of *L. curvatus* corresponds to the strain FBA2 (Nakano *et al.*, 2016). The chromosomes of all available genomes of *L. curvatus* range from 1.80 to 2.11 Mb and the percentage of GC content goes from 41.70 to 42.10.

As seen in Fig. 2.1, from 2015 to 2017, 25 new genomes of *L. sakei* and 11 genomes of *L. curvatus* were available in the NCBI database showing the impact of the accessibility of sequencing technologies.

Interestingly, despite the similarities between the genomes size of both species; the GC content of *L. curvatus* is slightly higher than the one of *L. sakei* (Fig. 2.2).

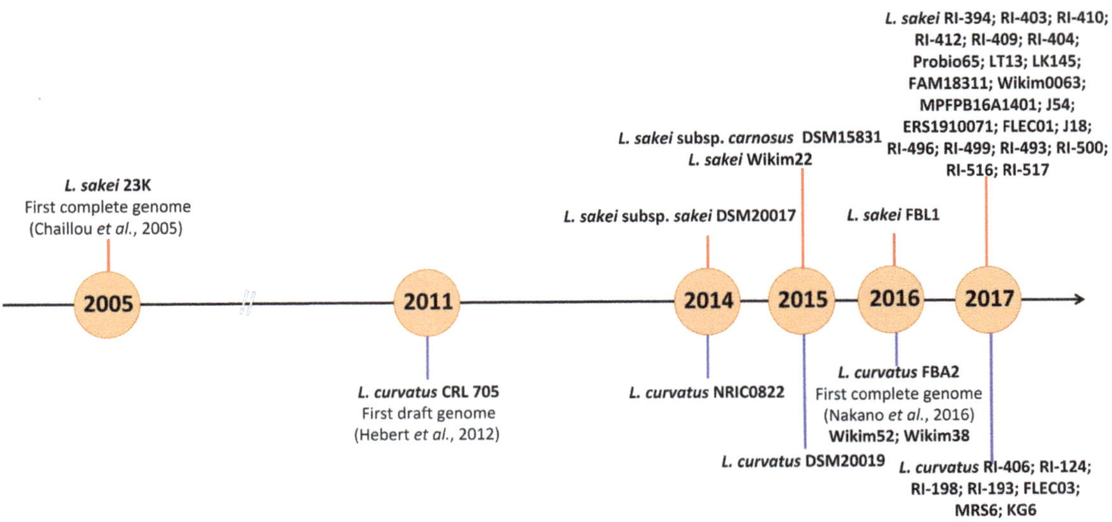

Figure 2.1 Time line of the incorporation of *L. sakei* and *L. curvatus* genome sequences in the GenBank database.

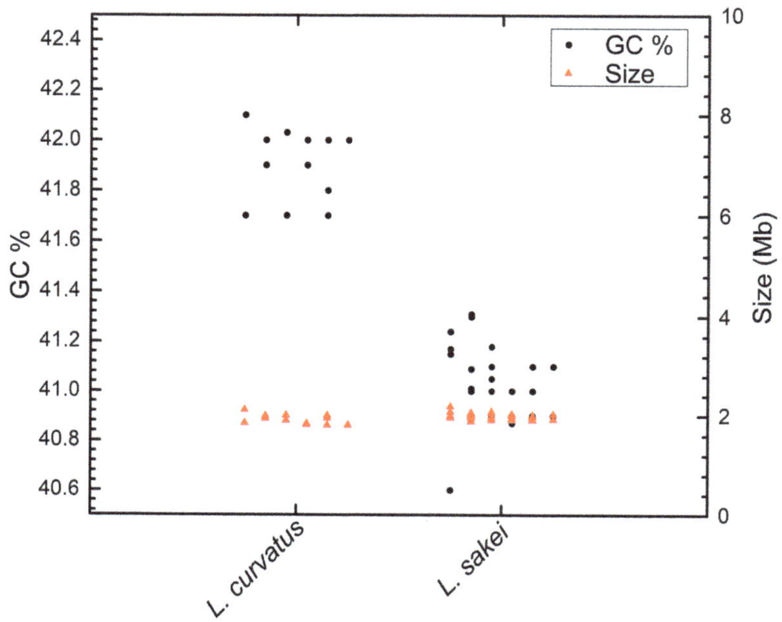

Figure 2.2 Size and GC content of the deposited genomes *L. sakei* and *L. curvatus* in the NCBI.

L. sakei and *L. curvatus* in meat products

L. sakei and *L. curvatus* are mainly associated with fermented meat products, vacuum-packaged refrigerated meat and fish (Berthier and Ehrlich, 1999; Lyhs *et al.*, 2002; Belfiore *et al.*, 2013). Both have been described as the most prevalent species found in meat products (Lyhs and Björkroth, 2008).

Meat is a challenging environment for the development of microorganisms with conditions such as low redox potential, low temperatures of storage and the presence of additives. These hard conditions act as hurdles and select microorganisms. In these environments *Lactobacillus* play an important role.

L. sakei 23K genome served as a model microorganism to reveal the adaptation of the species for growing in meat substrates (Chaillou *et al.*, 2005). The main fitness traits revealed from its genome were summarized by Eijsink and Axelsson (2005). These are (i) the ability to resist stressing conditions in meat production and meat processing (low temperature or oxidative stress, addition of curing salts); (ii) the ability of efficiently using nutrients and metabolites present in meat environments to ensure its growth; and (iii) the ability to compete with other bacteria in meat matrix by producing antagonistic molecules.

These adaptations for growing in such a challenging environment also includes: capacity of using glucose, ribose or nucleosides as a carbon source (Stentz *et al.*, 1997; Chaillou *et al.*, 2005; McLeod *et al.*, 2010); arginine-specific aminopeptidase activity (Sanz and Toldrá, 2002); pathway of arginine-deiminase as an energy source (Zúñiga *et al.*, 1998; Champomier Vergès *et al.*, 1999; Rimaux *et al.*, 2012) and a versatile redox metabolism (Chaillou *et al.*, 2005). The ability to resist salt and low-temperature stresses has also been studied (Champomier-Vergès *et al.*, 2002; Marceau *et al.*, 2004).

Energy sources

Meat is a poor source of carbohydrates. Ribose is one of the carbohydrates present in raw meat and *L. sakei* is capable of metabolizing it as a carbon source. There are some particularities of the uptake of ribose by *L. sakei* such as the presence of a permease encoded by *rbsU* instead of an ABC transporter (Stentz and Zagorec, 1999). This gene *rbsU* is also present in *L. curvatus* CRL 705, in place of those encoding the ABC transporter (*rbsABC*).

The growth of *L. sakei* on ribose was further studied by McLeod *et al.* (2010, 2011) and Nyquist *et al.* (2011). First using a comparative proteomic approach it was observed that *L. sakei* grew either using glucose or ribose as a carbon source, but one strain *L. sakei* LS25, showed a different pattern when growing in ribose characterized by a pronounced down regulation of the glycolytic pathway (McLeod *et al.*, 2010). Later using microarrays, a transcriptomic comparison of three strains showed that nucleoside catabolism was induced during the growth on ribose (McLeod *et al.*, 2011; Nyquist *et al.*, 2011). Experimentally probed in strain CTC494, it was shown that after transportation of adenosine and inosine into the cells, the ribose-5P moiety was used in the pentose-phosphate pathway leading to energy production while adenine and hypoxanthine were secreted (Rimaux *et al.*, 2011).

Other carbon sources present in meat as the N-acetyl neuraminic acid can be used by *L. sakei* (Anba-Mondoloni *et al.*, 2013). As this sugar it is also used by pathogenic bacteria, the ability of metabolizing it by *L. sakei* may reflect a competitive advantage to be used against pathogenic species in meat substrates.

There are two pathways for arginine catabolism, an abundant amino acid in meat (Zúñiga *et al.*, 1998; Champomier-Vergès *et al.*, 1999). The main pathway in all *L. sakei* is the arginine deiminase pathway (ADI) (Nyquist *et al.*, 2011). For its part, *L. curvatus* CRL 705 lacks this latest *arcABCT* gene cluster (Hebert *et al.*, 2012). The arginine degradation by the ADI pathway was considered a main taxonomical criterion for distinguishing the species *L. sakei* and *L. curvatus*. The mentioned pathway leads to the production of energy. The second pathway for the degradation of arginine that can coexist in *L. sakei* is the agmatine deiminase (Rimaux *et al.*, 2012).

Stress response

The presence of oxidoreductases helps the microorganisms cope with oxidative stress in meat by maintaining the redox balance. In *L. sakei* 23K and also in other strains of this species about 30 putative oxidoreductases were detected in genome sequences (Chaillou *et al.*, 2005; Nyquist *et al.*, 2011).

Among the features illustrating *L. sakei* adaptation to meat is its ability to use haem as revealed by gene sequences dedicated to haem/iron transport and its capacity for intracellular iron incorporation from haemoglobin, myoglobin and transferrin which are complex iron sources present in meat (Duhutrel *et al.*, 2010).

Besides the low temperatures (4°C) commonly used for meat refrigeration, microorganisms have to deal with different challenges of meat processing such as the addition of salt for curing (3–9% NaCl) (Leistner, 2000). The effect of NaCl was studied in *L. sakei* (Samapundo *et al.*, 2010) as well as in *L. curvatus*. For the latter species, the studies of the effect of NaCl on its growth are controversial. Indeed, studies referring to the description of *L. curvatus* as a species, were based on the phenotypic trait that strains were not able to grow with a final concentration of 10% NaCl (Torriani *et al.*, 1996). However, more recently, Belfiore *et al.* (2013) studied the saline stress of *L. sakei* and *L. curvatus* and showed their ability to grow in concentrations of NaCl between 10% and 18%.

On the other hand, it has been shown that the whole microbial community in pork sausages was modified following a reduction of NaCl concentration from 1.5% to 2% (Fougy *et al.*, 2016).

All *L. sakei* strains studied by Nyquist *et al.* (2011) presented the uptake systems for the incorporation of NaCl and a sodium-dependent symporter thought to drive the accumulation of osmo- and cryoprotective solutes such as betaine and carnitine, as well as genes encoding the putative cold stress proteins Csp1–4 and general stress proteins (Chaillou *et al.*, 2005).

Hüfner *et al.* (2007) identified 15 genes specifically induced during the fermentation of *L. sakei* 23K in meat substrates. By creating knock-out mutants, four of them were identified to have essential functions. These four genes correspond to a heat-shock regulator (*ctsR*), L-asparaginase (*asnA2*), the gene *LSA1065* related to post-transcriptional regulation and *LSA1194* that encodes a hypothetical protein of unknown function. As expected, these genes are also present in *L. curvatus* CRL 705.

Thus, *L. sakei* and *L. curvatus* have the potential fitness to overcome oxidative stress, the presence of additives and the low temperatures during the meat processing.

L. sakei and *L. curvatus*, as ubiquitous species

Although both species are clearly prepared for living in meat environments, they also grow in a wide diversity of niches as illustrated in Fig. 2.3. When the emended descriptions for both species were first published they were associated with plants as participants in the production of fermented plant material (silage, sauerkraut, mixed pickles) and also in their spoilage. They can be also be present in beverages (beer) and even in manure (Klein *et al.*, 1996).

L. sakei and *L. curvatus* may be inhabitants of a variety of different niches. Many of the genome sequences from these two species available in GenBank are issued from strains isolated from meat environments. However some of the sequenced strains have other origins. For instance strains Wikim22, Wikim0063 and Probio65 of *L. sakei* were isolated from kimchi, a Korean fermented cabbage. Wikim22 has 44 extra genes that are absent from the model strain *L. sakei* 23K (Lim *et al.*, 2014). This strain has a phosphotransferase system (PTS), for the uptake of the sugar, for utilizing beta-glucoside specific IIA, IIB, IIC, and BglG families (Lim *et al.*, 2014). Other *L. sakei* strains: RI-516 and RI-517 were isolated from cocoa-bean fermentation; while LT13 and LK145 are strains from 'sake' (Kato and Oikawa, 2017a; Kato and Oikawa, 2017b). Strain FLEC01 was isolated from human faeces, although the gastrointestinal tract is not a frequent environment of *L. sakei* (Dal Bello *et al.*, 2003; Chiaramonte *et al.*, 2009, 2010).

The strain NRIC0822 of *L. curvatus* was isolated from Japanese sushi (kabura-zushi). This strain presents the particular cluster of motility by flagella being the first record of this cluster in the *L. sakei* clade (Cousin *et al.*, 2015). Until that date, motility outside the *L. salivarius* clade for LAB had not been reported. *L. curvatus* NRIC0822 showed motility genotypically as well as phenotypically. This strain was motile during exponential growth phase under different conditions: under anaerobic and aerobic conditions, in the presence of 5% CO_2 or at temperatures ranging from 15°C to 37°C and with different carbon sources such as glucose, galactose, fructose, mannose and maltose (Cousin *et al.*, 2015).

The genome sequence of *L. curvatus* FBA2 isolated from radish and carrot pickled with rice bran and salt was the first uploaded as a complete genome (Nakano *et al.*, 2016). Some products with *L. curvatus* FBA2 strain are patented and being commercialized in Japan for skin improvement. Two other complete genomes are available, those of

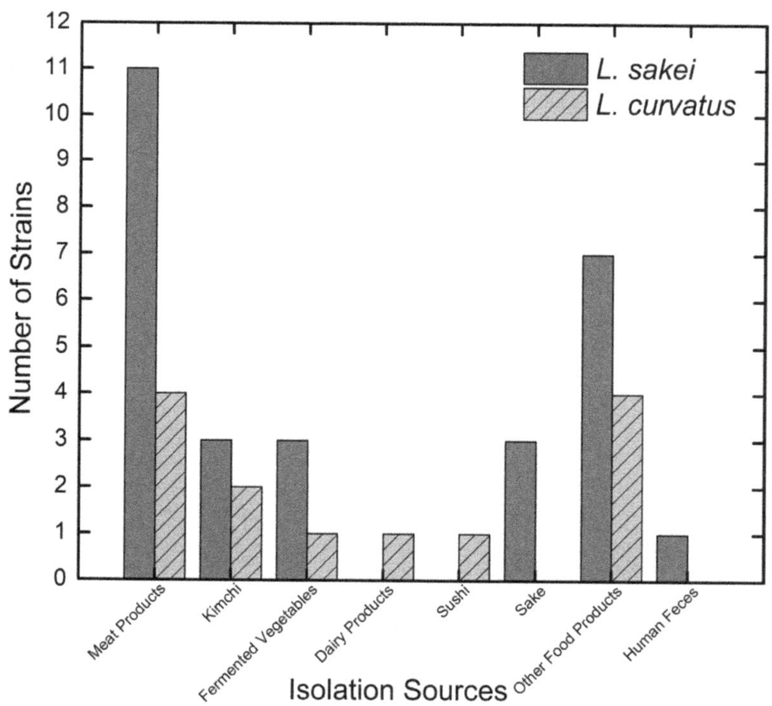

Figure 2.3 Isolation sources of *L. sakei* and *L. curvatus* strains whose genome sequences are publicly available in the GenBank database until November 2017.

Wikim52 and Wikim38, both isolated from kimchi (Lee *et al*., 2017). The latest has proven properties for the production of interleuquin IL-10 in dendritic cells (Jo *et al*., 2016).

L. curvatus FLEC03 was isolated from a slice of beef carpaccio after 7 days' storage at 8°C (Lucquin *et al*., 2012). Unlike other strains, *L. curvatus* FLEC03 is characterized by the presence of eight cell-surface multicomponent complexes (Terán *et al*., 2017) similar to those present in *L. sakei* 23K (Chaillou *et al*., 2005) and whose functions are related to bacterial adhesion and complex carbohydrate molecule scavenging (Siezen *et al*., 2006).

As part of a project of isolating *Lactobacillus* from food products, four other *L. curvatus* strains were sequenced. These strains were named RI-124, RI-193, RI-198 and RI-406 (Inglin *et al*., 2017). In October 2017, two additional *L. curvatus* complete genomes were incorporated in the GenBank database. These strains came from fermented meat products. Strain KG6 came from salami-type fermented meat and strain MRS6 came from traditional Swiss fermented sausage (Jans *et al*., 2017).

Bacteriocins

Bacteriocins are small, heat-stable peptides ribosomally synthesized by bacteria which are active against other bacteria. They can have a narrow- or a broad-target spectrum (Cotter *et al*., 2005). They can be used by the biotechnological and food industries as natural food preservatives against unwanted (spoilage or pathogenic) microorganisms.

Bacteriocins can be classified by their genetic and biochemical characteristics. There are different criteria for their classification. Klaenhammer (1993) proposed the classification of bacteriocins into four main groups while later on Cotter *et al*. (2005) proposed a classification with two major divisions: Class I (the lantibiotics) and Class II (the non-lantibiotic bacteriocins). Class II is found in both *L. sakei* and *L. curvatus*. Although the model strain 23K is a non bacteriocin producer (Chaillou *et al*., 2005), *L. sakei* and also *L. curvatus* are well known for their ability to produce bacteriocins (Champomier-Vergès *et al*., 2002; Woraprayote *et al*., 2016). Bacteriocins are frequently named after its species; the class II bacteriocins produced by some strains of *L. sakei* are called sakacins. The most frequent types are sakacins A, P, Q, T and X. The first sakacin described in *L. sakei* was sakacin A produced by *L. sakei* Lb 706 (Schillinger and Lücke, 1989). Sakacins usually have a narrow spectrum of action; they are active against other species of *Lactobacillus* and against *Listeria monocytogenes*, a ubiquitous Gram-positive foodborne pathogen. Sakacin P was described in many *L. sakei* strains such as Lb674 (Mathiesen *et al*., 2005) and CCUG 42687 (Møretrø *et al*., 2000), although not all strains of *L. sakei* produce sakacin P (Møretrø *et al*., 2005). The sakacin P operon consists of six genes: *sppK, sppR, sppA, spiA, sppT* and *sppE* (Hühne *et al*., 1996; Urso *et al*., 2006) which are mutated in some strains (Møretrø *et al*., 2005). However, frequently class II bacteriocin genes are found in plasmids (Drider *et al*., 2006).

Curvacin A was the first bacteriocin described from *L. curvatus* LTH1174 (Tichaczek *et al*., 1993). The structural gene is harboured by a 60 kb plasmid of *L. curvatus* LTH1174 (Tichaczek *et al*., 1993). Curvacin A, a class II bacteriocin, is active against *Listeria monocytogenes* and *Enterococcus faecalis* (Vogel *et al*., 1993).

The model strain *L. curvatus* CRL 705 produces a unique two-peptide bacteriocin, which was the first reported of this kind of bacteriocin from a meat-associated strain. This Lactocin 705 was shown to be antagonistic towards other LAB and *Brochothrix thermosphacta* when assayed in meat systems (Castellano *et al*., 2003; Castellano and Vignolo, 2006).

Bacteriocins can be applied in meat environments through direct inoculation, application of cell-free supernatant and with a certain degree of purification in the form of packaging (Woraprayote *et al*., 2016).

Biogenic amines

Biogenic amines are basic nitrogenous compounds of low molecular weight formed by the decarboxylation of amino acids or amination and transamination of aldehydes or ketones (Askar and Treptow, 1986). The biogenic amines are derived from bacterial decarboxylation of amino acids (Halász *et al*., 1994; Vidal-Carou *et al*., 2007). The catabolism of lysine, ornithine, tyrosine and histidine by decarboxylases ends in cadaverine, putrescine, tyramine and histamine, respectively. Some biogenic amines can be present in food products and they have been associated with spoilage.

Ingestion of these compounds in high concentrations can cause toxicological symptoms with negative health issues (Ladero *et al.*, 2010). As these biogenic amines can trigger health problems, the presence of genes encoding the mentioned decarboxylases, are not a desirable characteristic in microorganisms used as starter cultures (Coton *et al.*, 2010). The capacity of the bacteria to produce these compounds can be used as exclusion criteria for starter cultures (Buckenhüskes, 1993). *L. curvatus* have been more associated with the production of biogenic amines than *L. sakei* (Bover-Cid *et al.*, 2001). *L. curvatus* produced different biogenic amines with higher concentration of tyramine and also with less amounts putrescine, phenylethylamine, tryptamine and cadaverine (Bover-Cid *et al.*, 2001). It is not surprising then, how *L. sakei* acquired a greater biotechnological importance. Amine-negative strains of *L. sakei* have been used to reducing the accumulation of biogenic amines during the ripening of fermented sausages (Bover-Cid *et al.*, 2000). However, some *L. sakei* have an operon encoding an agmatine–deiminase pathway that produces putrescine and this helps with the spoilage in food products (Rimaux *et al.*, 2012).

Conclusions

L. sakei and *L. curvatus* are species belonging to the *L. sakei* clade of psychrotrophic *Lactobacillus*. Despite their ubiquity both species are well adapted to overcome the challenging environments of the meat matrix. These species also possess fitness traits that ensure their growth even in conditions of storage and fermentation. *L. sakei* with their well-studied model strain 23K and *L. curvatus* with their model strain CRL 705, currently under investigation, represent good alternatives for biopreservation purposes.

References

Anba-Mondoloni, J., Chaillou, S., Zagorec, M., and Champomier-Vergès, M.C. (2013). Catabolism of N-acetylneuraminic acid, as a fitness function for the food-borne lactic acid bacterium *Lactobacillus sakei*, involves two new proteins. Appl. Environ. Microbiol. *79*, 2012–2018. https://doi.org/10.1128/AEM.03301-12

Askar, A., and Treptow, H. (1986). Biogene Amine in Lebensmitteln: Vorkommen, Bedeutung und Bestimmung, 1st edn. (Eugen Ulmer Verlag, Germany). ISBN 10: 3800121328 ISBN 13: 9783800121328

Belfiore, C., Raya, R.R., and Vignolo, G.M. (2013). Identification, technological and safety characterization of *Lactobacillus sakei* and *Lactobacillus curvatus* isolated from Argentinean anchovies (Engraulis anchoita). Springerplus 2, 257. https://doi.org/10.1186/2193-1801-2-257

Berthier, F., and Ehrlich, S.D. (1999). Genetic diversity within *Lactobacillus sakei* and *Lactobacillus curvatus* and design of PCR primers for its detection using randomly amplified polymorphic DNA. Int. J. Syst. Bacteriol. *49*, 997–1007. https://doi.org/10.1099/00207713-49-3-997

Bover-Cid, S., Izquierdo-Pulido, M., and Vidal-Carou, M.C. (2000). Mixed starter cultures to control biogenic amine production in dry fermented sausages. J. Food Prot. *63*, 1556–1562. https://doi.org/10.4315/0362-028X-63.11.1556

Bover-Cid, S., Izquierdo-Pulido, M., and Vidal-Carou, M.C. (2001). Effectiveness of a *Lactobacillus sakei* starter culture in the reduction of biogenic amine accumulation as a function of the raw material quality. J. Food Prot. *64*, 367–373. https://doi.org/10.4315/0362-028X-64.3.367

Bredholt, S., Nesbakken, T., and Holck, A. (2001). Industrial application of an antilisterial strain of *Lactobacillus sakei* as a protective culture and its effect on the sensory acceptability of cooked, sliced, vacuum-packaged meats. Int. J. Food Microbiol. *66*, 191–196. https://doi.org/10.1016/S0168-1605(00)00519-5

Buckenhüskes, H.J. (1993). Selection criteria for lactic acid bacteria to be used as starter cultures for various food commodities. FEMS Microbiol. Rev. *12*, 253–271. https://doi.org/10.1016/0168-6445(93)90067-J

Castellano, P., Raya, R., and Vignolo, G. (2003). Mode of action of lactocin 705, a two-component bacteriocin from *Lactobacillus casei* CRL705. Int. J. Food Microbiol. *85*, 35–43. https://doi.org/10.1016/S0168-1605(02)00479-8

Castellano, P., and Vignolo, G. (2006). Inhibition of *Listeria innocua* and *Brochothrix thermosphacta* in vacuum-packaged meat by addition of bacteriocinogenic *Lactobacillus curvatus* CRL705 and its bacteriocins. Lett. Appl. Microbiol. *43*, 194–199. https://doi.org/10.1111/j.1472-765X.2006.01933.x

Chaillou, S., Champomier-Vergès, M.C., Cornet, M., Crutz-Le Coq, A.M., Dudez, A.M., Martin, V., Beaufils, S., Darbon-Rongère, E., Bossy, R., Loux, V., *et al.* (2005). The complete genome sequence of the meat-borne lactic acid bacterium *Lactobacillus sakei* 23K. Nat. Biotechnol. 23, 1527–1533. https://doi.org/10.1038/nbt1160

Chaillou, S., Christieans, S., Rivollier, M., Lucquin, I., Champomier-Vergès, M.C., and Zagorec, M. (2014). Quantification and efficiency of *Lactobacillus sakei* strain mixtures used as protective cultures in ground beef. Meat Sci. 97, 332–338. https://doi.org/10.1016/j.meatsci.2013.08.009

Chaillou, S., Daty, M., Baraige, F., Dudez, A.M., Anglade, P., Jones, R., Alpert, C.A., Champomier-Vergès, M.C., and Zagorec, M. (2009). Intraspecies genomic diversity and natural population structure of the meat-borne lactic acid bacterium *Lactobacillus sakei*. Appl. Environ. Microbiol. *75*, 970–980. https://doi.org/10.1128/AEM.01721-08

Chaillou, S., Lucquin, I., Najjari, A., Zagorec, M., and Champomier-Vergès, M.C. (2013). Population genetics of *Lactobacillus sakei* reveals three lineages with distinct evolutionary histories. PLOS ONE *8*, e73253. https://doi.org/10.1371/journal.pone.0073253

Champomier-Vergès, M.C., Chaillou, S., Cornet, M., and Zagorec, M. (2002). Erratum to '*Lactobacillus sakei*: recent developments and future prospects' [Research in Microbiology 152 (2001) 839.] Res. Microbiol. *153*, 115–123. https://doi.org/10.1016/S0923-2508(01)01296-7

Champomier Vergès, M.C., Zuñiga, M., Morel-Deville, F., Pérez-Martínez, G., Zagorec, M., and Ehrlich, S.D. (1999). Relationships between arginine degradation, pH and survival in *Lactobacillus sakei*. FEMS Microbiol. Lett. *180*, 297–304. https://doi.org/10.1016/S0378-1097(99)00492-9

Chiaramonte, F., Anglade, P., Baraige, F., Gratadoux, J.J., Langella, P., Champomier-Vergès, M.C., and Zagorec, M. (2010). Analysis of *Lactobacillus sakei* mutants selected after adaptation to the gastrointestinal tracts of axenic mice. Appl. Environ. Microbiol. *76*, 2932–2939. https://doi.org/10.1128/AEM.02451-09

Chiaramonte, F., Blugeon, S., Chaillou, S., Langella, P., and Zagorec, M. (2009). Behavior of the meat-borne bacterium *Lactobacillus sakei* during its transit through the gastrointestinal tracts of axenic and conventional mice. Appl. Environ. Microbiol. *75*, 4498–4505. https://doi.org/10.1128/AEM.02868-08

Coton, M., Romano, A., Spano, G., Ziegler, K., Vetrana, C., Desmarais, C., Lonvaud-Funel, A., Lucas, P., and Coton, E. (2010). Occurrence of biogenic amine-forming lactic acid bacteria in wine and cider. Food Microbiol. *27*, 1078–1085. https://doi.org/10.1016/j.fm.2010.07.012

Cotter, P.D., Hill, C., and Ross, R.P. (2005). Bacteriocins: developing innate immunity for food. Nat. Rev. Microbiol. *3*, 777–788. https://doi.org/10.1038/nrmicro1273

Cousin, F.J., Lynch, S.M., Harris, H.M., McCann, A., Lynch, D.B., Neville, B.A., Irisawa, T., Okada, S., Endo, A., and O'Toole, P.W. (2015). Detection and genomic characterization of motility in *Lactobacillus curvatus*: confirmation of motility in a species outside the *Lactobacillus salivarius* clade. Appl. Environ. Microbiol. *81*, 1297–1308. https://doi.org/10.1128/AEM.03594-14

Dal Bello, F., Walter, J., Hammes, W.P., and Hertel, C. (2003). Increased complexity of the species composition of lactic acid bacteria in human feces revealed by alternative incubation condition. Microb. Ecol. *45*, 455–463. https://doi.org/10.1007/s00248-003-2001-z

Drider, D., Fimland, G., Héchard, Y., McMullen, M., and Prévost, H. (2006). The continuing story of class IIa bacteriocins. Microbiol. Mol. Biol. Rev. *70*, 564–582. https://doi.org/10.1128/MMBR.00016-05

Duhutrel, P., Bordat, C., Wu, T.D., Zagorec, M., Guerquin-Kern, J.L., and Champomier-Vergès, M.C. (2010). Iron sources used by the nonpathogenic lactic acid bacterium *Lactobacillus sakei* as revealed by electron energy loss spectroscopy and secondary-ion mass spectrometry. Appl. Environ. Microbiol. *76*, 560–565. https://doi.org/10.1128/AEM.02205-09

Eijsink, V.G., and Axelsson, L. (2005). Bacterial lessons in sausage making. Nat. Biotechnol. *23*, 1494–1495. https://doi.org/10.1038/nbt1205-1494

Felis, G.E., and Dellaglio, F. (2007). Taxonomy of Lactobacilli and Bifidobacteria. Curr. Issues Intest. Microbiol. *8*, 44–61.

Fougy, L., Desmonts, M.H., Coeuret, G., Fassel, C., Hamon, E., Hézard, B., Champomier-Vergès, M.C., and Chaillou, S. (2016). Reducing salt in raw pork sausages increases spoilage and correlates with reduced bacterial diversity. Appl. Environ. Microbiol. *82*, 3928–3939. https://doi.org/10.1128/AEM.00323-16

Goris, J., Konstantinidis, K.T., Klappenbach, J.A., Coenye, T., Vandamme, P., and Tiedje, J.M. (2007). DNA–DNA hybridization values and their relationship to whole-genome sequence similarities. Int. J. Syst. Evol. Microbiol. *57*, 81–91. https://doi.org/10.1099/ijs.0.64483-0

Halász, A., Baráth, A., Simon-Sarkadi, S., and Holzapfel, W.H. (1994). Biogenic amines and their production by microorganisms in food. Trends Food Sci. Technol. *5*, 42–49. https://doi.org/10.1016/0924-2244(94)90070-1

Hebert, E.M., Saavedra, L., Taranto, M.P., Mozzi, F., Magni, C., Nader, M.E., Font de Valdez, G., Sesma, F., Vignolo, G., and Raya, R.R. (2012). Genome sequence of the bacteriocin-producing *Lactobacillus curvatus* strain CRL705. J. Bacteriol. *194*, 538–539. https://doi.org/10.1128/JB.06416-11

Huang, C.H., Liou, J.S., Huang, L., and Watanabe, K. (2017). Developing novel species-specific DNA markers for PCR-based species identification of the *Lactobacillus sakei* group. Lett. Appl. Microbiol. *66*, 138–144 https://doi.org/10.1111/lam.12825

Hüfner, E., Markieton, T., Chaillou, S., Crutz-Le Coq, A.M., Zagorec, M., and Hertel, C. (2007). Identification of *Lactobacillus sakei* genes induced during meat fermentation and their role in survival and growth. Appl. Environ. Microbiol. *73*, 2522–2531. https://doi.org/10.1128/AEM.02396-06

Hühne, K., Axelsson, L., Holck, A., and Kröckel, L. (1996). Analysis of the sakacin P gene cluster from *Lactobacillus sake* Lb674 and its expression in sakacin-negative *Lb. sake* strains. Microbiology *142*, 1437–1448. https://doi.org/10.1099/13500872-142-6-1437

Inglin, R.C., Meile, L., and Stevens, M.J.A. (2017). Draft genome sequences of 43 *Lactobacillus* strains from the species *L. curvatus*, *L. fermentum*, *L. paracasei*, *L. plantarum*, *L. rhamnosus*, and *L. sakei*, isolated from food products. Genome Announc. 5, e00632-17. https://doi.org/10.1128/genomeA.00632-17

Jans, C., Lagler, S., Lacroix, C., Meile, L., and Stevens, M.J.A. (2017). Complete genome sequences of *Lactobacillus curvatus* KG6, *L. curvatus* MRS6, and *Lactobacillus sakei* FAM18311, isolated from fermented meat products. Genome Announc. 5, e00915-17. https://doi.org/10.1128/genomeA.00915-17

Jo, S.G., Noh, E.J., Lee, J.Y., Kim, G., Choi, J.H., Lee, M.E., Song, J.H., Chang, J.Y., and Park, J.H. (2016). *Lactobacillus curvatus* WiKim38 isolated from kimchi induces IL-10 production in dendritic cells and alleviates DSS-induced colitis in mice. J. Microbiol. *54*, 503–509. https://doi.org/10.1007/s12275-016-6160-2

Katagiri, H., Kitahara, K., and Fukami, K. (1934). The characteristics of the lactic acid bacteria isolated from moto, yeast mashes for saké manufacture. IV.

Classification of the lactic acid bacteria. Bull. Agric. Chem. Soc. *10*, 156–157. https://doi.org/10.1080/03758397.1934.10857096

Kato, S., and Oikawa, T. (2017a). Genome sequence of *Lactobacillus sakei* LK-145 isolated from a Japanese Sake cellar as a high producer of d-amino acids. Genome Announc. *5*, e00656–17. https://doi.org/10.1128/genomeA.00656-17

Kato, S., and Oikawa, T. (2017b). Whole-genome sequence of *Lactobacillus sakei* LT-13 isolated from moto starter of sake. Genome Announc. *5*, e00651-17. https://doi.org/10.1128/genomeA.00651-17

Klaenhammer, T.R. (1993). Genetics of bacteriocins produced by lactic acid bacteria. FEMS Microbiol. Rev. *12*, 39–85. https://doi.org/10.1111/j.1574-6976.1993.tb00012.x

Klein, G., Dicks, L.M., Pack, A., Hack, B., Zimmermann, K., Dellaglio, F., and Reuter, G. (1996). Emended descriptions of *Lactobacillus sake* (Katagiri, Kitahara, and Fukami) and *Lactobacillus curvatus* (Abo-Elnaga and Kandler): Numerical classification revealed by protein figerprinting and identification based on biochemical patterns and DNA–DNA hybridizations. Int. J. Syst. Evol. Microbiol. *46*, 367–376. https://doi.org/10.1099/00207713-46-2-367

Ladero, V., Calles-Enriquez M., Fernández, M., Álvarez, M.A. (2010). Toxicological effects of dietary biogenic amines. Curr. Nutr. Food Sci. *6*, 145–156. https://doi.org/10.2174/157340110791233256

Lee, S.H., Jung, M.Y., Song, J.H., Lee, M., and Chang, J.Y. (2017). Complete genome sequence of *Lactobacillus curvatus* strain WiKim38 isolated from kimchi. Genome Announc. *5*, e00273-17. https://doi.org/10.1128/genomeA.00273-17.

Leisner, J.J., Pot, B., Christensen, H., Rusul, G., Olsen, J.E., Wee, B.W., Muhamad, K., and Ghazali, H.M. (1999). Identification of lactic acid bacteria from chili bo, a Malaysian food ingredient. Appl. Environ. Microbiol. *65*, 599–605.

Leistner, L. (2000). Basic aspects of food preservation by hurdle technology. Int. J. Food Microbiol. *55*, 181–186. https://doi.org/10.1016/S0168-1605(00)00161-6

Leroi, F., Cornet, J., Chevalier, F., Cardinal, M., Coeuret, G., Chaillou, S., and Joffraud, J.J. (2015). Selection of bioprotective cultures for preventing cold-smoked salmon spoilage. Int. J. Food Microbiol. *213*, 79–87. https://doi.org/10.1016/j.ijfoodmicro.2015.05.005

Lim, H.I., Lee, J., Jang, J.Y., Park, H.W., Choi, H.J., Kim, T.W., Kang, M.R., and Lee, J.H. (2014). Draft genome sequence of *Lactobacillus sakei* strain Wikim 22, isolated from kimchi in Chungcheong Province, South Korea. Genome Announc. *2*, e01296–14. https://doi.org/10.1128/genomeA.01296-14

Lucquin, I., Zagorec, M., Champomier-Vergès, M., and Chaillou, S. (2012). Fingerprint of lactic acid bacteria population in beef carpaccio is influenced by storage process and seasonal changes. Food Microbiol. *29*, 187–196. https://doi.org/10.1016/j.fm.2011.08.001

Lyhs, U., and Björkroth, J.K. (2008). *Lactobacillus sakei/curvatus* is the prevailing lactic acid bacterium group in spoiled maatjes herring. Food Microbiol. *25*, 529–533. https://doi.org/10.1016/j.fm.2007.11.007

Lyhs, U., Korkeala, H., and Björkroth, J. (2002). Identification of lactic acid bacteria from spoiled, vacuum-packaged 'gravad' rainbow trout using ribotyping. Int. J. Food Microbiol. 72, 147–153. https://doi.org/10.1016/S0168-1605(01)00634-1

Marceau, A., Zagorec, M., Chaillou, S., Méra, T., and Champomier-Vergès, M.C. (2004). Evidence for involvement of at least six proteins in adaptation of *Lactobacillus sakei* to cold temperatures and addition of NaCl. Appl. Environ. Microbiol. *70*, 7260–7268. https://doi.org/10.1128/AEM.70.12.7260-7268.2004

Mathiesen, G., Huehne, K., Kroeckel, L., Axelsson, L., and Eijsink, V.G. (2005). Characterization of a new bacteriocin operon in sakacin P-producing *Lactobacillus sakei*, showing strong translational coupling between the bacteriocin and immunity genes. Appl. Environ. Microbiol. *71*, 3565–3574. https://doi.org/10.1128/AEM.71.7.3565-3574.2005

McLeod, A., Snipen, L., Naterstad, K., and Axelsson, L. (2011). Global transcriptome response in *Lactobacillus sakei* during growth on ribose. BMC Microbiol. *11*, 145. https://doi.org/10.1186/1471-2180-11-145

McLeod, A., Zagorec, M., Champomier-Vergès, M.C., Naterstad, K., and Axelsson, L. (2010). Primary metabolism in *Lactobacillus sakei* food isolates by proteomic analysis. BMC Microbiol. *10*, 120. https://doi.org/10.1186/1471-2180-10-120

Møretrø, T., Aasen, I.M., Storrø, I., and Axelsson, L. (2000). Production of sakacin P by *Lactobacillus sakei* in a completely defined medium. J. Appl. Microbiol. *88*, 536–545. https://doi.org/10.1046/j.1365-2672.2000.00994.x

Møretrø, T., Naterstad, K., Wang, E., Aasen, I.M., Chaillou, S., Zagorec, M., and Axelsson, L. (2005). Sakacin P non-producing *Lactobacillus sakei* strains contain homologues of the sakacin P gene cluster. Res. Microbiol. *156*, 949–960. https://doi.org/10.1016/j.resmic.2005.05.005

Nakano, K., Shiroma, A., Tamotsu, H., Ohki, S., Shimoji, M., Ashimine, N., Shinzato, M., Minami, M., Nakanishi, T., Teruya, K., *et al.* (2016). First complete genome sequence of the skin-improving *Lactobacillus curvatus* strain FBA2, isolated from fermented vegetables, determined by PacBio single-molecule real-time technology. Genome Announc. *4*, e00884–16. https://doi.org/10.1128/genomeA.00884-16

Nyquist, O.L., McLeod, A., Brede, D.A., Snipen, L., Aakra, Å., and Nes, I.F. (2011). Comparative genomics of *Lactobacillus sakei* with emphasis on strains from meat. Mol. Genet. Genomics *285*, 297–311. https://doi.org/10.1007/s00438-011-0608-1

Richter, M., and Rosselló-Móra, R. (2009). Shifting the genomic gold standard for the prokaryotic species definition. Proc. Natl. Acad. Sci. U.S.A. *106*, 19126–19131. https://doi.org/10.1073/pnas.0906412106

Rimaux, T., Rivière, A., Illeghems, K., Weckx, S., De Vuyst, L., and Leroy, F. (2012). Expression of the arginine deiminase pathway genes in *Lactobacillus sakei* is strain dependent and is affected by the environmental pH. Appl. Environ. Microbiol. *78*, 4874–4883. https://doi.org/10.1128/AEM.07724-11

Rimaux, T., Vrancken, G., Vuylsteke, B., De Vuyst, L., and Leroy, F. (2011). The pentose moiety of adenosine and inosine is an important energy source for the fermented-meat starter culture *Lactobacillus sakei* CTC 494. Appl. Environ. Microbiol. 77, 6539–6550. https://doi.org/10.1128/AEM.00498-11

Salvetti, E., Torriani, S., and Felis, G.E. (2012). The genus *Lactobacillus*: a taxonomic update. Probiotics Antimicrob. Proteins *4*, 217–226. https://doi.org/10.1007/s12602-012-9117-8

Samapundo, S., Ampofo-Asiama, J., Anthierens, T., Xhaferi, R., Van Bree, I., Szczepaniak, S., Goemaere, O., Steen, L., Dhooge, M., Paelinck, H., *et al.* (2010). Influence of NaCl reduction and replacement on the growth of *Lactobacillus sakei* in broth, cooked ham and white sauce. Int. J. Food Microbiol. *143*, 9–16. https://doi.org/10.1016/j.ijfoodmicro.2010.07.014

Sanz, Y., and Toldrá, F. (2002). Purification and characterization of an arginine aminopeptidase from *Lactobacillus sakei*. Appl. Environ. Microbiol. *68*, 1980–1987. https://doi.org/10.1128/AEM.68.4.1980-1987.2002

Schillinger, U., and Lücke, F.K. (1989). Antibacterial activity of *Lactobacillus sake* isolated from meat. Appl. Environ. Microbiol. *55*, 1901–1906.

Siezen, R., Boekhorst, J., Muscariello, L., Molenaar, D., Renckens, B., and Kleerebezem, M. (2006). *Lactobacillus plantarum* gene clusters encoding putative cell-surface protein complexes for carbohydrate utilization are conserved in specific Gram-positive bacteria. BMC Genomics *7*, 126. https://doi.org/10.1186/1471-2164-7-126

Stentz, R., and Zagorec, M. (1999). Ribose utilization in *Lactobacillus sakei*: analysis of the regulation of the *rbs* operon and putative involvement of a new transporter. J. Mol. Microbiol. Biotechnol. *1*, 165–173.

Stentz, R., Lauret, R., Ehrlich, S.D., Morel-Deville, F., and Zagorec, M. (1997). Molecular cloning and analysis of the *ptsHI* operon in *Lactobacillus sakei*. Appl. Environ. Microbiol. *63*, 2111–2116.

Sun, Z., Harris, H.M., McCann, A., Guo, C., Argimón, S., Zhang, W., Yang, X., Jeffery, I.B., Cooney, J.C., Kagawa, T.F., *et al.* (2015). Expanding the biotechnology potential of lactobacilli through comparative genomics of 213 strains and associated genera. Nat. Commun. *6*, 8322. https://doi.org/10.1038/ncomms9322

Teran, L.C., Coeuret, G., Raya, R., Champomier-Vergès, M.C., and Chaillou, S. (2017). Draft genome sequence of *Lactobacillus curvatus* FLEC03, a meat-borne isolate from beef carpaccio packaged in a modified atmosphere. Genome Announc *5*, e00584-17. https://doi.org/10.1128/genomeA.00584-17

Tichaczek, P.S., Vogel, R.F., and Hammes, W.P. (1993). Cloning and sequencing of *curA* encoding curvacin A, the bacteriocin produced by *Lactobacillus curvatus* LTH1174. Arch. Microbiol. *160*, 279–283. https://doi.org/10.1007/BF00292077

Torriani, S., Van Reenen, G.A., Klein, G., Reuter, G., Dellaglio, F., and Dicks, L.M. (1996). *Lactobacillus curvatus* subsp. *curvatus* subsp. nov. and *Lactobacillus curvatus* subsp. *melibiosus* subsp. nov. and *Lactobacillus sake* subsp. *sake* subsp. nov. and *Lactobacillus sake* subsp. *carnosus* subsp. nov., new subspecies of *Lactobacillus curvatus* Abo-Elnaga and Kandler 1965 and *Lactobacillus sake* Katagiri, Kitahara, and Fukami 1934 (Klein *et al.* 1996, emended descriptions), respectively. Int. J. Syst. Bacteriol. *46*, 1158–1163. https://doi.org/10.1099/00207713-46-4-1158

Urso, R., Rantsiou, K., Cantoni, C., Comi, G., and Cocolin, L. (2006). Sequencing and expression analysis of the sakacin P bacteriocin produced by a *Lactobacillus sakei* strain isolated from naturally fermented sausages. Appl. Microbiol. Biotechnol. *71*, 480–485. https://doi.org/10.1007/s00253-005-0172-x

Vogel, R.F., Lohmann, M., Nguyen, M., Weller, A.N., and Hammes, W.P. (1993). Molecular characterization of *Lactobacillus curvatus* and *Lact. sake* isolated from sauerkraut and their application in sausage fermentations. J. Appl. Bacteriol. *74*, 295–300. https://doi.org/10.1111/j.1365-2672.1993.tb03029.x

Vidal-Carou, M.C., Latorre-Moratalla, M.L., Veciana-Nogués, M.T., and Bover-Cid S. (2007). Biogenic amines: Risks and control. In Handbook of Fermented Meat and Poultry, Toldrá, F., Hui, Y.H., Astiasarán, I., Nip, W.-K., Sebranek, J.G., Silveira, E.-T.F., Stahnke, L.H., and Talon, R., eds. (Blackwell Publishing, UK), pp. 455–468. https://doi.org/10.1002/9781118522653.ch47

Woraprayote, W., Malila, Y., Sorapukdee, S., Swetwiwathana, A., Benjakul, S., and Visessanguan, W. (2016). Bacteriocins from lactic acid bacteria and their applications in meat and meat products. Meat Sci. *120*, 118–132. https://doi.org/10.1016/j.meatsci.2016.04.004

Zagorec, M., and Champomier-Vergès, M.C. (2017). *Lactobacillus sakei*: a starter for sausage fermentation, a protective culture for meat products. Microorganisms *5*, E56. https://doi.org/10.3390/microorganisms5030056

Zheng, J., Ruan, L., Sun, M., and Gänzle, M. (2015). A genomic view of Lactobacilli and Pediococci demonstrates that phylogeny matches ecology and physiology. Appl. Environ. Microbiol. *81*, 7233–7243. https://doi.org/10.1128/AEM.02116-15

Zúñiga, M., Champomier-Vergès, M., Zagorec, M., and Pérez-Martínez, G. (1998). Structural and functional analysis of the gene cluster encoding the enzymes of the arginine deiminase pathway of *Lactobacillus sake*. J. Bacteriol. *180*, 4154–4159.

Complex Oligosaccharide Utilization Pathways in *Lactobacillus*

3

Manuel Zúñiga*, María Jesús Yebra and Vicente Monedero

Instituto de Agroquímica y Tecnología de Alimentos (IATA-CSIC), Valencia, Spain.

*Correspondence: manolo@iata.csic.es

https://doi.org/10.21775/9781910190890.03

Abstract

Lactobacillus is the bacterial genus that contains the highest number of characterized probiotics. Lactobacilli in general can utilize a great variety of carbohydrates. This characteristic is an essential trait for their survival in highly competitive environments such as the gastrointestinal tract of animals. In particular, the ability of some strains to utilize complex carbohydrates such as milk oligosaccharides as well as their precursor monosaccharides, confer upon lactobacilli a competitive advantage. For this reason, many of these carbohydrates are considered as prebiotics. Genome sequencing of many lactobacilli strains has revealed a great variety of genes involved in the metabolism of carbohydrates and some of them have already been characterized. In this chapter, the current knowledge at the biochemical and genetic levels of the catabolic pathways of complex carbohydrates utilized by lactobacilli will be summarized.

Introduction

All animals establish symbiotic associations with microbes. Paramount among these is the establishment of complex microbial communities in the gastrointestinal tract. These communities may comprise thousands of species and are particular to each individual (Eckburg *et al.*, 2005; Ley *et al.*, 2006; Turnbaugh *et al.*, 2009). The composition and distribution of microbiota changes along the gastrointestinal tract reflecting the differing physicochemical conditions and epithelial surfaces that the microbes find in the different compartments of the gastrointestinal tract (Gu *et al.*, 2013; Kim and Isaacson, 2015; Stearns *et al.*, 2011; Tropini *et al.*, 2017). For example, Lactobacillaceae is predominantly found in the stomach and small intestine of mice whereas anaerobes such as Bacteroidaceae, Prevotellaceae, Rikenellaceae, Lachnospiraceae and Ruminococcaceae are mainly found in the large intestine and faeces (Gu *et al.*, 2013). In the oral cavity of humans, a neutral pH and aerobic conditions prevail although anaerobic niches are also present. The environment in the stomach is acidic and microaerophilic whereas pH increases and oxygen availability decreases along the small intestine and colon. Nutrient availability and immune effectors also vary along the gastrointestinal tract. Furthermore, microbes are not uniformly distributed along the transverse axis of the gut (Tropini *et al.*, 2017). Species such as *Akkermansia muciniphila* and some *Bacteroides* spp. are thought to be predominantly associated with the mucus layer whereas closer to the mucosa, aerotolerant taxa such as Proteobacteria and Actinobacteria are more abundant due to the radial oxygen gradient established across the intestinal wall (Albenberg *et al.*, 2014).

Numerous studies have demonstrated that gut microbiota has a great influence on manifold host physiology aspects, from metabolism (Li *et al.*, 2008) to behaviour (Ezenwa *et al.*, 2012). Perturbation of the gut microbiota may markedly affect the host health (Claesson *et al.*, 2012; Cho and Blaser, 2012) and has been related to a number of diseases such as metabolic disorders (Sonnenburg and Bäckhed, 2016), inflammatory diseases (Blander *et al.*, 2017), diabetes (Membrez *et al.*, 2008; Vaarala *et al.*, 2008), coeliac disease (Collado *et al.*, 2007), etc. Although the microbiota associated with an individual may display some resilience to perturbation (Lozupone *et al.*, 2012; Sommer *et al.*, 2017), external environmental factors such as food intake (Kohl *et al.*, 2014) and composition (David *et al.*, 2014) can alter it substantially. This aspect is pivotal in the relationship between microbiota and host nutrition and health as the gastrointestinal microbiota provide important metabolic capabilities, including the ability to obtain energy from indigestible dietary polysaccharides (Goh and Klaenhammer, 2015). The human genome encodes only 17 glycosidases, enabling the utilization of a very limited set of polysaccharides (Cantarel *et al.*, 2012; Goh and Klaenhammer, 2015). Therefore, most complex carbohydrates are available to gut microbes able to utilize these compounds.

Carbohydrates, as oligo- and polysaccharides or as glycoconjugates constitute an amazingly diverse group of molecules due to the structural diversity and numerous bonding sites of their constituent monosaccharides that allow their assembly among themselves or to almost any other organic molecule in a wide array of architectures. Because of this structural versatility, carbohydrates fulfil a wide variety of functions in organisms as structural polymers, energy reserve, signalling, etc. Furthermore, as a major component of human diet, carbohydrates also have a determining influence on the interactions between host and associated microbiota (Hooper *et al.*, 2002). In addition to dietary carbohydrates, host-derived glycans constitute a secondary source of carbohydrates for gut microbiota (Hooper *et al.*, 2002).

Although the gut microbiota taken as a whole can utilize a wide variety of glycans, cells belonging to individual taxa can usually metabolize a limited set of them. From this fact stems the concept of prebiotic, defined as a non-digestible food ingredient that beneficially affects the host by selectively stimulating the growth and/or activity of one or a limited number of bacteria in the colon, and thus improves host health (Gibson and Roberfroid, 1995). Prebiotics were first thought of as a means to selectively enrich probiotic gut microbes, specifically *Lactobacillus* and *Bifidobacterium* (Gibson and Roberfroid, 1995; Goh and Klaenhammer, 2015). Although not limited by the definition, all prebiotics are glycans. The definition of prebiotic has been revised and the specific stimulatory effect of prebiotics on lactobacilli and bifidobacteria has been challenged (Hutkins *et al.*, 2016). Notwithstanding, the original idea gave a strong boost to the study of the glycan catabolic pathways of these organisms.

Lactobacillus is a large genus currently comprising over 200 species that have been isolated from a wide variety of habitats. They are Gram-positive, microaerophilic or anaerobic obligate fermentative organisms that produce lactic acid as the major end product of sugar fermentation. Together with genera *Paralactobacillus*, *Pediococcus* and *Sharpea* they constitute the family Lactobacillaceae within the order Lactobacillales. *Lactobacillus* strains play a major role in the production of a wide variety of fermented products. Others are naturally associated with mucosal surfaces of humans and animals and have been considered as probiotics (Tannock, 2004). Either as foodstuff fermenters or as probiotics, their capacity to utilize glycans is an important trait for their performance. This chapter focuses on our current knowledge on the pathways of complex glycan dissimilation identified in species of *Lactobacillus*.

Fructan and fructooligosaccharide catabolic pathways

Structural characteristics of fructans

Fructans are linear or branched fructose polymers and can be broadly divided into inulins (β-2,1-linked) and levans (β-2,6-linked) (Fuchs, 1991) (Fig. 3.1). Fructans are usually synthesized from sucrose by repeated fructosyl transfer so that they have a terminal glucose unit. Levans are produced by many microorganisms, including some lactobacilli (Bello *et al.*, 2001; Tieking *et al.*, 2003; Van Geel-Schutten *et al.*, 1999), and a few plant species (Öner *et al.*, 2016). In contrast, inulins are relatively common in plants, especially Asteraceae, but only a few bacterial species produce them, among them some *Lactobacillus* and *Leuconostoc* species (Anwar *et al.*, 2008; Olivares-Illana *et al.*, 2003; van Hijum *et al.*, 2002). While bacterial

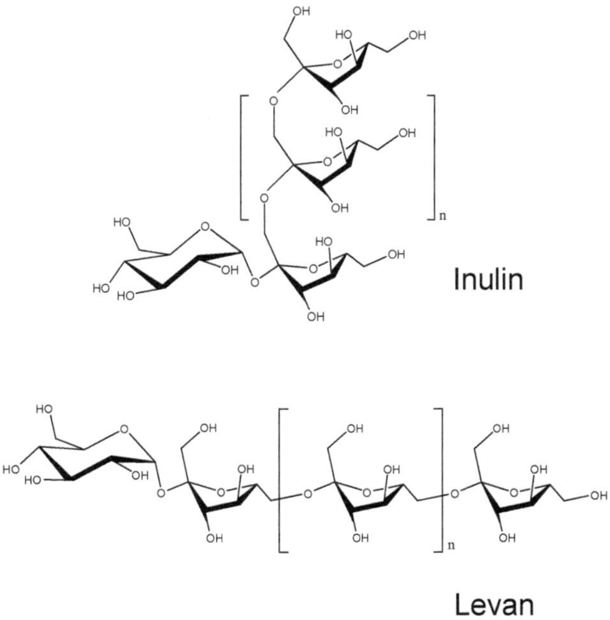

Inulin

Levan

Figure 3.1 Chemical structures of inulin and levan.

fructans have a very high degree of polymerization (DP) up to 10^5 fructose units, the DP of plant-derived fructans does generally not exceed DP 100. High DP of bacterial fructans is possibly related to their function as structural components of biofilms, but they also constitute an extracellular nutrient reservoir (Öner *et al.*, 2016). In plants, fructans serve essentially as reserve carbohydrates.

Synthesis of fructans is catalysed in bacteria by fructosyltransferases, named inulosucrases (EC 2.4.1.9) when they synthesize inulin, and levansucrases (EC 2.4.1.10) when they produce levan. Levansucrases can also form β-2,1-fructosyl-fructose linkages either to synthesize inulin-type fructooligosaccharides (FOS) or to create branching points that connect the β-2,6-linked chains of the polymer (Öner *et al.*, 2016). In contrast to bacteria, the biosynthesis of fructans in plants is catalysed by three different classes of enzymes: sucrose:sucrose 1-fructosyltransferase (EC 2.4.1.99) (1-SST), fructan:fructan 1-fructosyltransferase (EC 2.4.1.100) (1-FFT) and fructan exohydrolase (EC.3.2.1.153) (1-FEH) (van Arkel *et al.*, 2013). 1-SST primarily catalyses the synthesis of the trisaccharide 1-kestose, from two molecules of sucrose. In this reaction glucose is formed in equimolar amounts to 1-kestose (β-D-fructofuranosyl-(2→1)-β-D-fructofuranosyl-(2→1)-α-D-glucopyranoside). 1-FFT catalyses the transfer of fructosyl units from 1-kestose and any other fructan molecule onto 1-kestose and higher DP fructan molecules. 1-FEH, catalyses the degradation of inulin by hydrolysing terminal fructosyl units, which results in the formation of fructose and lower DP inulin (van Arkel *et al.*, 2013).

Metabolic pathways for fructans utilization

The interest on fructans in relation to the intestinal microbiota stemmed from the search of carbohydrate sources that reached the large intestine and were selectively used by beneficial microbes such as bifidobacteria. Several studies noted that fructooligosaccharides derived from inulin hydrolysis (usually referred to as FOSs) were not hydrolysed by the host endogenous enzymes but efficiently used by bifidobacteria (Hidaka *et al.*, 1986; Yazawa and Tamura, 1982). FOS naturally occur in many kinds of plants but they are commercially produced from the hydrolysis of inulin or synthesized from sucrose by transfructosylation by β-fructofuranosidases or β-D-fructosyltransferases (Goh and Klaenhammer, 2015). The most utilized natural source of inulins is chicory and depending on the method of extraction the product obtained may be almost exclusively FOS of the GF_n type (a glucose monomer linked α-1,2 to two or more β-2,1-linked fructosyl units) or a mixture of GF_n and F_m type (two or more β-2,1-linked fructosyl units) oligomers (Roberfroid *et al.*, 1998).

However, the first studies that reported the utilization of fructans by lactobacilli were focused on the usage of lactic acid bacteria for ensilage and showed that some lactobacilli could utilize fructans for growth (Kleeberger and Kühbauch, 1976). In a later study, it was shown that 16 strains out of 712 were able to degrade levan and eight of them could also degrade inulin (Müller and Lier, 1994). Subsequent studies confirmed the ability of some lactobacilli to degrade fructans (Merry *et al.*, 1995; Müller and Steller, 1995; Winters *et al.*, 1998) although the enzymes and pathways involved were not determined. Other studies also established that some lactobacilli could grow on FOSs as carbon sources (Kaplan and Hutkins, 2000; Sghir *et al.*, 1998). The characterization of an extracellular fructan hydrolase from *Lactobacillus paracasei* subsp. *paracasei* P 4134 provided the first clues on the fructan degradative pathways of lactobacilli (Müller and Seyfarth, 1997). The enzyme hydrolysed β-2,6-linked fructans more rapidly than β-2,1 linked fructans and the main product of hydrolysis was fructose, suggesting that the enzyme is an exofructanase (Müller and Seyfarth, 1997). Later studies on other *L. casei/paracasei* have confirmed these conclusions (Kuzuwa *et al.*, 2012; Velikova *et al.*, 2017). Subsequently, another extracellular fructanhydrolase purified from *Lactobacillus pentosus* B235 was characterized (Paludan-Müller *et al.*, 2002). The purified enzyme had the highest activity for levan, but also hydrolysed garlic extract, a β-2,1-linked fructan with β-2,6-linked fructosyl sidechains, 1,1,1-kestose, 1,1-kestose, 1-kestose, inulin and sucrose at 60, 45, 39, 12, 9 and 3%, respectively, of the activity observed for levan (Paludan-Müller *et al.*, 2002).

Genome sequencing and transcriptomic analyses paved the way for the identification and characterization of genes involved in the utilization of fructans. The studies carried out so far have characterized three different fructan utilization pathways in *Lactobacillus acidophilus, Lactobacillus casei* and *Lactobacillus plantarum* that differ in their transport systems and fructofuranosidase enzymes (Fig. 3.2).

In 2003, Barrangou *et al.* (2003) identified an operon (*msm*) involved in FOS utilization by *Lactobacillus acidophilus* NCFM. The *msm* operon consisted of genes encoding for a transcriptional regulator of the LacI family (*msmR*), an ABC transport system (*msmEFGK*), a fructosidase (*bfrA*) and a sucrose phosphorylase (*gtfA*; Fig. 3.3) (Barrangou *et al.*, 2003). Similar operons can be detected in a limited number of strains of other species of lactobacilli (Fig. 3.3). In *Lactobacillus delbrueckii, Lactobacillus perolens, Lactobacillus saniviri* and *Lactobacillus concavus* strains, a gene encoding a putative fructokinase is associated with the *msm*

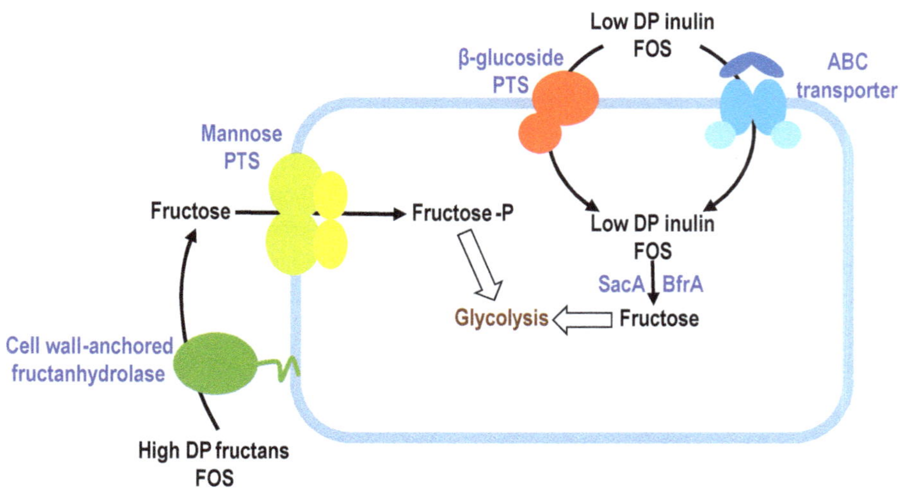

Figure 3.2 Fructan utilization pathways characterized in LAB.

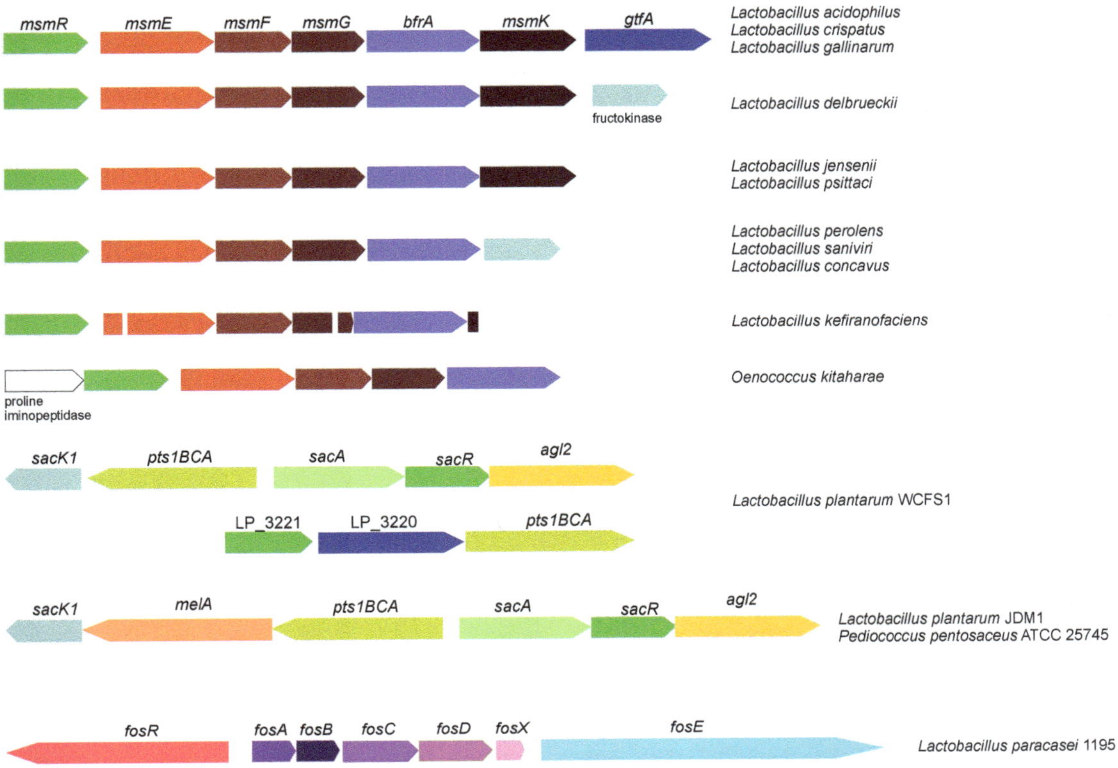

Figure 3.3 Schematic representation of fructan utilization gene clusters present in selected lactobacilli. Colours indicate homologous genes. Arrows indicate directionality. Interrupted arrows indicate translational frameshifts.

cluster (Fig. 3.3). The transcriptional analysis of the *msm* operon of *L. acidophilus* NCFM showed that all genes were transcribed in a single transcriptional unit (Barrangou *et al.*, 2003). Sucrose and oligofructose (both GF_n and F_n types) induced the expression of the operon whereas glucose and fructose did not. The expression of the operon was repressed by glucose suggesting that it is subjected to carbon catabolite repression (CCR). This hypothesis was further supported by the presence of several CRE-like sites in the *msm* promoter region. The functionality of the operon was shown by inactivation *msmE* and *bfrA* that resulted in defective growth on FOS-F_n (Barrangou *et al.*, 2003).

A comparative analysis of *L. ruminis* strains of human and bovine origin led to the identification of an operon possibly involved in FOS utilization consisting of a β-fructan hydrolase and an oligosaccharide H^+ symporter (O'Donnell *et al.*, 2011) although experimental evidence is still lacking.

In *Lactobacillus plantarum* WCFS1, a gene cluster consisting of a putative fructokinase (*sacK1*), a putative phosphoenolpyruvate-dependent phosphotransferase transport system (PTS) of the β-glucoside family (*pts1BCA*), a β-fructofuranosidase (*sacA*), a LacI family transcriptional regulator (*sacR*) and a putative α-glucosidase (*agl2*) (Fig. 3.3), was induced when this strain was grown in the presence of low molecular weight FOS (Saulnier *et al.*, 2007). The biochemical characterization of the *L. plantarum* ST-III SacA demonstrated that the enzyme has exofructofuranosidase activity with preference for β-2,1 linkages between two fructose moieties in fructans with low DP (Chen *et al.*, 2014). The heterologous expression of SacA in *Lactobacillus rhamnosus* GG, an organism that can utilize fructose but not FOS, enabled this strain to grow on FOSs, thus demonstrating the functional role of this enzyme (Chen *et al.*, 2014). Furthermore, inactivation of *sacA* in *L. plantarum* ST-III severely impaired the growth of this strain on FOSs (Chen *et al.*, 2015). In contrast, a mutant defective in *pts1BCA* still could grow with FOSs although at a lower growth rate than the wild-type strain (Chen *et al.*, 2015). A transcriptomic analysis of this strain had detected a second putative *sac* (*sacPTS26*) gene cluster constituted by a β-glucoside PTS (PTS26), an α-glucosidase (Agl4), and a transcriptional regulator (SacR2), that was also induced in the presence of FOSs (Chen *et al.*, 2015). A double mutant *pts1BCA/pts26* was unable to grow on FOS, indicating that both transporters are required for optimal FOS uptake and utilization (Chen *et al.*, 2015). This second *sac* cluster is also present in *L. plantarum* WCFS1 (Fig. 3.3) but it was not detected as up-regulated in the presence of FOS in this strain (Saulnier *et al.*, 2007) and its involvement in FOS utilization in WCFS1 remain to be determined (Chen *et al.*, 2015). The presence of α-glucosidase-encoding genes in *sac* clusters is also intriguing. The functional role of these genes remains to be established.

As mentioned above, *L. paracasei* utilizes an extracellular fructanhydrolase. Goh *et al.* (2006) determined that the genes required for FOS utilization by *L. paracasei* 1195 are organized in a cluster (*fosRABCDXE*) encoding a putative mannose family PTS transporter (*fosABCDX*), a β-fructosidase (*fosE*) and, divergently transcribed, a transcriptional antiterminator (*fosR*) (Fig. 3.3). Homologous clusters are found in other *L. casei/paracasei* but considerable variability is observed. For example, *L. paracasei* ATCC 334 *fosE* homologue (LSEI_0564) lacks the C-terminal part including the cell wall anchor motif whereas this gene is absent in strain BL23 (Fig. 3.3). This variability may account for the conflicting observations of extracellular or cell wall anchored fructanhydrolase activity in different *L. paracasei* strains. Inactivation of *fosE* led to the loss of the ability to grow on sucrose, FOS, oligofructose (FF_n type), inulin and levan, thus demonstrating the functionality of the operon (Goh *et al.*, 2006). Furthermore, introduction of *fosE* into *Lactobacillus rhamnosus* GG enabled this strain to utilize FOS (Goh *et al.*, 2007). The analysis of the FosE encoding sequence revealed an N-terminal signal peptide sequence and an LPQAG cell wall anchor motif at the C-terminal region, suggesting its localization at the cell wall. Cell fractionation assays confirmed this hypothesis as FOS hydrolysis activity was present exclusively in the cell wall extract of *L. paracasei* previously grown on FOS (Goh *et al.*, 2007). In agreement with a previous biochemical characterization of a fructanhydrolase of *L. paracasei* (Müller and Seyfarth, 1997), the analysis of the degradation products of *L. paracasei* 1195 FosE indicated that it is an exofructanhydrolase (Goh *et al.*, 2007). The transcriptional regulation of the *fos* cluster has also been studied. Expression of *fos* genes is induced in the presence of FOS, inulin and, to a lesser extent, sucrose and fructose, but repressed by glucose (Goh *et al.*, 2007; Goh *et al.*, 2006). A CRE sequence is present in the *lev* promoter region, suggesting that the operon is subjected to CCR via the P-Ser-HPr/CcpA complex (Goh *et al.*, 2006).

The role of FosR has not been addressed in *L. paracasei* 1195; however, the functional role of the homologous LevR of strain BL23 has been studied (Mazé *et al.*, 2004). The intergenic regions of *fosR–fosA* and *levR–levA* in strain BL23 differ only in a single nucleotide and the FosABCD proteins shared more than 99% identity with their BL23 counterparts (Goh *et al.*, 2006). LevR is homologous to the *Bacillus subtilis* LevR transcriptional regulator, which controls the expression of a mannose-class PTS transporter and a levanase involved in levan utilization. *B. subtilis* LevR interacts with the σ^{54} factor and its activity is modulated via phosphorylation by P-His-HPr and P-$EIIB^{Lev}$ (Martin-Verstraete *et al.*, 1998). In contrast, BL23 strain LevR do not require σ^{54} although the regulation of its activity by phosphorylation still occurs by dual PTS-catalysed phosphorylation at conserved histidine residues in the EIIA and PRD2 domains of LevR by P-His-HPr and P-His-$EIIB^{Lev}$, respectively (Mazé *et al.*, 2004). When the PTS^{Lev} transporter is active, P-His-$EIIB^{Lev}$ preferably donates its phosphoryl group to the transported sugar, leading to dephosphorylation of LevR at His-776 by P-His-$EIIB^{Lev}$ and LevR activation and thereby induction of the *lev* PTS. On the other hand, when metabolically preferred PTS sugars, such as glucose, are present, the phosphoryl group of P-His-HPr is used for sugar phosphorylation. Poor phosphorylation at His-488 by P-His-HPr renders LevR less active and down regulates expression of the *lev* PTS.

The different strategies of fructan utilization by lactobacilli possibly determine their abilities to utilize different fructans. Internalization and subsequent hydrolysis possibly limits the ability of fructan utilization to low DP oligosaccharides whereas extracellular degradation would enable the utilization of high DP fructans. Experimental evidence available supports this view. Makras *et al.* (2005) assayed the capacity of ten strains of lactobacilli to degrade inulin-type fructans, observing that *L. acidophilus* only degraded oligofructose whereas *L. paracasei* could also degrade long-chain inulin. *L. plantarum* could utilize short-chain fructooligosaccharides but grew poorly with FOS (Saulnier *et al.*, 2007). It has been proposed that the different strategies of fructan utilization in lactobacilli may respond to different ecological strategies. Internalization and subsequent degradation eliminates cross-feeding while conferring an advantage in a nutrient-competitive environment. On the other hand, extracellular degradation let other organisms profit of the hydrolytic products thereby allowing the establishment of symbiotic relationships with other members of the community (Goh and Klaenhammer, 2015).

Metabolism of glucans and glucooligosaccharides

Structural characteristics of glucans

Lactobacillus species are equipped with the enzymatic machinery to utilize multiple glucan structures, which consist of glucose homopolymers with different linkages and branching types. However, the presence of these capacities is species- and strain-specific and usually linked to particular niche adaptations. Glucans can be classified into α- and β-glucans depending on the type of glycosidic bond present in the molecules (Fig. 3.4). Starch is the main example of α-glucan and it represents the major carbon storage polysaccharide in plants. It is made of linear glucose chains with α-1,4 linkages with high DP (amylose) and shorter chains which in addition to α-1,4 bonds possess around 5% of α-1,6 branching of 18 to 25 glucoses (amylopectin). Other α-glucans can be linear or branched and carry diverse bonds (α-1,2; α-1,3) in addition to α-1,4 and α-1,6. Most of these glucans are produced by bacteria and fungi, such as dextrans (α-1,6 with α-1,3 branching; Fig. 3.4) produced by strains of lactobacilli such as *Leuconostoc mesenteroides* (Chen *et al.*, 2016), *Weisella cibaria* (Malang *et al.*, 2015) or *Lactobacillus sakei* (Nácher-Vázquez *et al.*, 2017), reuteran (α-1,4 with α-1,6 branching), synthesized by *Lactobacillus reuteri* (Chen *et al.*, 2016) or pullulan (maltotriose units linked by α-1,6 bonds; Fig. 3.5), produced by *Aureobasidium pullulans* (Cheng *et al.*, 2011). Isomaltooligosaccharides (IMO) are a type of α-glucans that are gaining interest due to their prebiotic effects (Ketabi *et al.*, 2011; Leemhuis *et al.*, 2014; Yen *et al.*, 2011). They are short α-1,6-linked glucans (e.g. isomaltose [α-D-glucopyranosyl-(1→6)-D-glucopyranoside], panose [α-D-glucopyranosyl-(1→6)-α-D-glucopyranosyl-(1→4)-D-glucopyranoside], isomaltotriose, isomaltotetraose and isomaltopentaose) that are present in some foods or can be commercially prepared from starch. Their prebiotic effect derives from the fact that humans and other monogastric animals generally lack IMO-degrading enzymes. Glycogen is an α-glucan equivalent to starch but present in animals and it is characterized by being more extensively branched. Many bacteria, including lactobacilli (Goh and Klaenhammer, 2013), can synthesize and utilize this molecule as carbon storage.

Cellulose is the most abundant β-glucan present in nature. It is a key component of plant cell walls so that it accounts for the major proportion of fixed carbon in living organisms. Plant as well as fungal cell walls may contain other β-glucans with β-1,3 or multiple alternating β-1,3 and β-1,4 linkages and β-1,6 branching linkages. Similar to α-glucans, some β-glucans can be produced by bacteria, including lactobacilli. Examples of these are the β-glucans produced by *Lactobacillus suebicus* (β-1,3 linked) (Garai-Ibabe *et al.*, 2010) and *L. brevis* (Fraunhofer *et al.*, 2017), where the glycosyl transferases catalysing the synthetic process have been characterized.

α-Glucan metabolic pathways in *Lactobacillus*

The α-glucans are the substrates of a wide variety of hydrolytic enzymes. The α-amylases cleave α-1,4 linkages at any part of the polymers, whereas β-amylases act at the non-reducing ends liberating maltose [α-D-glucopyranosyl-(1→4)-D-glucopyranoside]. Pullulanases and amylopullulanases degrade α-1,6 linkages and α-1,6 as well as α-1,4 linkages, respectively. Among pullulanases, isopullulanases and neopullulanases can be distinguished because they produce isopanose or panose from pullulan, respectively (Fig. 3.5).

Lactobacilli usually encode multiple α-glycosidases in their genomes, mainly from the glycosyl hydrolase (GH) 13 family (CAZy, Carbohydrate Active enZYmes classification; www.cazy.org), the family to which α-amylases belong,

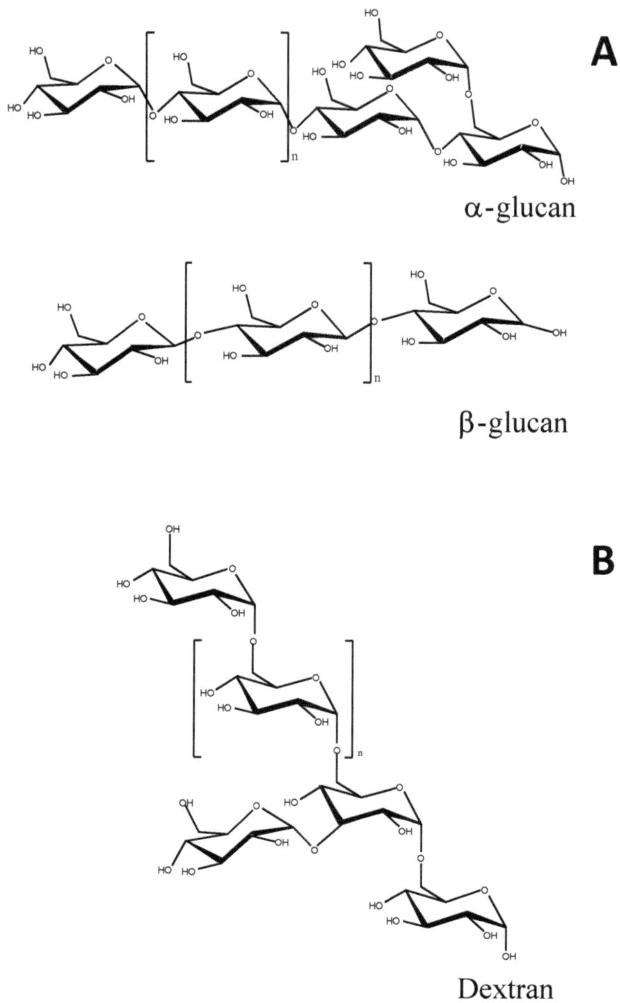

Figure 3.4 A. Chemical structures of α- and β-glucans. An α-1,6 ramification is represented in the structure of α-glucan. B. Chemical structure of dextran. An α-1,3 ramification is represented.

although their specificities remain to be investigated in most cases. Among α-glucans, starch degradation by lactobacilli is relatively well characterized. The ability to degrade starch was noticed after the isolation and phenotypic characterization of new *Lactobacillus* strains from waste maize fermentations such as *Lactobacillus amylophilus* (Nakamura and Crowell, 1979) and *Lactobacillus amylovorus* (Nakamura, 1981). However, extracellular amylases are not common in this genus (less than 2% of all the GH13 glycosyl hydrolases present in lactobacilli). They are mainly concentrated in five species: *L. acidophilus, L. amylovorus, Lactobacillus fermentum, L. plantarum* and *Lactobacillus manihotivorans* (Petrova *et al.*, 2013), although some particular strains belonging to different species can also degrade starch (e.g. *Lactobacillus paracasei* B41 (Petrova and Petrov, 2012)). In addition to amylases, pullulanases and amylopullulanases have been found in lactobacilli. A thermostable pullulanase (endo-α-1,6-glucosidase, GH13_14 subfamily) encoded by the LBA_1710 gene from *L. acidophilus* NCFM has been characterized. This enzyme preferentially acts on β-limit dextrins (amylopectins digested by β-amylases) over amylopectin (Møller *et al.*, 2017). The products of LBA_1710 can also act on the linear polymer pullulan [(maltotriose-α1,6-maltotriose)$_n$] and possess a very low K_m and very high specific activity for this polysaccharide. Notwithstanding, pullulan and amylopectin do not support growth of this strain. This would reflect the lack of α-amylase activity and

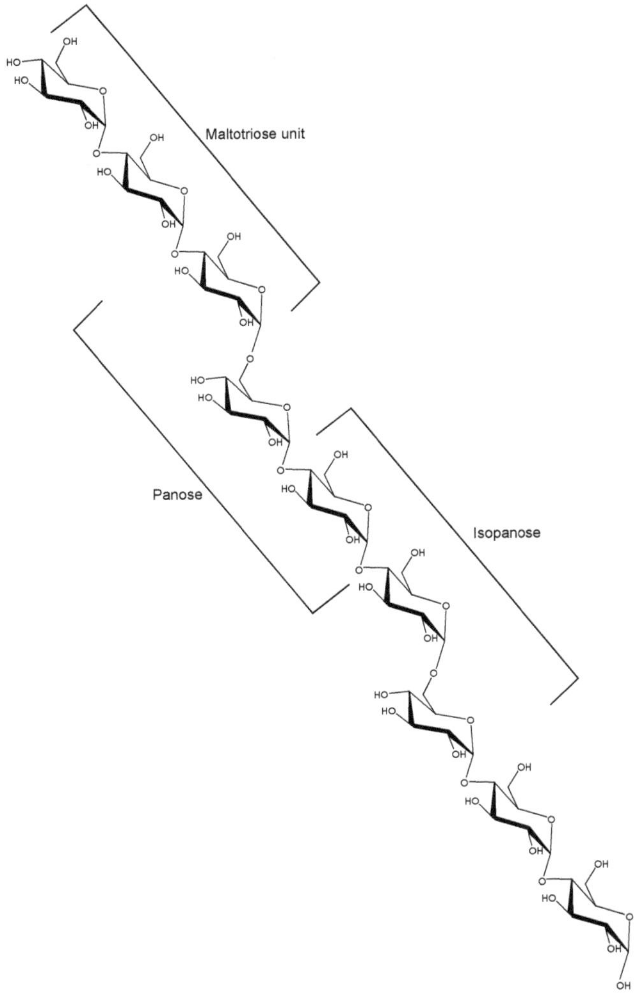

Figure 3.5 Chemical structure of pullulan. Maltotriose, isopanose and panose moieties are indicated.

a non-efficient transport system for the resulting oligosaccharides after pullulanase digestion in this strain. The enzyme is secreted due to presence of a signal peptide and possesses two starch-binding modules (CBM48), a catalytic domain (GH13_14 subfamily) and an additional surface-layer associated domain (SLAP) which warrants its retention at the cell surface. Homologues of this enzyme are present in many lactobacilli from intestinal origin. The enzyme may act on short-branched α-glucans derived from the degradation of dietary starch and glycogen by the human enzymes, supporting that these debranching enzymes play a role in the adaptation of these lactobacilli to the gut niche (Møller *et al.*, 2017). An extracellular amylopullulanase from *L. plantarum* L137, an isolate from a traditional fermented food containing fish and rice with high hydrolytic activity towards starch, has also been characterized. The enzyme degrades soluble starch to maltotriose and maltotreaose whereas it produces only maltotriose from pullulan (Kim *et al.*, 2008). The encoding gene (*apuA*) is located in the endogenous plasmid pLTK13 (Kim *et al.*, 2008). The enzyme contains a number of amino acid repeats at the N- and C-terminus that are derived from the same repeated DNA sequence (5′-ACCGACGCAGCCAACTCA-3′) but translated in different frames. The C-terminal repeats are similar to mucin-binding domains present in bacterial peptidoglycan-bound proteins and most probably participate in substrate binding (Kim *et al.*, 2008). Domains with amino acid repeats are typically found in carbohydrate-degrading enzymes. The amylases from *L. amylovorus*, *L.*

plantarum and *L. manihotivorans* carry a C-terminal starch-binding domain (SBD) of almost 500 amino acids consisting of tandem repeat units of 91 amino acids in variable numbers (Morlon-Guyot *et al.*, 2001). SBDs promote attachment to the substrate and allow degradation of non-soluble starch (Rodriguez Sanoja *et al.*, 2000). A neopullulanase has been cloned and characterized from *Lactobacillus mucosae* LM1 (Balolong *et al.*, 2016). A homologous gene had been previously shown to be induced in the amylolytic *L. plantarum* A6 strain during pearl millet fermentation and neopullulanase activity detected (Humblot *et al.*, 2014).

A number enzymes involved in the degradation of IMOs encoded by lactobacilli have also been characterized. A glucan-α-1,6-glucosidase (GH13_31 subfamily) encoded by the *L. acidophilus* NCFM gene LBA_0264 is involved in the catabolism of IMO (Møller *et al.*, 2012). The participation of glucan-α1,6-glucosidase and maltose phosphorylase (see below) have also been implicated in the catabolism of IMO in *L. brevis* (Hu *et al.*, 2013). The *L. acidophilus* NCFM enzyme is induced after growth in IMO mixtures, displays a high activity on panose and prefers IMO with more than two glucoses, as it acts preferentially on isomaltotriose and isomaltotetraose compared to isomaltose. The LBA_0264 genes and its homologues in other lactobacilli are not clustered with other sugar catabolism or sugar transporter genes except for *Lactobacillus johnsonii* ATCC 33200 and *Lactobacillus gasseri* JV-V03, which belong to the *acidophilus* group and where the IMO-utilizing gene forms part of the maltodextrin operon (Møller *et al.*, 2012), and in other *Lactobacillus* species such as *Lactobacillus casei* (Monedero et al., 2008). Another GH13_13 enzyme has been characterized in *L. plantarum* LL441. It displays activity on isomaltose and isomaltulose [α-D-glucopyranosyl-(1→6)-D-fructofuranoside], but not on panose or isomaltotriose, and its gene is clustered with genes encoding EIIABCD components of a mannose-class PTS (Delgado *et al.*, 2017). Whether the enzyme might be acting on phosphorylated disaccharides needs to be proven.

Metabolism of maltodextrins

Utilization of maltodextrins (linear oligosaccharides derived from starch hydrolysis) is a characteristic more extended in lactobacilli. Species adapted to starch-rich environments (e.g. plant material fermentations where the endogenous plant amylases release maltose and maltodextrins) such as *Lactobacillus sanfranciscensis*, participating in sourdough fermentations during bread making, are particularly efficient in maltose and maltodextrin utilization. The enzymatic machinery for the utilization of these carbohydrates in lactobacilli soon attracted attention and it has been thoroughly investigated. In the 90´s maltose catabolic proteins were partially characterized in *L. sanfranciscensis* DSM20451 (Ehrmann and Vogel, 1998) and *Lactobacillus brevis* ATCC 8287 (Hüwel *et al.*, 1997). These microorganisms relied on an intracellular maltose phosphorylase (MapA [EC 2.4.1.8]) non requiring pyridoxal 5′-phosphate and belonging to the GH65 family, which also includes trehalose phosphorylases (EC 2.4.1.64), and kojibiose phosphorylases (EC 2.4.1.230), for catalysing a phosphorolysis of the disaccharide that involves an inversion of the anomeric configuration of the C-1 atom, giving β-glucose 1-phosphate and glucose. β-glucose 1-phosphate is further converted to glucose 6-phosphate for its incorporation into glycolysis via β-phosphoglucomutase (Pgm), whose gene is co-localized in the chromosome with *mapA* (Fig. 3.6). MapA displays a high specificity for maltose, but it is not active on maltodextrins such as maltotriose or maltotetraose and it cannot phosphorolyse disaccharides with other α-linkage configurations such as isomaltose (α-1,6), nigerose (α-1,3), kojibiose (α-1,2) or trehalose (α-1,1). Furthermore, it differs in sequence from other maltose (maltodextrin) phosphorylases such as that of *E. coli* (GT35 family). MapA is a dimer in solution and its structure has been solved for the *L. brevis* enzyme (Egloff *et al.*, 2001). The structure consists of a β-sandwich domain linked to an $(\alpha/\alpha)_6$ barrel catalytic domain, and a C-terminal β-sheet domain. The $(\alpha/\alpha)_6$ barrel domain displays striking structural and functional similarities to the catalytic domain of a glucoamylase from *Aspergillus awamori*, suggesting evolution from a common ancestor (Egloff *et al.*, 2001). In the structures of these enzymes the catalytic conserved Glu residue from the glucoamylase superposes onto a conserved Glu of MapA that likely acts as the acid catalytic residue that promotes the nucleophilic attack of phosphate on the glycosidic bond. Modelling of the *L. acidophilus* MapA protein with different substrates

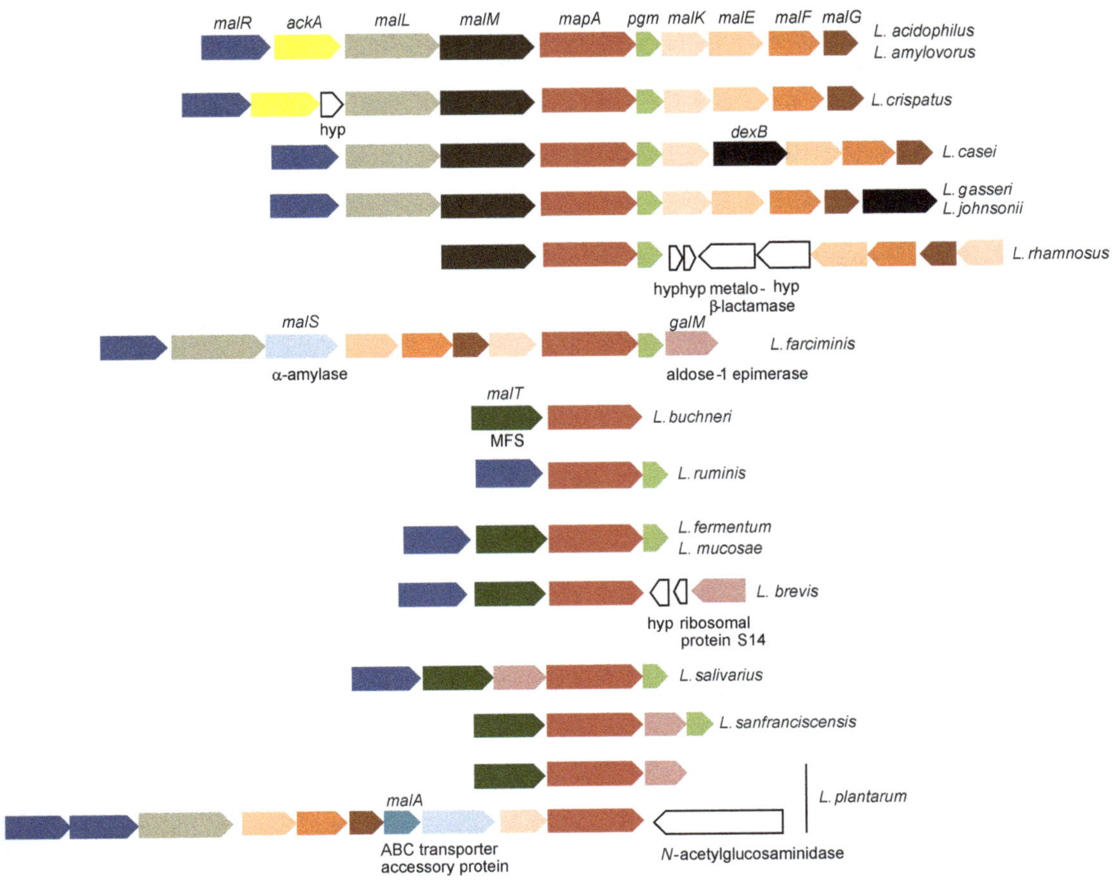

Figure 3.6 Comparison of maltose/maltodextrin operons in *Lactobacillus* species. The genetic organization of the operons encoding the enzymes and transporters for maltose/maltodextrin utilization in selected lactobacilli is depicted. The maltose/maltodextrin locus in *Lactobacillus* species carry genes for a specific ABC transporter, maltose phosphorylase and β-phosphoglucomutase together with several α-glycosidases or a MFS permease, maltose phosphorylase, β-phosphoglucomutase and, generally, an aldose 1-epimerase. In *L. plantarum* strains both types of *mal* operons are simultaneously present. Colours indicate homologous genes. Arrows indicate directionality.

allowed understanding why this enzyme does not accommodate maltotriose or maltotetraose in its active site (Nakai *et al.*, 2009). The protein adopts a configuration where loop His413–Glu421 between α3 and α4 of the $(\alpha/\alpha)_6$ barrel domain blocks the binding of longer malto-oligosaccharides. However, this enzyme has been applied in reverse phosphorolysis reactions for the synthesis of α-1,4-linked disaccharides with β-glucose 1-phosphate as donor and glucose, glucosamine, *N*-acetyl glucosamine, L-fucose, mannose or xylose as acceptors; being unable to use other sugars with axial hydroxyls at C-3 and C-4 positions or disaccharides/trisaccharides (Nakai *et al.*, 2009).

A genome survey of 38 *Lactobacillus* strains revealed that the presence of a set of genes for maltodextrins utilization, including *mapA* and *pgm*, is widespread in lactobacilli. These gene clusters usually include different α-glycosidases, although the presence of amylases is scarce (Gänzle and Follador, 2012). The maltodextrin operon which contain the *mapA* and *pgm* genes have been genetically characterized in *L. acidophilus* (Nakai *et al.*, 2009) and *L. casei* (Monedero *et al.*, 2008), showing that in these species maltodextrin transport is carried out by an ABC transporter ($MalEFGK_2$) homologous to that of the well-studied maltose/maltodextrin transporter of *E. coli* (Fig. 3.6). In *L. casei* BL23, ten *mal* genes are clustered and cotranscribed in a single mRNA whose expression is regulated by MalR, a transcriptional regulator of the LacI/GalR family and repressed by the presence of glucose via the general CCR mechanism mediated by the global regulator CcpA. According to studies conducted with *Streptococcus pneumoniae* MalR, this regulator acts as a transcriptional repressor and maltose strongly inhibits its DNA-binding capacity (Puyet *et al.*, 1993). *Cis*-acting sequences recognized by MalR can be identified adjacent to the −10 and −35 promoters of the *L. acidophilus* and *Lactococcus lactis mal* operons (Nakai *et al.*, 2009). In addition to the maltose phosphorylase and phosphoglucomutase genes, three α-glucosidase-encoding genes (GH13) are clustered together with the *L. casei* maltose-catabolic genes: *malL*, encoding a putative oligo-α1,6-glucosidase (putatively acting on short IMO like isomaltose); *malM*, encoding a maltogenic α-amylase (cuts α-1,4 linkages from dextrins yielding maltose) and *dexB*, coding for a second α-1,6 glucosidase (Monedero *et al.*, 2008). This last enzyme belongs to the GH13_31 subfamily (glucan-α-1,6-glucosidase) and, as described before, based on the studies of its *L. acidophilus* counterpart is able to degrade IMO. Therefore, it is postulated that in addition to maltodextrins, the ABC transporter $MalEFGK_2$ would be able to transport IMO. A second operon located in opposite direction encodes an additional ABC transporter with high homology to $MalEFGK_2$ but its function is unknown and mutations in their genes have no phenotypic effects on maltose or maltotriose growth. In other *L. casei* strains, such as ATCC334, maltodextrin utilization is impaired by a large deletion in the *mal* cluster (Monedero *et al.*, 2008).

The *mal* operon of *L. acidophilus* NCFM consists of nine genes and a *malR* regulator (Nakai *et al.*, 2009). In addition to the maltodextrin ABC transporter, MapA and Pgm, *L. acidophilus* encodes two glycosidases: a maltogenic α-amylase (MalM) and an oligo α-1,6-glucosidase (MalL) that are homologous to their *L. casei* counterparts. The *L. acidophilus mal* operon also carries an acetate kinase gene (*ackA*) involved in pyruvate metabolism during glycolysis. Although ABC transporters can be identified as the main maltodextrin transport systems in lactobacilli, in some species that efficiently use maltose as a carbon source a genetic association of a gene encoding a permease of the major facilitator superfamily (MFS) is found with *mapA* and *pgm* genes (Fig. 3.6). This suggests that these species (e.g. *L. sanfranciscensis*, *L. salivarius*, *L. brevis*) make use of a maltose-H^+ symport system for the uptake of the disaccharide. In contrast, no PTS systems for the transport of maltose similar to those described in *Bacillus subtilis* have been identified in lactobacilli. Studies on the maltose uptake in *L. sanfranciscensis* LTH2581, a strain which only ferments maltose and glucose, confirmed the presence of a maltose-H^+ symport system. When maltose is taken up by this strain the intracellularly generated glucose exceeds the metabolic capacity of the cells, which results in glucose expulsion through a glucose uniport system (Neubauer *et al.*, 1994). In lactobacilli utilizing maltose through a MFS permease, a genetic association of the *mal* cluster with a gene encoding an aldose 1-epimerase (EC 5.1.3.3) can be found (Fig. 3.6). This enzyme is involved in the anomeric conversion of D-glucose between the α and β forms, which possibly speeds up the entry of glucose into the glycolytic pathway via its phosphorylation by glucokinase. Strains of *L. plantarum* are remarkable by the fact that they carry two *mapA* and *pgm* genes. One couple is linked to an ABC transporter ($MalEFGK_2$) and α-glucosidases (*mal* cluster 1), whereas the other forms a cluster (*mal* cluster 2) with a MFS permease (Fig. 3.6). Therefore, strains of this species possess the capacity to use maltose and maltodextrins by using two separated sets of genes. Unlike the rest of lactobacilli, the *L. plantarum mal* cluster 1 contains two transcriptional regulators (genes Lp_0172 and Lp_0173) with homology to MalR (Muscariello *et al.*, 2011). Expression of *malE* from this cluster is induced by maltose and repressed by glucose via CcpA. Mutational analysis suggested that only the product of Lp_0173 participated in *malE* regulation. A mutant in this gene showed a glucose-insensitive expression of *malE* together with a lack of induction by maltose (Muscariello *et al.*, 2011). Remarkably, an *in silico* approach for the study of LacI-GalR transcriptional regulators in *L. plantarum* WCFS1 identified five operons putatively controlled by the products of Lp_0172 and Lp_0173 which include the *mal*1 and *mal*2 clusters, an operon for β-glucosides utilization, an operon carrying the genes for teichoic acid synthesis *tagB1* and *tagB2* and an amino acid permease (Francke *et al.*, 2008).

Transport studies with ^{14}C-maltose has revealed an unusually high K_m for the *L. casei* maltose ABC transporter (around 0.3 mM; K_m for ABC transporters and their substrates are usually in the μM range), which suggested that maltose is not the preferred substrate and points to maltodextrins as the natural oligosaccharides taken up by this transporter (Monedero *et al.*, 2008). This notion is further substantiated by the fact that the three glycosidase enzymes encoded by the *mal* cluster are intracellular. Therefore, in this microorganism maltodextrins are preferentially metabolized over maltose, and they are hydrolysed in the cytoplasm to render maltose and glucose. This characteristic is probably shared by the rest of lactobacilli harbouring maltodextrin clusters with ABC transporters. *In vitro* studies on the binding capacity of the solute-binding component of the maltodextrin ABC transporter from *L. casei* (MalE) revealed that it is able to interact with maltotriose, maltotetraose, maltopentaose and with α, β and γ-cyclodextrins, which carry six, seven and eight glucose molecules, respectively, with K_d values that were in the μM range, albeit showing a preference for linear maltodextrins over cyclodextrins (Homburg *et al.*, 2017). However, contrarily to *E. coli* MalE, the solute-binding component of the *L. casei* maltose system does not interact with maltose. The fact that mutants in the *L. casei malK* gene are not able to ferment maltose (Monedero *et al.*, 2008) suggests that other solute-binding proteins may be responsible to recognize and deliver maltose intracellularly via the MalFG permease component of the ABC system. Alternatively, it cannot be excluded that the MalK ATPase component of the transporter can be shared by an as yet unidentified and incomplete maltose-specific ABC system lacking a cognate ATPase unit (Homburg *et al.*, 2017). Crystallographic data of *L. casei* MalE complexed with maltotriose, maltotetraose and cyclodextrins provided structural clues for its lack of interaction with maltose (Homburg *et al.*, 2017). The globular carbohydrate-binding protein MalE consists of two N- and C-terminal domains which form a ligand-binding pocket situated between them, a characteristic shared by other MalE homologues. However, in *L. casei* MalE three aromatic residues from the C-terminal domain (W234, Y164 and W353) stack against a specific glucose moiety of the bound substrate and create three distinct subpockets, where the position of three glycosidic moieties is fixed with additional hydrogen bonds from the N-terminal MalE domain. Thus, the disaccharide maltose cannot adequately accommodate the three subpockets, preventing MalE to adopt the closed conformation that is achieved after interaction with linear and cyclic dextrins. This observation is also confirmed by the fact that, contrarily to maltotetraose, maltose does not stimulate ATPase activity of the MalEFGK$_2$ complex (Homburg *et al.*, 2017).

Transcriptomic data under laboratory or natural fermentation conditions has shed some light on the regulation of the expression of glucans/starch utilizing enzymes and their concerted action during the degradation of these carbohydrates by lactobacilli. Experiments with *L. acidophilus* NCFM show that this strain induces preferentially the expression of PTS transport systems in the presence of prebiotic glucans such as cellobiose [β-D-glucopyranosyl-(1→4)-D-glucopyranoside], isomaltose, panose or gentiobiose [β-D-glucopyranosyl-(1→6)-D-glucopyranoside], whereas polydextrose (a synthetic glucose polymer consisting of a mixture of different α-glycosidic linkages) induces ABC transporters (Andersen *et al.*, 2012). Transport via the PTS results in intracellular phosphorylated sugars that can be cleaved by different phosphoglucosidases. As expected, MapA was also induced by polydextrose. Of note, MalH, a isomaltose 6-phosphate hydrolase (GH4) encoded by LBA_1689 was induced by isomaltose, isomaltulose, panose and polydextrose. This enzyme participates in the formation of glucose 6-phosphate and glucose from isomaltose 6-phosphate but also glucose 6-phosphate plus fructose from isomaltulose internalized via a PTS encoded by the LBA_0606-LBA_0609 locus (Andersen *et al.*, 2012). MalH has been found in lactobacilli associated with the gastrointestinal tract and it could be a good indicator of prebiotic activity of α-1,6 glucosides such as panose and polydextrose that can in addition be degraded by the activity of the product of LBA_0264, a glucan-α-1,6-glucosidase (GH13_31) which is induced by IMO (Andersen *et al.*, 2012). Expression of genes involved in starch metabolism in *L. plantarum* A6 has been studied during a natural fermentation of pearl millet porridge (Humblot *et al.*, 2014). This highly amylolytic strain expresses α-amylases (intracellular and extracellular), α-glucosidase, neopullulanase, amylopectin phosphorylase and MapA when growing in this natural substrate. The ability of this strain to liquefy the pearl millet gruel compared to other *L. plantarum* strains that do not grow in this substrate is attributable to the presence of the extracellular α-amylase (*amyA*). Metatranscriptomic analyses during the spontaneous fermentation by natural microbial consortia of different sourdoughs (wheat and spelt) with back-slopping for ten days have been carried out (Weckx *et al.*, 2011). By using a DNA microarray carrying genes from several lactic acid bacteria, a high expression of glycolytic enzymes was observed during these fermentations. However, the expression of *mapA* and maltose/maltodextrins ABC transporter encoding genes was low in both sourdough types and it corresponded mainly to the *L. plantarum* and *Lactococcus lactis* genes, respectively; even although the microarray also carried genes from *L. sakei*, *L. curvatus*, *L. brevis*, and *L. fermentum* (Weckx *et al.*, 2011). These expression levels were related to the concentration of maltose throughout the back-slopping process. The sourdough microbiota was capable of degrading other carbohydrates important in sourdough (e.g. saccharose and fructose) and their metabolism could cause CCR of the utilization of other carbon sources mediated by CcpA, whose gene is highly expressed during sourdough fermentations (Weckx *et al.*, 2011).

Metabolism of glycogen

Lactobacilli carry in their genomes gene clusters for the biosynthesis of the storage polysaccharide glycogen. In some lactobacilli that dwell in specific mucosal niches, such as the vagina, glycogen metabolism has been associated with their proliferation (Miller *et al.*, 2016). Glycogen metabolic genes have been characterized in *L. acidophilus* and it is postulated that they play a role in the persistence of this bacterium in the gut (Goh and Klaenhammer, 2013). The bacterial glycogen synthesis starts by the synthesis of ADP-glucose from glucose 1-phosphate by GlgC or GlgD enzymes. Then the glycogen synthase (GlgA) transfers glucose from ADP-glucose to a chain of α-1,4-glucan, whereas GlgB is involved in the formation of the α-1,6 branching points. For its catabolism, glycogen phosphorylase (GlgP) catalyses the breakdown of α-1,4 linkages and GlgX participates in the debranching at the α-1,6 bonds in dextrins that cannot be further processed by GlgP. In *L. acidophilus* NCFM the glycogen cluster encompasses 11.7 kb and carry *glgBCDAP* together with two genes coding for a α-amylase and β-phosphoglucomutase (Goh and Klaenhammer, 2013). Similar genetic structures are found in approximately one third of the sequenced *Lactobacillus*, being mainly present in strains associated with the gastrointestinal tract of mammals and other animals. As an example, although the cluster is present in human intestinal isolates of *L. bulgaricus* and *L. helveticus*, it is not found in dairy isolates of these species (Goh and Klaenhammer, 2013). Expression of the *L. acidophilus glg* genes depends on the carbon source and the growth phase, showing maximal expression with raffinose [α-D-galactopyranosyl-(1→6)-α-D-glucopyranosyl-(1→2)-β-D-fructofuranoside] and repression by glucose. As the enzymes for the synthesis and degradation are produced in parallel, the regulation of intracellular glycogen levels depending on the carbon sources may rely on the carbon fluxes. Mutants impaired in *glgA* or *glgB* present a reduced growth on raffinose and a mutant in *glgB* and in the gene encoding the catabolic glycogen phosphorylase (*glgP*) grows slower in MRS medium (containing glucose) and are less resistant to simulated gastrointestinal conditions (Goh and Klaenhammer, 2013). This suggests that the accumulation of α-glucan polymers with sequestered glucose that cannot be further metabolized results in impaired growth. This highlights the need for co-ordinated glycogen synthesis and degradation for retrieving glucose from glycogen storage under normal and stress conditions. *In vivo* experiments in a germ-free mice model in which wild-type *L. acidophilus* and its *glgA* mutant were delivered to animals by intragastric gavage showed that the wild-type strain was able to compete and to displace the *glgA* mutant in monocolonized mice. This demonstrated the role of glycogen synthesis and utilization in lactobacilli in competitive fitness in the gut (Goh and Klaenhammer, 2014).

β-Glucan metabolic pathways in *Lactobacillus*

Despite the fact that lactobacilli are usually associated with the microbiota of plant decaying material, cellulases (endo-β-1,4-D-glucanases) have not been described for this genus and in general, for the lactic acid bacteria. Notwithstanding, the prebiotic activity of some β-glucans has been established but the metabolism of these polymers by lactobacilli has been rarely demonstrated. Very few examples of growth stimulation effects of β-glucans in lactobacilli are found in the literature, and they have mainly proved by using β-glucan hydrolysates (Dong *et al.*, 2017).

Important efforts have been made to use lactobacilli for the conversion of lignocellulosic material by applying saccharolytic processes prior fermentation or by the use of engineered strains expressing β-D-glucanases from other microbial sources (Moraïs *et al.*, 2014; Okano *et al.*, 2010; Overbeck *et al.*, 2016). Similarly, lactobacilli expressing β-glucanases from other sources have been engineered for fermentation or health promoting effects (Liu *et al.*, 2005; Wang *et al.*, 2014a).

Metabolism of xylooligosaccharides

Structural characteristics of xylooligosaccharides

Xylooligosaccharides (XOS) are plant-derived oligosaccharides with β-1,4 linkages between xylose (a pentose) monomers that can be decorated with residues of the pentose arabinose. These residues can be linked by α-1,2 or α-1,3 bonds to xylose molecules along the chain (arabinoxylan oligosaccharides) where one or two arabinose residues can be found per xylose. These polymers are abundant in plant cell walls and together with other heteropolysaccharides form the hemicellulose component in plants.

Xylooligosaccharides metabolic pathways in *Lactobacillus*

In vitro growth assays and studies of the microbiota of humans concluded that these polysaccharides possess a prebiotic effect that stimulates the growth of bifidobacteria (Childs *et al.*, 2014; Lin *et al.*, 2016). This is supported by the characterization of multiple ABC transporters for XOS and XOS-degrading enzymes in species of *Bifidobacterium* (Ejby *et al.*, 2013). *In vitro* growth assays and human trials which explored changes in gut microbiota composition after XOS intake also pointed to XOS as prebiotic polysaccharides that stimulate growth of certain lactobacilli (Lin *et al.*, 2016). Notwithstanding, the capacity to ferment XOS by lactobacilli seems to be limited (Ananieva *et al.*, 2014). In accordance to this, the information about enzymes degrading XOS and arabinoxylans in lactobacilli is scarce. Two different enzymatic activities are needed to completely degrade arabinoxylans: arabinofuranosidase (liberating the arabinose residues that decorate the xylooligosaccharide backbone) and β-xylosidase (acting on the β-1,4 linkage between xylose molecules). Enzymes with this activity are classified into GH43 and GH51 glycosyl hydrolase families. *L. brevis* is thus far the only *Lactobacillus* species in which these activities have been studied (Michlmayr *et al.*, 2013; Michlmayr *et al.*, 2011). This species can be found in a wide variety of habitats including fermentations of hemicellulose-rich plant materials. Three GH43 β-xylosidases and two GH51 arabinofuranosidases have been found during the study of the genomes of several strains. The sequences of these enzymes show a high level of amino acid identity to enzymes from typical intestinal bacteria (e.g. bifidobacteria), suggesting events of horizontal gene transfer at the intestinal niche. The GH43 enzymes from *L. brevis* DSM 20054, annotated as β-xylosidases, have been thoroughly characterized (Michlmayr *et al.*, 2013). The β-xylosidase encoded by LVIS_0375 (*xynB1*) gene exhibited activity towards β-1,4-xylobiose and β-1,4-xylotriose. LVIS_2285 (*xynB2*) showed low activity with *p*-nitrophenyl-β-D-xylopyranoside and no activity with β-1,4-xylooligosaccharides, whereas the β-xylosidase encoded by LVIS_1748 (*abf3*) exhibited activity for α-1,5-arabinooligosaccharides. XynB1 and XynB2 are 32% identical and are also present in strains of *Lactobacillus buchneri*, *L. fermentum*, *Lactobacillus hilgardii*, *L. pentosus* and *L. reuteri*. These species belong to the group of heterofermentative lactobacilli so that it has been postulated that the capacity to degrade XOS and arabinoxylans is restricted to this particular group within lactobacilli (Michlmayr *et al.*, 2013). Unlike the arabinofuranosidases Abf1 and Abf2 characterized from *L. brevis* DSM 20054 (GH53) (Michlmayr *et al.*, 2011), Abf3 (and XynB1 and XynB2) cannot release arabinose from arabinoxylans with a composition of 65% α-1,3-linked arabinose, 8% α-1,2-linked arabinose and 26% doubly substituted xylose (two arabinose linkages per xylose monomer) indicating that these enzymes do not act as arabinofuranosidases. Furthermore, Abf1 and Abf2 are selective for α-1,3-linked arabinose residues of monosubstituted xylose (Michlmayr *et al.*, 2011).

The degradation of other hemicelluloses such as xyloglucan (a β-1,4 glucan backbone with xylose residues linked to glucose via α-1,6 bonds) by lactobacilli has received less attention. *L. pentosus* MD353, isolated from cucumber fermentation, carries a xylose operon (*xylAB*) involved in the metabolism of this pentose and encoding a xylose isomerase (*xylA*) and xylulose kinase (*xylB*) (Lokman *et al.*, 1991). These enzymes convert cytoplasmic D-xylose to D-xylulose 5-phosphate, an intermediate of the pentose phosphate pathway. Adjacent to this operon two genes are present, *xylPQ*, which are also induced by xylose via the xylose repressor XylR. XylQ is a GH31 α-xylosidase which has been demonstrated to act on isoprimeverose [α-D-xylopyranosyl-(1→6)-D-glucopyranoside] and it is also able to liberate with very low efficiency small amounts of xylose from xyloglucan oligosaccharides with different linkage configurations (Chaillou *et al.*, 1998). Isoprimeverose is usually released from xyloglucan by cellulolytic microorganisms producing endoglucanases and it can be taken up by *L. pentosus* via the product of *xylP*. This gene encodes a galactoside-pentoside-hexuronide family transporter that catalyses the transport of isoprimeverose, but not xylose, through a proton motive force-driven process (Heuberger *et al.*, 2001).

Metabolism of galactooligosaccharides

Structural characteristics of galactooligosaccharides

β-Galacto-oligosaccharides (GOS) are non-digestible carbohydrates usually composed of lactose at the reducing end and one to ten galactose units linked by β-1,3, β-1,4 or β-1,6 bonds (Macfarlane *et al.*, 2008). They can be acquired naturally through the diet from the degradation of galactan side chains of the rhamnogalacturonan I fraction of pectin (Jones *et al.*, 1997). In addition, they are also incorporated as prebiotics after their synthesis by the transgalactosylation activity of β-galactosidases on lactose, which acts both as donor and as acceptor of the galactose moiety (Vera *et al.*, 2016). Analysis of some GOS mixtures revealed the presence of oligosaccharides with galactose at the reducing end instead of glucose, and they may also contain disaccharides different from lactose that are considered GOS as well. Indeed, commercial GOS are typically mixed-length galactosylated compounds with a DP ranging from 2 to 12 (Coulier *et al.*, 2009). The type of linkage, mostly β-1,3, β-1,4 and/or β-1,6, and to a lesser extent, β-1,2, is determined by the enzyme source. Commercial enzymes used for GOS synthesis belong to the CAZy glycosyl hydrolase family 2 (GH2) and they are obtained from *Bifidobacterium bifidum* and *Bacillus circulans* for GOS with β-1,3 and β-1,4 bonds, respectively, and from *Kluyveromyces lactis* and *Aspergillus oryzae* to obtain β-1,6-linked GOS (Rodriguez-Colinas *et al.*, 2011). Most of the published studies use the term GOS when referring to β-GOS, but there are also GOS with α-configuration, which are produced by transgalactosylation reactions with α-galactosidases (Wang *et al.*, 2014b). The oligosaccharides 3′-, 4′- and 6′-galactosyllactose have been found in colostrum and human milk, and they constitute the only oligosaccharides contained in common between human milk oligosaccharides (HMOs) and GOS.

GOS are metabolized by specific bacteria in the gastrointestinal tract, and they have been found to modulate the gut microbiota by stimulation of beneficial bacteria such as bifidobacteria and lactobacilli, and inhibition of pathogenic bacteria (Macfarlane *et al.*, 2008; Rastall *et al.*, 2005). The fermentation of GOS in the gastrointestinal tract leads to an increased production of specific short-chain fatty acids (SCFA), which are known for their health benefits including reduction of the risk of developing cancer and intestinal disorders (Cardelle-Cobas *et al.*, 2009; Sangwan *et al.*, 2011).

GOS metabolic pathways in *Lactobacillus*

Lactobacillus species in general can efficiently utilize GOS, although their utilization is a strain-dependent character and it may also vary depending on GOS DP (Endo *et al.*, 2016; Thongaram *et al.*, 2017). A study on GOS utilization by different species of *Lactobacillus* showed that 9 out of 10 tested species metabolized the galactosyllactose fraction (35%) of a GOS mixture to a different degree (Endo *et al.*, 2016). All tested strains of *L. delbrueckii*, *L. plantarum*, *L. fermentum*, *L. reuteri*, *L. johnsoni* and *L. acidophilus* metabolized galactosyllactose whereas only some strains of *L. rhamnosus*,

L. paracasei and *L. sakei* did it. As with fructans, the availability of genomic sequences allowed the determination of the genetic basis of GOS metabolism by lactobacilli. In this way, transcriptomic analyses of *L. acidophilus* NCFM with whole-genome DNA microarrays, revealed that GOS induce the *lac–gal* gene cluster, which encodes a galactoside-pentose-hexuronide permease (LacS), two β-galactosidases belonging to the GH family 42 (LacA) and GH family 2 (LacLM), and enzymes of the Leloir Pathway (GalM, GalT, GalK and GalE) involved in the metabolism of galactose (Andersen *et al.*, 2011). Inactivation of LacS impaired growth on lactose, lactitol and GOS (Andersen *et al.*, 2011). Phylogenetic analysis showed that *lacS* is mainly found in human gut-associated *Lactobacillus* species, suggesting that transport and catabolism of those carbohydrates could be a significant energy source for lactobacilli in the gut (Andersen *et al.*, 2011). Possibly, GOS are transported into the cells by the LacS permease, hydrolysed by the β-galactosidases LacA and LacLM into galactose and glucose, which would be directed to the Leloir Pathway and glycolysis, respectively. Interestingly, the *lac–gal* cluster is also induced by bile acids (Barrangou *et al.*, 2006; Pfeiler *et al.*, 2007), suggesting that bile may act as a location signal in the gut environment where GOS and related carbohydrates would be readily available.

L. fermentum is able to utilize α-GOS as carbon source for growth in soymilk (LeBlanc *et al.*, 2004). This capability relies in the expression of the gene *melA*, that encodes an α-galactosidase with activity on α-1,6 linkages (Carrera-Silva *et al.*, 2006). α-GOS such as raffinose (α-D-galactopyranosyl-(1→6)-α-D-glucopyranosyl-(1→2)-β-D-fructofuranoside), and stachyose (α-D-galactopyranosyl-(1→6)-α-D-galactopyranosyl-(1→6)-α-D-glucopyranosyl-(1→2)-β-D-fructofuranoside), are abundant in vegetables and they are substrates for MelA. Other α-galactosidases that hydrolyse those α-galactosides were also identified in *L. plantarum* (Silvestroni *et al.*, 2002) and *L. reuteri* (Tzortzis *et al.*, 2003). The growth of *L. acidophilus* NCFM in the presence of stachyose induced a cluster of genes encoding an ATP-binding cassette (ABC) transporter, a GH36 α-galactosidase and enzymes from the Leloir Pathway. Inactivation of the α-galactosidase-encoding gene resulted in the loss of the ability to grow on raffinose, stachyose and the disaccharide melibiose (α-D-galactopyranosyl-(1→6)-D-glucopyranoside) (Andersen *et al.*, 2012). These results suggested that *L. acidophilus* NCFM transports α-galactosides into the cytoplasm via an ABC system and then, they are hydrolysed by the action of the GH36 α-galactosidase into galactose and sucrose or galactose and glucose.

Metabolism of human milk oligosaccharides (HMOs)

Structural characteristics and functional properties of HMOs

HMOs constitute a group of non-conjugated and structurally diverse carbohydrates that represent the third largest solid component of human milk after lactose and lipids (Kunz *et al.*, 2000; Thurl *et al.*, 2010). They consist of combinations of five monosaccharides: D-glucose, D-galactose, *N*-acetylglucosamine (GlcNAc), L-fucose and sialic acid (Sia). The only form of Sia found in human milk is *N*-acetylneuraminic acid (Neu5Ac), while oligosaccharides present in milk of other mammals may contain too *N*-glycolylneuraminic acid (Neu5Gc) (Urashima *et al.*, 2013). All HMOs contain a lactose unit [β-D-galactopyranosyl-(1→4)-D-glucopyranoside] at their reducing end, that can be further elongated by the addition of β-1,3-linked lacto-*N*-biose [LNB; β-D-galactopyranosyl-(1→3)-*N*-acetyl-D-glucosamine; type-1 core] and/or β1-3/6-linked N-acetyllactosamine (LacNAc; β-D-galactopyranosyl-(1→4)-N-acetyl-D-glucosamine; type-2 chain). The basic core structures can be modified by L-Fuc with an α1-2, α1-3 or α1-4 linkage and/or Sia with an α2-3 or α2-6 linkage. HMOs can be simple trisaccharides as 2'/3'-fucosyllactose (2'/3'FL) and 3'/6'-syalyllactose (6'SL) or complex oligosaccharides with several LNB and LacNAc repeat units (Bode, 2012). These disaccharides have also been found in free form in human milk (Balogh *et al.*, 2015). Currently, over a 100 structurally distinct HMOs have been identified, including neutral (non-sialylated) and acidic (sialylated) compounds (Kobata, 2010). Most of the neutral HMOs are fucosylated, with concentrations ranging from 50% to 80%, whereas sialylated HMOs ranged from 10% to 20%. In contrast, only about 1% of the oligosaccharides are fucosylated in bovine milk (Bode, 2012). The overall concentration of HMOs varies during lactation, colostrum contains about 20–25 g/l HMOs and this amount diminishes to approximately 5–20 g/l in mature milk (Bode, 2012; Thurl *et al.*, 2010). In addition, the HMOs composition and quantity is variable among mothers, since their synthesis is correlated with the Secretor and Lewis blood group characteristics, which depend on the expression of FUT2 and FUT3 fucosyltransferases, respectively, within the mammary gland (Kunz *et al.*, 2017; Thurl *et al.*, 2010).

Human milk also contains highly glycosylated proteins, including mucins, that have attached oligosaccharide moieties with structures that resemble those of the free HMOs (Liu and Newburg, 2013). Both HMOs and the glycan moieties of the proteins are synthesized by the same glycosyltransferases. *N*-glycans are linked to an asparagine residue through an GlcNAc, that is elongated by an additional GlcNAc residue through a β-1,4 linkage and three mannose residues. The GlcNAc residue linked to Asn can be modified via α-1,6-fucosylation and the mannose residues with other monosaccharides, including L-fucose and Sia, and it becomes a complex structure (Nwosu *et al.*, 2012). *O*-glycans usually contain an *N*-acetylgalactosamine (GalNAc) linked to a serine or threonine residue. In the type-1 sugar chain found in mucins, the GalNAc is extended with Gal, linked via a β-1,3 bond, forming the disaccharide galacto-*N*-biose [GNB; β-D-galactopyranosyl-(1→3)-*N*-acetyl-D-galactosamine] (Fig. 3.7). GNB is also present in glycosphingolipids and in bioactive sugar structures like the T-antigen disaccharide (Liu and Newburg, 2013; Moran *et al.*, 2011).

In the last years, many studies have suggested that HMOs act as anti-adhesins against pathogens. HMOs are structurally similar to host receptors for pathogens since they are synthesized by the same glycosyltransferases that synthesize cell surface glycoproteins and glycolipids. As soluble receptor analogues, HMOs can act as decoys protecting infants against infections. Some HMOs inhibit the attachment of norovirus and bacterial pathogens such as *Listeria monocytogenes* and pathogenic *E. coli* strains (Newburg *et al.*, 2005), *Campylobacter jejuni* (Ruiz-Palacios *et al.*, 2003), *Helicobacter pylori* (Mysore *et al.*, 1999) and parasites such as *Entamoeba histolytica* (Jantscher-Krenn *et al.*, 2012), explaining the fact that breast-fed infants are at lower risk to acquire *E. histolytica* infections than formula-fed infants (Islam *et al.*, 1988). Two disaccharides, α-L-fucosyl-(1→3)-*N*-acetyl-D-glucosamine (3FN) and α-L-fucosyl-(1→6)-*N*-acetyl-D-glucosamine (6FN) (Rodríguez-Díaz *et al.*, 2013) that form part of the structure of many HMOs, either free or

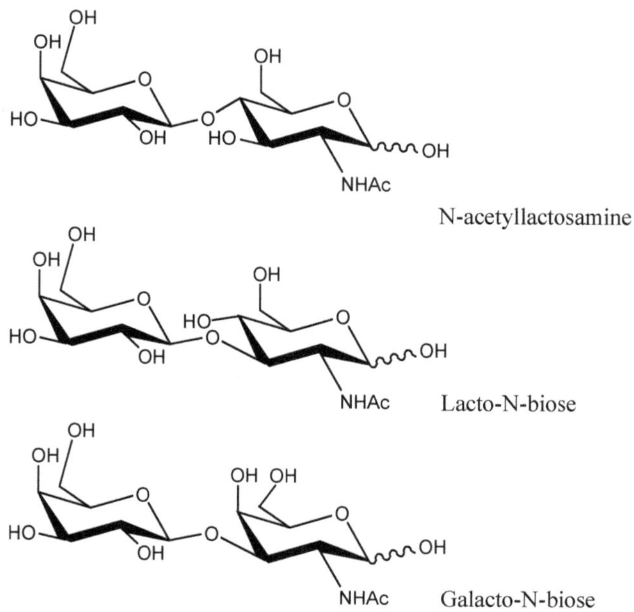

Figure 3.7 Chemical structures of *N*-acetyllactosamine, lacto-*N*-biose, and galacto-*N*-biose.

glycoconjugated to proteins, possess anti-adhesive properties against the enteropathogenic *E. coli* (O86). 6FN was also able to block the binding of the enteropathogenic *E. coli* (O127a) to HT29 cells (Becerra *et al.*, 2015a). Recently, it has been shown that HMOs function as antimicrobial and antibiofilm agents against *Streptococcus agalactiae*, an invasive pathogen in both children and adults (Ackerman *et al.*, 2017; Lin *et al.*, 2017). Interestingly, specific neutral HMOs directly inhibit the growth of this bacterium and a mutant impaired in a putative glycosyltransferase is resistant to those HMOs (Lin *et al.*, 2017).

Metabolic pathways for HMOs in *Lactobacillus*

Convincing evidence supports that HMOs favour the growth of beneficial bacteria present in the gastrointestinal tract of breastfed infants. HMOs were first identified as the prebiotic 'bifidus factor' described for human milk. Bifidobacteria constitute a considerable proportion of the intestinal microbiota of infants (Gomez-Llorente *et al.*, 2013), and they are highly adapted to use HMOs as a carbon source (Garrido *et al.*, 2013). Genome analyses have revealed that strains of *Bifidobacterium longum* subsp. *infantis, Bifidobacterium longum* subsp. *longum, Bifidobacterium breve* and *B. bifidum* encode a battery of enzymes involved in HMOs catabolism (Kwak *et al.*, 2016; LoCascio *et al.*, 2010). Unlike *Bifidobacterium*, species of the genera *Lactobacillus*, that are often isolated from breast-fed infant faeces (Albesharat *et al.*, 2011; Martín *et al.*, 2007; Rubio *et al.*, 2014), usually showed a limited capacity for HMOs utilization. The only exception is represented by members of the *L. casei/ paracasei/rhamnosus* group, which contain several genes encoding enzymes involved in the hydrolysis of fucosyl-oligosaccharides (Rodríguez-Díaz *et al.*, 2011) and in the metabolism of the type-1 (Bidart *et al.*, 2014) and type-2 core structures from HMOs (Bidart *et al.* unpublished) (Fig. 3.8). Three α-L-fucosidases (AlfA, AlfB and AlfC) encoded in the *L. casei* BL23 genome have been characterized and they were able to hydrolyse *in vitro* fucosylated HMOs (Rodríguez-Díaz *et al.*, 2011). All three enzymes are possibly intracellularly located as they lack secretion signals, suggesting that *L. casei* must transport the fucosylated substrates into the cytoplasm before their hydrolysis. This notion was demonstrated for the disaccharide 3FN that is transported into the cells by the mannose-class PTS encoded by the genes *alfEFG*, without being phosphorylated: L-fucose is a 6-deoxy-galactose and therefore lacks a phosphorylatable hydroxyl group at the carbon in the sixth position. These genes are divergently oriented from the gene cluster *alfBR*, encoding the α-L-fucosidase AlfB and the transcriptional repressor AlfR (Rodríguez-Díaz *et al.*, 2012). AlfB digested the disaccharide within the cells into L-fucose and GlcNAc. The latter is metabolized by *L. casei*, whereas the L-fucose moiety is excreted to the medium (Fig. 3.8) because, in contrast to *L. rhamnosus* GG (Becerra *et al.*, 2015b), *L. casei* lacks L-fucose catabolic genes.

The release of L-fucose and Sia from the non-reducing ends is the first step to degrade the HMOs core structures. *Bifidobacterium* spp. and *Bacteroides* spp. generally are good consumers of fucosylated and sialylated HMOs as they usually possess fucosidase and sialidase activities. The last activity is not a common feature among lactobacilli, although *L. delbrueckii* ATCC7830 showed sialidase activity when cultured in the present of 6′-sialyllactose (Yu *et al.*, 2013). A number of lactobacilli contain genes enabling the use of L-fucose or Sia moieties released from HMOs. *L. rhamnosus* GG can utilize L-fucose since it contains an operon encoding a specific catabolic pathway similar to that of *E. coli* (Becerra *et al.*, 2015b). *L. sakei* 23K contains two gene clusters, *nanTEAR* and *nanKMP* involved in the catabolism of Neu5Ac (Anba-Mondoloni *et al.*, 2013), and some strains of *L. plantarum, L. salivarius* (Almagro-Moreno and Boyd, 2009) and *L. paracasei* (Hammer *et al.*, 2017) also contain genes for Sia metabolism. Extracellular sialidase and fucosidase activities have not been described in those species; therefore, the utilization of these carbohydrates as well as the glycan core structures by lactobacilli probably depends on their release from HMOs and mucin by other members of the intestinal microbiota. This is the case of *L. casei*, which has a complete machinery to metabolize HMOs and *O*-glycans core structures such as LNB, GNB (Bidart *et al.*, 2014), LacNAc (Bidart *et al.*, unpublished) and lacto-*N*-triose II (LNTII; β-*N*-acetyl-D-glucosamine-(1→3)-β-D-galactopyranosyl-(1→4)-D-glucopyranoside) (Bidart *et al.*, 2016) (Fig. 3.8). LNB and GNB utilization relies on the *gnb* operon, which contains genes encoding a transcriptional repressor (*gnbR*), a

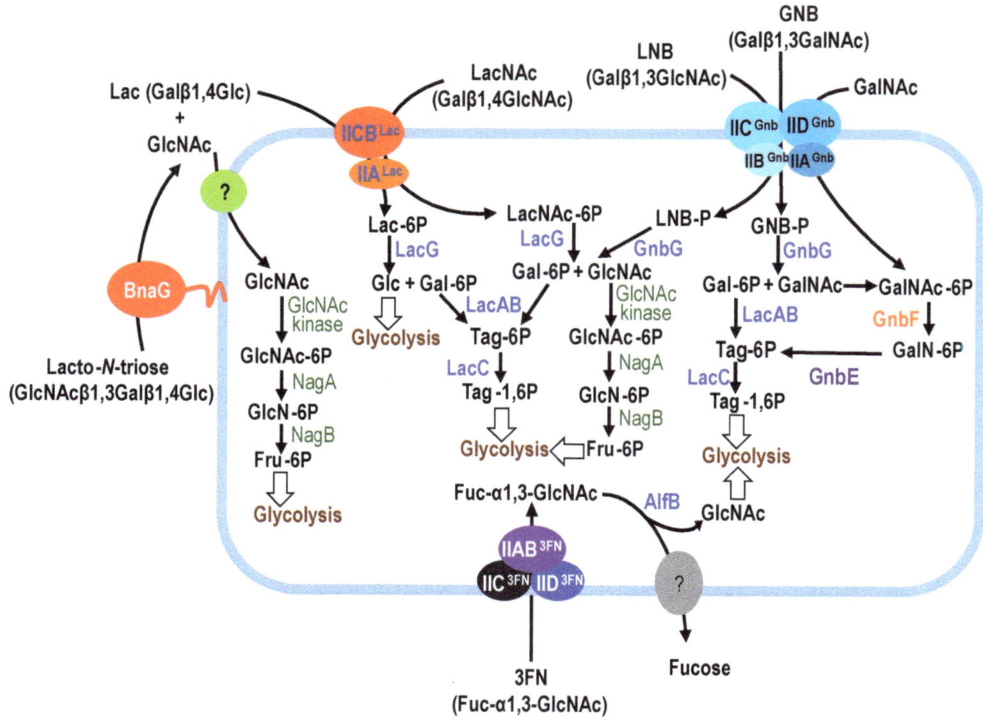

Figure 3.8 HMOs catabolic pathways identified from *Lactobacillus casei*. LNB, lacto-*N*-biose; GNB, galacto-*N*-biose; LacNAc, *N*-acetyllactosamine; Lac, lactose; GalNAc, *N*-acetylgalactosamine; GalN, galactosamine; Gal, galactose; Glc, glucose, GlcNAc; *N*-acetylglucosamine; GlcN, glucosamine; Tag, tagatose; 3FN, fucosyl-α1,3-*N*-acetylglucosamine; Fuc, fucose; $IICB^{Lac}$ and IIA^{Lac}, lactose-specific domains of the phosphoenolpyruvate: phosphotransferase system (PTS); IIC^{Gnb}, IID^{Gnb}, IIA^{Gnb} and IIB^{Gnb}, LNB/GNB/GalNAc-specific domains of the PTS; IIC^{3FN}, IID^{3FN} and $IIAB^{3FN}$, $3FN^{3FN}$ 3FN-specific domains of the PTS; BnaG, beta-*N*-acetylglucosaminidase; LacG, phospho-β-galactosidase; GnbG, phospho-β-galactosidase; GnbF, *N*-acetylgalactosamine 6-phosphate deacetylase; GnbE, galactosamine 6-phosphate isomerase/deaminase; LacAB, galactose 6-phosphate isomerase; LacC, tagatose 6-phosphate kinase; NagA, *N*-acetylglucosamine 6-phosphate deacetylase; NagB, glucosamine 6-phosphate deaminase.

galactosamine 6-phosphate isomerizing deaminase (*gnbE*), a GalNAc 6-phosphate deacetylase (*gnbF*), a phospho-β-galactosidase (*gnbG*) and four genes (*gnbBCDA*) encoding the EIIB, EIIC, EIID and EIIA components of a mannose-class PTS system (PTS^{Gnb}) (Bidart *et al.*, 2014). LNB, GNB and also GalNAc are transported and phosphorylated by the PTS^{Gnb} and then, both disaccharides are hydrolysed by the specific β-1,3-galactosidase GnbG (GH family 35) into galactose 6-phosphate and the corresponding *N*-acetylhexosamines (GlcNAc and GalNAc). Galactose 6-phosphate is metabolized through the tagatose 6-phosphate pathway, whereas GlcNAc and GalNAc would be phosphorylated by as yet unknown kinases before entering different catabolic routes (Fig. 3.8). GlcNAc 6-phosphate is converted by the NagA deacetylase to glucosamine 6-phosphate, which enters the glycolysis pathway via conversion to fructose 6-phosphate by the NagB deaminase. GalNAc 6-phosphate would be deacetylated and deaminated to tagatose 6-phosphate by the products of the genes *gnbF* and *gnbE*. Therefore, all *gnb* genes would participate in GNB and GalNAc metabolism while LNB utilization by *L. casei* would not require GnbE and GnbF activities. It is worth noting that *gnb* genes are highly induced by GNB and GalNAc which relieve repression by GnbR, whereas the presence of LNB barely induces the *gnb* operon (Bidart *et al.*, 2014). According to this, the *gnb* operon would be primarily adapted to catabolize GNB and GalNAc. The coexistence of these sugars with LNB in environments as the gastrointestinal tract might account for the utilization of LNB by this pathway although this sugar would not induce the expression of *gnb* genes. The *gnb* gene cluster is conserved in the *L. casei/paracasei/*

rhamnosus/zeae group, and it has been shown that both GNB and LNB are fermented by several strains of *L. casei*, *L. rhamnosus* and *L. zeae* species (Bidart *et al.*, 2017). Strains belonging to *L. gasseri* and *L. johnsonii* species are also consumers of LNB and GNB, although they do not have a *gnb* operon (Bidart *et al.*, 2017). Therefore, at least another catabolic system for those disaccharides remains to be discovered in lactobacilli.

Lactose metabolism in lactic acid bacteria has been widely studied due to the economic relevance of lactose fermentation in the dairy industry (Cavanagh *et al.*, 2015; de Vos and Vaughan, 1994; Lapierre *et al.*, 2002; Stefanovic *et al.*, 2017). Lactose can be transported by lactose/galactose antiport permeases, proton symport permeases or through a PTS transporter (Alpert and Chassy, 1990; de Vos and Vaughan, 1994; Francl *et al.*, 2012; Gosalbes *et al.*, 1997; Leong-Morgenthaler *et al.*, 1991). Recently, it has been shown that the *lac* operon from *L. casei* is also responsible of the utilization of LacNAc (Bidart *et al.* unpublished). This carbohydrate is transported and phosphorylated by the PTS^{Lac} (Gosalbes *et al.*, 2002; Gosalbes *et al.*, 1997; Gosalbes *et al.*, 1999), and then is hydrolysed by the phospho-β-galactosidase LacG (GH family 1) into galactose 6-phosphate and GlcNAc. In fact, the *lac* operon of *L. casei* showed higher induction levels in the presence of LacNAc than with lactose, suggesting that LacNAc may be the preferential substrate of this transporter. Indeed, this carbohydrate is present in the human gastrointestinal tract through all stages of life (Marionneau *et al.*, 2001; Moran *et al.*, 2011) whereas lactose would be only present during the lactating period since the introduction of dairy farming is a very recent event in the evolutionary history of humankind. Genome sequence analyses (http://www.ncbi.nlm.nih.gov/genomes) showed that many species belonging to the genus *Lactobacillus* contain genes encoding PTS transporters homologous to the PTS^{Lac} from *L. casei* BL23, suggesting that lactose-specific PTS transporters are quite common among lactobacilli. Lactose is also the product resulting from the metabolism of LNTII by the action of the exoglycosidase β-*N*-acetylglucosaminidase BnaG (GH family 20) from *L. casei* (Bidart *et al.*, 2016). Unlike the other glycosidases active on HMOs characterized in this species, BnaG is a cell wall -anchored extracellular protein. This enzyme shows high specificity for *N*-acetylhexosaminyl-β-1,3-linked sugars as it releases GlcNAc not only from LNTII but also from β-*N*-acetyl-D-glucosamine-(1→3)-D-mannopyranoside, a disaccharide that forms part of glycoproteins (Garrido *et al.*, 2012). As well, BnaG liberates GalNAc from β-*N*-acetyl-D-galactosamine-(1→3)-D-galactopyranoside, which forms part of globotetraose, a glycan moiety of human glycosphingolipids present at cell surfaces (Schnaar *et al.*, 2009). The oligosaccharide part of these lipids have recently been described as substrates for lacto-*N*-biosidases isolated from *B. longum* subsp. *longum*, which, in contrast to BnaG, are endoglycosidases and release GNB (Gotoh *et al.*, 2015). The possession of cell wall-attached glycosidases may provide a competitive advantage by allowing cleavage and consumption of complex-linked sugars. In addition, there is also evidence suggesting that these enzymes might also modulate the activity of host glycoproteins and pathogen–host receptor interactions through modification of the surface-exposed host glycans structures (Garbe and Collin, 2012; Garrido *et al.*, 2012; Kobata, 2013).

The presence of α-L-fucosidases and catabolic pathways for the utilization of LNB, GNB, LacNAc and LNTII in *L. casei* shows the capacity of this species for the exploitation of human milk and mucosa-associated glycans. This feature probably constitutes an adaptation of these bacteria to survive in the gastrointestinal tract of breastfed infants.

Concluding remarks

Lactobacilli play a major role in the production of many fermented foods and, in human and animal health as components of the microbial communities associated with different mucosal surfaces or as health-improving food supplements. These organisms rely on sugar utilization for growth so that knowing their sugar utilization pathways is a must to understand their role in microbial communities and their performance in food production. Due to their economic relevance, some species of lactobacilli have been extensively studied (e.g. specific strains of *L. plantarum*, *L. casei* or *L. acidophilus*) and the pathways of utilization of many monosaccharides and disaccharides have been elucidated. However, there are still great gaps in our knowledge of the utilization of complex glycans, especially host mucosal glycans, even in well-characterized strains of lactobacilli. Also, information is partial or simply lacking for many other species that inhabit mucosal and food/feed niches. Genomic sequencing in lactobacilli has revealed an enormous variety of putative glycan transport systems and glycosyl hydrolases belonging to different glycosyl hydrolase (GH) families. However, the function of most of these transporters and enzymes remains to be elucidated. Furthermore, most studies have been carried out using pure cultures in laboratory conditions. However, utilization of complex glycans in food matrixes or microbial communities is possibly a key factor for growth and survival in these environments. In the natural niches where lactobacilli dwell, complex ecological relationships are found and cross-feeding is established between different microbial groups, where sequential and cooperative degradation of complex glycans probably takes place. Unravelling these associations will require the use of different 'omic' technologies that include metagenomics, transcriptomics and metabolomics. This constitutes a future challenge in the study of carbohydrate metabolism in lactobacilli and its functional and ecological relevance.

References

Ackerman, D.L., Craft, K.M., and Townsend, S.D. (2017). Infant food applications of complex carbohydrates: structure, synthesis, and function. Carbohydr. Res. *437*, 16–27. https://doi.org/10.1016/j.carres.2016.11.007

Albenberg, L., Esipova, T.V., Judge, C.P., Bittinger, K., Chen, J., Laughlin, A., Grunberg, S., Baldassano, R.N., Lewis, J.D., Li, H., *et al.* (2014). Correlation between intraluminal oxygen gradient and radial partitioning of intestinal microbiota. Gastroenterology *147*, 1055–1063.e8. https://doi.org/10.1053/j.gastro.2014.07.020

Albesharat, R., Ehrmann, M.A., Korakli, M., Yazaji, S., and Vogel, R.F. (2011). Phenotypic and genotypic analyses of lactic acid bacteria in local fermented food, breast milk and faeces of mothers and their babies. Syst. Appl. Microbiol. *34*, 148–155. https://doi.org/10.1016/j.syapm.2010.12.001

Almagro-Moreno, S., and Boyd, E.F. (2009). Insights into the evolution of sialic acid catabolism among bacteria. BMC Evol. Biol. *9*, 118. https://doi.org/10.1186/1471-2148-9-118

Alpert, C.A., and Chassy, B.M. (1990). Molecular cloning and DNA sequence of *lacE*, the gene encoding the lactose-specific enzyme II of the phosphotransferase system of *Lactobacillus casei*. Evidence that a cysteine residue is essential for sugar phosphorylation. J. Biol. Chem. *265*, 22561–22568.

Ananieva, M., Tzenova, M., Iliev, I., and Ivanova, I. (2014). Gene expression of enzymes involved in utilization of xylooligosaccharides by *Lactobacillus* strains. Biotechnol. Biotechnol. Equip. *28*, 941–948. https://doi.org/10.1080/13102818.2014.948257

Anba-Mondoloni, J., Chaillou, S., Zagorec, M., and Champomier-Vergès, M.C. (2013). Catabolism of N-acetylneuraminic acid, a fitness function of the food-borne lactic acid bacterium *Lactobacillus sakei*, involves two newly characterized proteins. Appl. Environ. Microbiol. *79*, 2012–2018. https://doi.org/10.1128/AEM.03301-12

Andersen, J.M., Barrangou, R., Abou Hachem, M., Lahtinen, S., Goh, Y.J., Svensson, B., and Klaenhammer, T.R. (2011). Transcriptional and functional analysis of galactooligosaccharide uptake by *lacS* in *Lactobacillus acidophilus*. Proc. Natl. Acad. Sci. U.S.A. *108*, 17785–17790. https://doi.org/10.1073/pnas.1114152108

Andersen, J.M., Barrangou, R., Hachem, M.A., Lahtinen, S.J., Goh, Y.J., Svensson, B., and Klaenhammer, T.R. (2012). Transcriptional analysis of prebiotic uptake and catabolism by *Lactobacillus acidophilus* NCFM. PLOS ONE *7*, e44409. https://doi.org/10.1371/journal.pone.0044409

Anwar, M.A., Kralj, S., van der Maarel, M.J., and Dijkhuizen, L. (2008). The probiotic *Lactobacillus johnsonii* NCC 533 produces high-molecular-mass inulin from sucrose by using an inulosucrase enzyme. Appl. Environ. Microbiol. *74*, 3426–3433. https://doi.org/10.1128/AEM.00377-08

Balogh, R., Jankovics, P., and Béni, S. (2015). Qualitative and quantitative analysis of N-acetyllactosamine and lacto-N-biose, the two major building blocks of human milk oligosaccharides in human milk samples by high-performance liquid chromatography-tandem mass spectrometry using a porous graphitic carbon column. J. Chromatogr. A *1422*, 140–146. https://doi.org/10.1016/j.chroma.2015.10.006

Balolong, M.P., Chae, J.P., and Kang, D.K. (2016). Expression and characterisation of neopullulanase from *Lactobacillus mucosae*. Biotechnol. Lett. *38*, 1753–1760. https://doi.org/10.1007/s10529-016-2152-x

Barrangou, R., Altermann, E., Hutkins, R., Cano, R., and Klaenhammer, T.R. (2003). Functional and comparative genomic analyses of an operon involved in fructooligosaccharide utilization by *Lactobacillus acidophilus*. Proc. Natl. Acad. Sci. U.S.A. *100*, 8957–8962. https://doi.org/10.1073/pnas.1332765100

Barrangou, R., Azcarate-Peril, M.A., Duong, T., Conners, S.B., Kelly, R.M., and Klaenhammer, T.R. (2006). Global analysis of carbohydrate utilization by *Lactobacillus acidophilus* using cDNA microarrays. Proc. Natl. Acad. Sci. U.S.A. *103*, 3816–3821. https://doi.org/10.1073/pnas.0511287103

Becerra, J.E., Coll-Marqués, J.M., Rodríguez-Díaz, J., Monedero, V., and Yebra, M.J. (2015a). Preparative scale purification of fucosyl-N-acetylglucosamine disaccharides and their evaluation as potential prebiotics and antiadhesins. Appl. Microbiol. Biotechnol. *99*, 7165–7176. https://doi.org/10.1007/s00253-015-6666-2

Becerra, J.E., Yebra, M.J., and Monedero, V. (2015b). An L-Fucose operon in the probiotic *Lactobacillus rhamnosus* GG is involved in adaptation to gastrointestinal conditions. Appl. Environ. Microbiol. *81*, 3880–3888. https://doi.org/10.1128/AEM.00260-15

Bello, F.D., Walter, J., Hertel, C., and Hammes, W.P. (2001). *In vitro* study of prebiotic properties of levan-type exopolysaccharides from lactobacilli and non-digestible carbohydrates using denaturing gradient gel electrophoresis. Syst. Appl. Microbiol. *24*, 232–237. https://doi.org/10.1078/0723-2020-00033

Bidart, G.N., Rodríguez-Díaz, J., Monedero, V., and Yebra, M.J. (2014). A unique gene cluster for the utilization of the mucosal and human milk-associated glycans galacto-N-biose and lacto-N-biose in *Lactobacillus casei*. Mol. Microbiol. *93*, 521–538. https://doi.org/10.1111/mmi.12678

Bidart, G.N., Rodríguez-Díaz, J., Palomino-Schätzlein, M., Monedero, V., and Yebra, M.J. (2017). Human milk and mucosal lacto- and galacto-N-biose synthesis by transgalactosylation and their prebiotic potential in *Lactobacillus* species. Appl. Microbiol. Biotechnol. *101*, 205–215. https://doi.org/10.1007/s00253-016-7882-0

Bidart, G.N., Rodríguez-Díaz, J., and Yebra, M.J. (2016). The extracellular wall-bound β-N-acetylglucosaminidase from *Lactobacillus casei* is involved in the metabolism of the human milk oligosaccharide lacto-N-triose. Appl. Environ. Microbiol. *82*, 570–577. https://doi.org/10.1128/AEM.02888-15

Blander, J.M., Longman, R.S., Iliev, I.D., Sonnenberg, G.F., and Artis, D. (2017). Regulation of inflammation by microbiota interactions with the host. Nat. Immunol. *18*, 851–860. https://doi.org/10.1038/ni.3780

Bode, L. (2012). Human milk oligosaccharides: every baby needs a sugar mama. Glycobiology *22*, 1147–1162. https://doi.org/10.1093/glycob/cws074

Cantarel, B.L., Lombard, V., and Henrissat, B. (2012). Complex carbohydrate utilization by the healthy human microbiome. PLOS ONE *7*, e28742. https://doi.org/10.1371/journal.pone.0028742

Cardelle-Cobas, A., Fernández, M., Salazar, N., Martínez-Villaluenga, C., Villamiel, M., Ruas-Madiedo, P., and de los Reyes-Gavilán, C.G. (2009). Bifidogenic effect and stimulation of short chain fatty acid production in human faecal slurry cultures by oligosaccharides derived from lactose and lactulose. J. Dairy Res. *76*, 317–325. https://doi.org/10.1017/S0022029909004063

Carrera-Silva, E.A., Silvestroni, A., LeBlanc, J.G., Piard, J.C., Savoy de Giori, G., and Sesma, F. (2006). A thermostable alpha-galactosidase from *Lactobacillus fermentum* CRL722: genetic characterization and main properties. Curr. Microbiol. *53*, 374–378. https://doi.org/10.1007/s00284-005-0442-y

Cavanagh, D., Fitzgerald, G.F., and McAuliffe, O. (2015). From field to fermentation: the origins of *Lactococcus lactis* and its domestication to the dairy environment. Food Microbiol. *47*, 45–61. https://doi.org/10.1016/j.fm.2014.11.001

Claesson, M.J., Jeffery, I.B., Conde, S., Power, S.E., O'Connor, E.M., Cusack, S., Harris, H.M., Coakley, M., Lakshminarayanan, B., O'Sullivan, O., *et al.* (2012). Gut microbiota composition correlates with diet and health in the elderly. Nature *488*, 178–184. https://doi.org/10.1038/nature11319

Collado, M.C., Calabuig, M., and Sanz, Y. (2007). Differences between the fecal microbiota of coeliac infants and healthy controls. Curr. Issues Intest. Microbiol. *8*, 9–14.

Coulier, L., Timmermans, J., Bas, R., Van Den Dool, R., Haaksman, I., Klarenbeek, B., Slaghek, T., and Van Dongen, W. (2009). In-depth characterization of prebiotic galacto-oligosaccharides by a combination of analytical techniques. J. Agric. Food Chem. *57*, 8488–8495. https://doi.org/10.1021/jf902549e

Chaillou, S., Lokman, B.C., Leer, R.J., Posthuma, C., Postma, P.W., and Pouwels, P.H. (1998). Cloning, sequence analysis, and characterization of the genes involved in isoprimeverose metabolism in *Lactobacillus pentosus*. J. Bacteriol. *180*, 2312–2320.

Chen, C., Zhao, G., Chen, W., and Guo, B. (2015). Metabolism of fructooligosaccharides in *Lactobacillus plantarum* ST-III via differential gene transcription and alteration of cell membrane fluidity. Appl. Environ. Microbiol. *81*, 7697–7707. https://doi.org/10.1128/AEM.02426-15

Chen, C., Zhou, F., Ren, J., Ai, L., Dong, Y., Wu, Z., Liu, Z., Chen, W., and Guo, B. (2014). Cloning, expression and functional validation of a β-fructofuranosidase from *Lactobacillus plantarum*. Process Biochem. *49*, 758–767. https://doi.org/10.1016/j.procbio.2014.02.013

Chen, X.Y., Levy, C., and Gänzle, M.G. (2016). Structure-function relationships of bacterial and enzymatically produced reuterans and dextran in sourdough bread baking application. Int. J. Food Microbiol. *239*, 95–102. https://doi.org/10.1016/j.ijfoodmicro.2016.06.010

Cheng, K.C., Demirci, A., and Catchmark, J.M. (2011). Pullulan: biosynthesis, production, and applications. Appl. Microbiol. Biotechnol. *92*, 29–44. https://doi.org/10.1007/s00253-011-3477-y

Childs, C.E., Röytiö, H., Alhoniemi, E., Fekete, A.A., Forssten, S.D., Hudjec, N., Lim, Y.N., Steger, C.J., Yaqoob, P., Tuohy, K.M., *et al.* (2014). Xylo-oligosaccharides alone or in synbiotic combination with *Bifidobacterium animalis* subsp. *lactis* induce bifidogenesis and modulate markers of immune function in healthy adults: a double-blind, placebo-controlled, randomised, factorial cross-over study. Br. J. Nutr. *111*, 1945–1956. https://doi.org/10.1017/S0007114513004261

Cho, I., and Blaser, M.J. (2012). The human microbiome: at the interface of health and disease. Nat. Rev. Genet. *13*, 260–270. https://doi.org/10.1038/nrg3182

David, L.A., Maurice, C.F., Carmody, R.N., Gootenberg, D.B., Button, J.E., Wolfe, B.E., Ling, A.V., Devlin, A.S., Varma, Y., Fischbach, M.A., *et al.* (2014). Diet rapidly and reproducibly alters the human gut microbiome. Nature *505*, 559–563. https://doi.org/10.1038/nature12820

de Vos, W.M., and Vaughan, E.E. (1994). Genetics of lactose utilization in lactic acid bacteria. FEMS Microbiol. Rev. *15*, 217–237.

Delgado, S., Flórez, A.B., Guadamuro, L., and Mayo, B. (2017). Genetic and biochemical characterization of an oligo-α-1,6-glucosidase from *Lactobacillus plantarum*. Int. J. Food Microbiol. *246*, 32–39. https://doi.org/10.1016/j.ijfoodmicro.2017.01.021

Dong, J.L., Yu, X., Dong, L.E., and Shen, R.L. (2017). *In vitro* fermentation of oat β-glucan and hydrolysates by fecal microbiota and selected probiotic strains. J. Sci. Food Agric. *97*, 4198–4203. https://doi.org/10.1002/jsfa.8292

Eckburg, P.B., Bik, E.M., Bernstein, C.N., Purdom, E., Dethlefsen, L., Sargent, M., Gill, S.R., Nelson, K.E., and Relman, D.A. (2005). Diversity of the human intestinal microbial flora. Science *308*, 1635–1638. https://doi.org/10.1126/science.1110591

Egloff, M.P., Uppenberg, J., Haalck, L., and van Tilbeurgh, H. (2001). Crystal structure of maltose phosphorylase from *Lactobacillus brevis*: unexpected evolutionary relationship with glucoamylases. Structure *9*, 689–697. https://doi.org/10.1016/S0969-2126(01)00626-8

Ehrmann, M.A., and Vogel, R.F. (1998). Maltose metabolism of *Lactobacillus sanfranciscensis*: cloning and heterologous expression of the key enzymes, maltose phosphorylase and phosphoglucomutase. FEMS Microbiol. Lett. *169*, 81–86. https://doi.org/10.1111/j.1574-6968.1998.tb13302.x

Ejby, M., Fredslund, F., Vujicic-Zagar, A., Svensson, B., Slotboom, D.J., and Abou Hachem, M. (2013). Structural basis for arabinoxylo-oligosaccharide capture by the probiotic *Bifidobacterium animalis* subsp. *lactis* Bl-04. Mol. Microbiol. *90*, 1100–1112. https://doi.org/10.1111/mmi.12419

Endo, A., Nakamura, S., Konishi, K., Nakagawa, J., and Tochio, T. (2016). Variations in prebiotic oligosaccharide fermentation by intestinal lactic acid bacteria. Int. J. Food Sci. Nutr. *67*, 125–132. https://doi.org/10.3109/09637486.2016.1147019

Ezenwa, V.O., Gerardo, N.M., Inouye, D.W., Medina, M., and Xavier, J.B. (2012). Animal behavior and the microbiome. Science *338*, 198–199. https://doi.org/10.1126/science.1227412

Francke, C., Kerkhoven, R., Wels, M., and Siezen, R.J. (2008). A generic approach to identify Transcription Factor-specific operator motifs; Inferences for LacI-family mediated regulation in *Lactobacillus plantarum* WCFS1. BMC Genomics *9*, 145. https://doi.org/10.1186/1471-2164-9-145

Francl, A.L., Hoeflinger, J.L., and Miller, M.J. (2012). Identification of lactose phosphotransferase systems in *Lactobacillus gasseri* ATCC 33323 required for lactose utilization. Microbiology *158*, 944–952. https://doi.org/10.1099/mic.0.052928-0

Fraunhofer, M.E., Geissler, A.J., Wefers, D., Bunzel, M., Jakob, F., and Vogel, R.F. (2017). Characterization of β-glucan formation by *Lactobacillus brevis* TMW 1.2112 isolated from slimy spoiled beer. Int. J. Biol. Macromol. *107*, 874–881. https://doi.org/10.1016/j.ijbiomac.2017.09.063

Fuchs, A. (1991). Current and potential food and non-food applications of fructans. Biochem. Soc. Trans. *19*, 555–560. https://doi.org/10.1042/bst0190555

Gänzle, M.G., and Follador, R. (2012). Metabolism of oligosaccharides and starch in lactobacilli: a review. Front. Microbiol. *3*, 340. https://doi.org/10.3389/fmicb.2012.00340

Garai-Ibabe, G., Dueñas, M.T., Irastorza, A., Sierra-Filardi, E., Werning, M.L., López, P., Corbí, A.L., and Fernández de Palencia, P. (2010). Naturally occurring 2-substituted (1,3)-beta-D-glucan producing *Lactobacillus suebicus* and *Pediococcus parvulus* strains with potential utility in the production of functional foods. Bioresour. Technol. *101*, 9254–9263. https://doi.org/10.1016/j.biortech.2010.07.050

Garbe, J., and Collin, M. (2012). Bacterial hydrolysis of host glycoproteins - powerful protein modification and efficient nutrient acquisition. J. Innate Immun. *4*, 121–131. https://doi.org/10.1159/000334775

Garrido, D., Dallas, D.C., and Mills, D.A. (2013). Consumption of human milk glycoconjugates by infant-associated bifidobacteria: mechanisms and implications. Microbiology *159*, 649–664. https://doi.org/10.1099/mic.0.064113-0

Garrido, D., Nwosu, C., Ruiz-Moyano, S., Aldredge, D., German, J.B., Lebrilla, C.B., and Mills, D.A. (2012). Endo-β-N-acetylglucosaminidases from infant gut-associated bifidobacteria release complex N-glycans from human milk glycoproteins. Mol. Cell Proteomics *11*, 775–785. https://doi.org/10.1074/mcp.M112.018119

Gibson, G.R., and Roberfroid, M.B. (1995). Dietary modulation of the human colonic microbiota: introducing the concept of prebiotics. J. Nutr. *125*, 1401–1412. https://doi.org/10.1093/jn/125.6.1401

Goh, Y.J., and Klaenhammer, T.R. (2013). A functional glycogen biosynthesis pathway in *Lactobacillus acidophilus*: expression and analysis of the *glg* operon. Mol. Microbiol. *89*, 1187–1200. https://doi.org/10.1111/mmi.12338

Goh, Y.J., and Klaenhammer, T.R. (2014). Insights into glycogen metabolism in *Lactobacillus acidophilus*: impact on carbohydrate metabolism, stress tolerance and gut retention. Microb. Cell Fact. *13*, 94. https://doi.org/10.1186/s12934-014-0094-3

Goh, Y.J., and Klaenhammer, T.R. (2015). Genetic mechanisms of prebiotic oligosaccharide metabolism in probiotic microbes. Annu. Rev. Food Sci. Technol. *6*, 137–156. https://doi.org/10.1146/annurev-food-022814-015706

Goh, Y.J., Lee, J.H., and Hutkins, R.W. (2007). Functional analysis of the fructooligosaccharide utilization operon in *Lactobacillus paracasei* 1195. Appl. Environ. Microbiol. *73*, 5716–5724. https://doi.org/10.1128/aem.00805-07

Goh, Y.J., Zhang, C., Benson, A.K., Schlegel, V., Lee, J.H., and Hutkins, R.W. (2006). Identification of a putative operon involved in fructooligosaccharide utilization by *Lactobacillus paracasei*. Appl. Environ. Microbiol. *72*, 7518–7530. https://doi.org/10.1128/aem.00877-06

Gomez-Llorente, C., Plaza-Diaz, J., Aguilera, M., Muñoz-Quezada, S., Bermudez-Brito, M., Peso-Echarri, P., Martinez-Silla, R., Vasallo-Morillas, M.I., Campaña-Martin, L., Vives-Piñera, I., *et al.* (2013). Three main factors define changes in fecal microbiota associated with feeding modality in infants. J. Pediatr. Gastroenterol. Nutr. *57*, 461–466. https://doi.org/10.1097/MPG.0b013e31829d519a

Gosalbes, M.J., Esteban, C.D., and Pérez-Martínez, G. (2002). *In vivo* effect of mutations in the antiterminator

LacT in *Lactobacillus casei*. Microbiology *148*, 695–702. https://doi.org/10.1099/00221287-148-3-695

Gosalbes, M.J., Monedero, V., Alpert, C.A., and Pérez-Martínez, G. (1997). Establishing a model to study the regulation of the lactose operon in *Lactobacillus casei*. FEMS Microbiol. Lett. *148*, 83–89. https://doi.org/10.1016/S0378-1097(97)00017-7

Gosalbes, M.J., Monedero, V., and Pérez-Martínez, G. (1999). Elements involved in catabolite repression and substrate induction of the lactose operon in *Lactobacillus casei*. J. Bacteriol. *181*, 3928–3934.

Gotoh, A., Katoh, T., Sugiyama, Y., Kurihara, S., Honda, Y., Sakurama, H., Kambe, T., Ashida, H., Kitaoka, M., Yamamoto, K., *et al.* (2015). Novel substrate specificities of two lacto-N-biosidases towards β-linked galacto-N-biose-containing oligosaccharides of globo H, Gb5, and GA1. Carbohydr. Res. *408*, 18–24. https://doi.org/10.1016/j.carres.2015.03.005

Gu, S., Chen, D., Zhang, J.N., Lv, X., Wang, K., Duan, L.P., Nie, Y., and Wu, X.L. (2013). Bacterial community mapping of the mouse gastrointestinal tract. PLOS ONE *8*, e74957. https://doi.org/10.1371/journal.pone.0074957

Hammer, A.J., Walters, A., Carroll, C., Newell, P.D., and Chaston, J.M. (2017). Draft genome sequence of *Lactobacillus paracasei* DmW181, a bacterium isolated from wild *Drosophila*. Genome Announc. *5*, e00545-17 https://doi.org/10.1128/genomeA.00545-17

Heuberger, E.H., Smits, E., and Poolman, B. (2001). Xyloside transport by XylP, a member of the galactoside-pentoside-hexuronide family. J. Biol. Chem. *276*, 34465–34472. https://doi.org/10.1074/jbc.M105460200

Hidaka, H., Eida, T., Takizawa, T., Tokunaga, T., and Tashiro, Y. (1986). Effects of fructooligosaccharides on intestinal flora and human health. Bifidobact. Microfl. *5*, 37–50.

Homburg, C., Bommer, M., Wuttge, S., Hobe, C., Beck, S., Dobbek, H., Deutscher, J., Licht, A., and Schneider, E. (2017). Inducer exclusion in Firmicutes: insights into the regulation of a carbohydrate ATP binding cassette transporter from *Lactobacillus casei* BL23 by the signal transducing protein P-Ser46-HPr. Mol. Microbiol. *105*, 25–45. https://doi.org/10.1111/mmi.13680

Hooper, L.V., Midtvedt, T., and Gordon, J.I. (2002). How host-microbial interactions shape the nutrient environment of the mammalian intestine. Annu. Rev. Nutr. 22, 283–307. https://doi.org/10.1146/annurev.nutr.22.011602.092259

Hu, Y., Ketabi, A., Buchko, A., and Gänzle, M.G. (2013). Metabolism of isomalto-oligosaccharides by *Lactobacillus reuteri* and bifidobacteria. Lett. Appl. Microbiol. *57*, 108–114. https://doi.org/10.1111/lam.12076

Humblot, C., Turpin, W., Chevalier, F., Picq, C., Rochette, I., and Guyot, J.P. (2014). Determination of expression and activity of genes involved in starch metabolism in *Lactobacillus plantarum* A6 during fermentation of a cereal-based gruel. Int. J. Food Microbiol. *185*, 103–111. https://doi.org/10.1016/j.ijfoodmicro.2014.05.016

Hutkins, R.W., Krumbeck, J.A., Bindels, L.B., Cani, P.D., Fahey, G., Goh, Y.J., Hamaker, B., Martens, E.C., Mills, D.A., Rastal, R.A., *et al.* (2016). Prebiotics: why definitions matter. Curr. Opin. Biotechnol. *37*, 1–7. https://doi.org/10.1016/j.copbio.2015.09.001

Hüwel, S., Haalck, L., Conrath, N., and Spener, F. (1997). Maltose phosphorylase from *Lactobacillus brevis*: purification, characterization, and application in a biosensor for ortho-phosphate. Enzyme Microb. Technol. *21*, 413–420. https://doi.org/10.1016/S0141-0229(97)00014-8

Islam, A., Stoll, B.J., Ljungstrom, I., Biswas, J., Nazrul, H., and Huldt, G. (1988). The prevalence of *Entamoeba histolytica* in lactating women and in their infants in Bangladesh. Trans. R. Soc. Trop. Med. Hyg. *82*, 99–103. https://doi.org/10.1016/0035-9203(88)90276-3

Jantscher-Krenn, E., Lauwaet, T., Bliss, L.A., Reed, S.L., Gillin, F.D., and Bode, L. (2012). Human milk oligosaccharides reduce *Entamoeba histolytica* attachment and cytotoxicity *in vitro*. Br. J. Nutr. *108*, 1839–1846. https://doi.org/10.1017/S0007114511007392

Jones, L., Seymour, G.B., and Knox, J.P. (1997). Localization of pectic galactan in tomato cell walls using a monoclonal antibody specific to (1[->]4)-[beta]-D-galactan. Plant Physiol. *113*, 1405–1412. https://doi.org/10.1104/pp.113.4.1405

Kaplan, H., and Hutkins, R.W. (2000). Fermentation of fructooligosaccharides by lactic acid bacteria and bifidobacteria. Appl. Environ. Microbiol. *66*, 2682–2684. https://doi.org/10.1128/aem.66.6.2682-2684.2000

Ketabi, A., Dieleman, L.A., and Gänzle, M.G. (2011). Influence of isomalto-oligosaccharides on intestinal microbiota in rats. J. Appl. Microbiol. *110*, 1297–1306. https://doi.org/10.1111/j.1365-2672.2011.04984.x

Kim, H.B., and Isaacson, R.E. (2015). The pig gut microbial diversity: understanding the pig gut microbial ecology through the next generation high throughput sequencing. Vet. Microbiol. *177*, 242–251. https://doi.org/10.1016/j.vetmic.2015.03.014

Kim, J.H., Sunako, M., Ono, H., Murooka, Y., Fukusaki, E., and Yamashita, M. (2008). Characterization of gene encoding amylopullulanase from plant-originated lactic acid bacterium, *Lactobacillus plantarum* L137. J. Biosci. Bioeng. *106*, 449–459. https://doi.org/10.1263/jbb.106.449

Kleeberger, A., and Kühbauch, W. (1976). Über den Abbau von Grasfruktosanen durch Laktobazillen aus Silage. Zentralblatt für Bakteriologie, Parasitenkunde, Infektionskrankheiten und Hygiene. Zweite Naturwissenschaftliche Abteilung: Allgemeine, Landwirtschaftliche und Technische Mikrobiologie *131*, 398–404. https://doi.org/10.1016/S0044-4057(76)80018-4

Kobata, A. (2010). Structures and application of oligosaccharides in human milk. Proc. Jpn. Acad., Ser. B, Phys. Biol. Sci. *86*, 731–747. https://doi.org/10.2183/pjab.86.731

Kobata, A. (2013). Exo- and endoglycosidases revisited. Proc. Jpn. Acad., Ser. B, Phys. Biol. Sci. *89*, 97–117. https://doi.org/10.2183/pjab.89.97

Kohl, K.D., Amaya, J., Passement, C.A., Dearing, M.D., and McCue, M.D. (2014). Unique and shared responses of the gut microbiota to prolonged fasting: a comparative study across five classes of vertebrate hosts. FEMS Microbiol. Ecol. *90*, 883–894. https://doi.org/10.1111/1574-6941.12442

Kunz, C., Meyer, C., Collado, M.C., Geiger, L., García-Mantrana, I., Bertua-Ríos, B., Martínez-Costa, C., Borsch, C., and Rudloff, S. (2017). Influence of gestational age, secretor, and Lewis blood group status on the oligosaccharide content of human milk. J. Pediatr. Gastroenterol. Nutr. *64*, 789–798. https://doi.org/10.1097/MPG.0000000000001402

Kunz, C., Rudloff, S., Baier, W., Klein, N., and Strobel, S. (2000). Oligosaccharides in human milk: structural, functional, and metabolic aspects. Annu. Rev. Nutr. *20*, 699–722. https://doi.org/10.1146/annurev.nutr.20.1.699

Kuzuwa, S., Yokoi, K.J., Kondo, M., Kimoto, H., Yamakawa, A., Taketo, A., and Kodaira, K. (2012). Properties of the inulinase gene levH1 of *Lactobacillus casei* IAM 1045; cloning, mutational and biochemical characterization. Gene *495*, 154–162. https://doi.org/10.1016/j.gene.2011.12.004

Kwak, M.J., Kwon, S.K., Yoon, J.K., Song, J.Y., Seo, J.G., Chung, M.J., and Kim, J.F. (2016). Evolutionary architecture of the infant-adapted group of Bifidobacterium species associated with the probiotic function. Syst. Appl. Microbiol. *39*, 429–439. https://doi.org/10.1016/j.syapm.2016.07.004

Lapierre, L., Mollet, B., and Germond, J.E. (2002). Regulation and adaptive evolution of lactose operon expression in *Lactobacillus delbrueckii*. J. Bacteriol. *184*, 928–935. https://doi.org/10.1128/jb.184.4.928-935.2002

LeBlanc, J.G., Garro, M.S., and Savoy de Giori, G. (2004). Effect of pH on *Lactobacillus fermentum* growth, raffinose removal, alpha-galactosidase activity and fermentation products. Appl. Microbiol. Biotechnol. *65*, 119–123. https://doi.org/10.1007/s00253-003-1532-z

Leemhuis, H., Dobruchowska, J.M., Ebbelaar, M., Faber, F., Buwalda, P.L., van der Maarel, M.J., Kamerling, J.P., and Dijkhuizen, L. (2014). Isomalto/malto-polysaccharide, a novel soluble dietary fiber made via enzymatic conversion of starch. J. Agric. Food Chem. *62*, 12034–12044. https://doi.org/10.1021/jf503970a

Leong-Morgenthaler, P., Zwahlen, M.C., and Hottinger, H. (1991). Lactose metabolism in *Lactobacillus bulgaricus*: analysis of the primary structure and expression of the genes involved. J. Bacteriol. *173*, 1951–1957. https://doi.org/10.1128/jb.173.6.1951-1957.1991

Ley, R.E., Turnbaugh, P.J., Klein, S., and Gordon, J.I. (2006). Microbial ecology: human gut microbes associated with obesity. Nature *444*, 1022–1023. https://doi.org/doi:10.1038/4441022a

Li, M., Wang, B., Zhang, M., Rantalainen, M., Wang, S., Zhou, H., Zhang, Y., Shen, J., Pang, X., Zhang, M., *et al.* (2008). Symbiotic gut microbes modulate human metabolic phenotypes. Proc. Natl. Acad. Sci. U.S.A. *105*, 2117–2122. https://doi.org/10.1073/pnas.0712038105

Lin, A.E., Autran, C.A., Szyszka, A., Escajadillo, T., Huang, M., Godula, K., Prudden, A.R., Boons, G.J., Lewis, A.L., Doran, K.S., *et al.* (2017). Human milk oligosaccharides inhibit growth of group B *Streptococcus*. J. Biol. Chem. *292*, 11243–11249. https://doi.org/10.1074/jbc.M117.789974

Lin, S.H., Chou, L.M., Chien, Y.W., Chang, J.S., and Lin, C.I. (2016). Prebiotic effects of xylooligosaccharides on the improvement of microbiota balance in human subjects. Gastroenterol. Res. Pract. *2016*, 5789232. https://doi.org/10.1155/2016/5789232

Liu, B., and Newburg, D.S. (2013). Human milk glycoproteins protect infants against human pathogens. Breastfeed. Med. *8*, 354–362. https://doi.org/10.1089/bfm.2013.0016

Liu, J.R., Yu, B., Liu, F.H., Cheng, K.J., and Zhao, X. (2005). Expression of rumen microbial fibrolytic enzyme genes in probiotic *Lactobacillus reuteri*. Appl. Environ. Microbiol. *71*, 6769–6775. https://doi.org/10.1128/AEM.71.11.6769-6775.2005

LoCascio, R.G., Desai, P., Sela, D.A., Weimer, B., and Mills, D.A. (2010). Broad conservation of milk utilization genes in *Bifidobacterium longum* subsp. *infantis* as revealed by comparative genomic hybridization. Appl. Environ. Microbiol. *76*, 7373–7381. https://doi.org/10.1128/AEM.00675-10

Lokman, B.C., van Santen, P., Verdoes, J.C., Krüse, J., Leer, R.J., Posno, M., and Pouwels, P.H. (1991). Organization and characterization of three genes involved in D-xylose catabolism in *Lactobacillus pentosus*. Mol. Gen. Genet. *230*, 161–169.

Lozupone, C.A., Stombaugh, J.I., Gordon, J.I., Jansson, J.K., and Knight, R. (2012). Diversity, stability and resilience of the human gut microbiota. Nature *489*, 220–230. https://doi.org/10.1038/nature11550

Macfarlane, G.T., Steed, H., and Macfarlane, S. (2008). Bacterial metabolism and health-related effects of galacto-oligosaccharides and other prebiotics. J. Appl. Microbiol. *104*, 305–344. https://doi.org/10.1111/j.1365-2672.2007.03520.x

Makras, L., Van Acker, G., and De Vuyst, L. (2005). *Lactobacillus paracasei* subsp. *paracasei* 8700:2 degrades inulin-type fructans exhibiting different degrees of polymerization. Appl. Environ. Microbiol. *71*, 6531–6537. https://doi.org/10.1128/aem.71.11.6531-6537.2005

Malang, S.K., Maina, N.H., Schwab, C., Tenkanen, M., and Lacroix, C. (2015). Characterization of exopolysaccharide and ropy capsular polysaccharide formation by *Weissella*. Food Microbiol. *46*, 418–427. https://doi.org/10.1016/j.fm.2014.08.022

Marionneau, S., Cailleau-Thomas, A., Rocher, J., Le Moullac-Vaidye, B., Ruvoën, N., Clément, M., and Le Pendu, J. (2001). ABH and Lewis histo-blood group antigens, a model for the meaning of oligosaccharide diversity in the face of a changing world. Biochimie *83*, 565–573. https://doi.org/10.1016/S0300-9084(01)01321-9

Martin-Verstraete, I., Charrier, V., Stülke, J., Galinier, A., Erni, B., Rapoport, G., and Deutscher, J. (1998). Antagonistic effects of dual PTS-catalysed phosphorylation on the *Bacillus subtilis* transcriptional activator LevR. Mol. Microbiol. *28*, 293–303. https://doi.org/10.1046/j.1365-2958.1998.00781.x

Martín, R., Heilig, G.H., Zoetendal, E.G., Smidt, H., and Rodríguez, J.M. (2007). Diversity of the *Lactobacillus* group in breast milk and vagina of healthy women and potential role in the colonization of the infant gut. J. Appl. Microbiol. *103*, 2638–2644. https://doi.org/10.1111/j.1365-2672.2007.03497.x

Mazé, A., Boël, G., Poncet, S., Mijakovic, I., Le Breton, Y., Benachour, A., Monedero, V., Deutscher, J., and Hartke, A. (2004). The *Lactobacillus casei* ptsHI47T mutation causes overexpression of a LevR-regulated but RpoN-independent operon encoding a mannose class phosphotransferase system. J. Bacteriol. *186*, 4543–4555. https://doi.org/10.1128/JB.186.14.4543-4555.2004

Membrez, M., Blancher, F., Jaquet, M., Bibiloni, R., Cani, P.D., Burcelin, R.G., Corthesy, I., Macé, K., and Chou, C.J. (2008). Gut microbiota modulation with norfloxacin and ampicillin enhances glucose tolerance in mice. FASEB J. *22*, 2416–2426. https://doi.org/10.1096/fj.07-102723

Merry, R.J., Winters, A.L., Thomas, P.I., Müller, M., and Müller, T. (1995). Degradation of fructans by epiphytic and inoculated lactic acid bacteria and by plant enzymes during ensilage of normal and sterile hybrid ryegrass. J. Appl. Bacteriol. *79*, 583-591. https://doi.org/10.1111/j.1365-2672.1995.tb00941.x

Michlmayr, H., Hell, J., Lorenz, C., Böhmdorfer, S., Rosenau, T., and Kneifel, W. (2013). Arabinoxylan oligosaccharide hydrolysis by family 43 and 51 glycosidases from *Lactobacillus brevis* DSM 20054. Appl. Environ. Microbiol. *79*, 6747–6754. https://doi.org/10.1128/AEM.02130-13

Michlmayr, H., Schümann, C., Kulbe, K.D., and del Hierro, A.M. (2011). Heterologously expressed family 51 alpha-L-arabinofuranosidases from *Oenococcus oeni* and *Lactobacillus brevis*. Appl. Environ. Microbiol. *77*, 1528–1531. https://doi.org/10.1128/AEM.01385-10

Miller, E.A., Beasley, D.E., Dunn, R.R., and Archie, E.A. (2016). Lactobacilli dominance and vaginal pH: why is the human vaginal microbiome unique? Front. Microbiol. *7*, 1936. https://doi.org/10.3389/fmicb.2016.01936

Møller, M.S., Fredslund, F., Majumder, A., Nakai, H., Poulsen, J.C., Lo Leggio, L., Svensson, B., and Abou Hachem, M. (2012). Enzymology and structure of the GH13_31 glucan 1,6-α-glucosidase that confers isomaltooligosaccharide utilization in the probiotic *Lactobacillus acidophilus* NCFM. J. Bacteriol. *194*, 4249–4259. https://doi.org/10.1128/JB.00622-12

Møller, M.S., Goh, Y.J., Rasmussen, K.B., Cypryk, W., Celebioglu, H.U., Klaenhammer, T.R., Svensson, B., and Abou Hachem, M. (2017). An extracellular cell-attached pullulanase confers branched α-glucan utilization in human gut *Lactobacillus acidophilus*. Appl. Environ. Microbiol. *83*, e00402-17. https://doi.org/10.1128/AEM.00402-17

Monedero, V., Yebra, M.J., Poncet, S., and Deutscher, J. (2008). Maltose transport in *Lactobacillus casei* and its regulation by inducer exclusion. Res. Microbiol. *159*, 94–102. https://doi.org/10.1016/j.resmic.2007.10.002

Moraïs, S., Shterzer, N., Lamed, R., Bayer, E.A., and Mizrahi, I. (2014). A combined cell-consortium approach for lignocellulose degradation by specialized *Lactobacillus plantarum* cells. Biotechnol. Biofuels 7, 112. https://doi.org/10.1186/1754-6834-7-112

Moran, A.P., Gupta, A., and Joshi, L. (2011). Sweet-talk: role of host glycosylation in bacterial pathogenesis of the gastrointestinal tract. Gut *60*, 1412–1425. https://doi.org/10.1136/gut.2010.212704

Morlon-Guyot, J., Mucciolo-Roux, F., Rodriguez Sanoja, R., and Guyot, J.P. (2001). Characterization of the *L. manihotivorans* alpha-amylase gene. DNA Seq. *12*, 27–37. https://doi.org/10.3109/10425170109042048

Müller, M., and Lier, D. (1994). Fermentation of fructans by epiphytic lactic acid bacteria. J. Appl. Bacteriol. *76*, 406–411. https://doi.org/10.1111/j.1365-2672.1994.tb01647.x

Müller, M., and Seyfarth, W. (1997). Purification and substrate specificity of an extracellular fructanhydrolase from *Lactobacillus paracasei* ssp. *paracasei* P 4134. New Phytol. *136*, 89–96. https://doi.org/10.1046/j.1469-8137.1997.00725.x

Müller, M., and Steller, J. (1995). Comparative studies of the degradation of grass fructan and inulin by strains of *Lactobacillus paracasei* subsp. *paracasei* and *Lactobacillus plantarum*. J. Appl. Bacteriol. *78*, 229–236. https://doi.org/10.1111/j.1365-2672.1995.tb05021.x

Muscariello, L., Vastano, V., Siciliano, R.A., Sacco, M., and Marasco, R. (2011). Expression of the *Lactobacillus plantarum malE* gene is regulated by CcpA and a MalR-like protein. J. Microbiol. *49*, 950–955. https://doi.org/10.1007/s12275-011-0495-5

Mysore, J.V., Wigginton, T., Simon, P.M., Zopf, D., Heman-Ackah, L.M., and Dubois, A. (1999). Treatment of *Helicobacter pylori* infection in rhesus monkeys using a novel antiadhesion compound. Gastroenterology *117*, 1316–1325. https://doi.org/10.1016/S0016-5085(99)70282-9

Nácher-Vázquez, M., Iturria, I., Zarour, K., Mohedano, M.L., Aznar, R., Pardo, M.Á., and López, P. (2017). Dextran production by *Lactobacillus sakei* MN1 coincides with reduced autoagglutination, biofilm formation and epithelial cell adhesion. Carbohydr. Polym. *168*, 22–31. https://doi.org/10.1016/j.carbpol.2017.03.024

Nakai, H., Baumann, M.J., Petersen, B.O., Westphal, Y., Schols, H., Dilokpimol, A., Hachem, M.A., Lahtinen, S.J., Duus, J.Ø., and Svensson, B. (2009). The maltodextrin transport system and metabolism in *Lactobacillus acidophilus* NCFM and production of novel alpha-glucosides through reverse phosphorolysis by maltose phosphorylase. FEBS J. *276*, 7353–7365. https://doi.org/10.1111/j.1742-4658.2009.07445.x

Nakamura, L.K. (1981). *Lactobacillus amylovorus*, a new starch-hydrolyzing species from cattle waste-corn fermentations. Int. J. Syst. Evol. Microbiol. *31*, 56–63. https://doi.org/doi:10.1099/00207713-31-1-56

Nakamura, L.K., and Crowell, C.D. (1979). *Lactobacillus amylophilus*, a new starch-hydrolyzing species from swine waste-corn fermentation.. Dev. Ind. Microbiol. *20*, 532–540.

Neubauer, H., Glaasker, E., Hammes, W.P., Poolman, B., and Konings, W.N. (1994). Mechanism of maltose uptake and glucose excretion in *Lactobacillus sanfrancisco*. J. Bacteriol. *176*, 3007–3012. https://doi.org/10.1128/jb.176.10.3007-3012.1994

Newburg, D.S., Ruiz-Palacios, G.M., and Morrow, A.L. (2005). Human milk glycans protect infants against enteric pathogens. Annu. Rev. Nutr. *25*, 37–58. https://doi.org/10.1146/annurev.nutr.25.050304.092553

Nwosu, C.C., Aldredge, D.L., Lee, H., Lerno, L.A., Zivkovic, A.M., German, J.B., and Lebrilla, C.B. (2012). Comparison of the human and bovine milk N-glycome via high-performance microfluidic chip liquid chromatography and tandem mass spectrometry. J. Proteome Res. *11*, 2912–2924. https://doi.org/10.1021/pr300008u

O'Donnell, M.M., Forde, B.M., Neville, B., Ross, P.R., and O'Toole, P.W. (2011). Carbohydrate catabolic flexibility in the mammalian intestinal commensal *Lactobacillus ruminis* revealed by fermentation studies aligned to genome annotations. Microb. Cell Fact. 10 (Suppl. 1), S12. https://doi.org/10.1186/1475-2859-10-S1-S12

Okano, K., Zhang, Q., Yoshida, S., Tanaka, T., Ogino, C., Fukuda, H., and Kondo, A. (2010). D-lactic acid production from cellooligosaccharides and beta-glucan using L-LDH gene-deficient and endoglucanase-secreting *Lactobacillus plantarum*. Appl. Microbiol. Biotechnol. *85*, 643–650. https://doi.org/10.1007/s00253-009-2111-8

Olivares-Illana, V., López-Munguía, A., and Olvera, C. (2003). Molecular characterization of inulosucrase from *Leuconostoc citreum*: a fructosyltransferase within a glucosyltransferase. J. Bacteriol. *185*, 3606–3612. https://doi.org/10.1128/JB.185.12.3606-3612.2003

Öner, E.T., Hernández, L., and Combie, J. (2016). Review of levan polysaccharide: from a century of past experiences to future prospects. Biotechnol. Adv. *34*, 827–844. https://doi.org/10.1016/j.biotechadv.2016.05.002

Overbeck, T., Steele, J.L., and Broadbent, J.R. (2016). Fermentation of de-oiled algal biomass by *Lactobacillus casei* for production of lactic acid. Bioprocess Biosyst. Eng. *39*, 1817–1823. https://doi.org/10.1007/s00449-016-1656-z

Paludan-Müller, C., Gram, L., and Rattray, F.P. (2002). Purification and characterisation of an extracellular fructan beta-fructosidase from a *Lactobacillus pentosus* strain isolated from fermented fish. Syst. Appl. Microbiol. *25*, 13–20. https://doi.org/10.1078/0723-2020-00101

Petrova, P., and Petrov, K. (2012). Direct starch conversion into L-(+)-lactic acid by a novel amylolytic strain of *Lactobacillus paracasei* B41. Starch *64*, 10–17. https://doi.org/10.1002/star.201100074

Petrova, P., Petrov, K., and Stoyancheva, G. (2013). Starch-modifying enzymes of lactic acid bacteria - structures, properties and applications. Starch *65*, 34–47. https://doi.org/10.1002/star.201200192

Pfeiler, E.A., Azcarate-Peril, M.A., and Klaenhammer, T.R. (2007). Characterization of a novel bile-inducible operon encoding a two-component regulatory system in *Lactobacillus acidophilus*. J. Bacteriol. *189*, 4624–4634. https://doi.org/10.1128/JB.00337-07

Puyet, A., Ibáñez, A.M., and Espinosa, M. (1993). Characterization of the *Streptococcus pneumoniae* maltosaccharide regulator MalR, a member of the LacI-GalR family of repressors displaying distinctive genetic features. J. Biol. Chem. *268*, 25402–25408.

Rastall, R.A., Gibson, G.R., Gill, H.S., Guarner, F., Klaenhammer, T.R., Pot, B., Reid, G., Rowland, I.R., and Sanders, M.E. (2005). Modulation of the microbial ecology of the human colon by probiotics, prebiotics and synbiotics to enhance human health: an overview of enabling science and potential applications. FEMS Microbiol. Ecol. *52*, 145–152. https://doi.org/10.1016/j.femsec.2005.01.003

Roberfroid, M.B., Van Loo, J.A., and Gibson, G.R. (1998). The bifidogenic nature of chicory inulin and its hydrolysis products. J. Nutr. *128*, 11–19. https://doi.org/10.1093/jn/128.1.11

Rodriguez-Colinas, B., de Abreu, M.A., Fernandez-Arrojo, L., de Beer, R., Poveda, A., Jimenez-Barbero, J., Haltrich, D., Ballesteros Olmo, A.O., Fernandez-Lobato, M., and Plou, F.J. (2011). Production of galacto-oligosaccharides by the β-galactosidase from *Kluyveromyces lactis*: comparative analysis of permeabilized cells versus soluble enzyme. J. Agric. Food Chem. *59*, 10477–10484. https://doi.org/10.1021/jf2022012

Rodríguez-Díaz, J., Carbajo, R.J., Pineda-Lucena, A., Monedero, V., and Yebra, M.J. (2013). Synthesis of fucosyl-N-acetylglucosamine disaccharides by transfucosylation using α-L-fucosidases from *Lactobacillus casei*. Appl. Environ. Microbiol. *79*, 3847–3850. https://doi.org/10.1128/AEM.00229-13

Rodríguez-Díaz, J., Monedero, V., and Yebra, M.J. (2011). Utilization of natural fucosylated oligosaccharides by three novel alpha-L-fucosidases from a probiotic *Lactobacillus casei* strain. Appl. Environ. Microbiol. *77*, 703–705. https://doi.org/10.1128/AEM.01906-10

Rodríguez-Díaz, J., Rubio-del-Campo, A., and Yebra, M.J. (2012). *Lactobacillus casei* ferments the N-Acetylglucosamine moiety of fucosyl-α-1,3-N-acetylglucosamine and excretes L-fucose. Appl. Environ. Microbiol. *78*, 4613–4619. https://doi.org/10.1128/AEM.00474-12

Rodriguez Sanoja, R., Morlon-Guyot, J., Jore, J., Pintado, J., Juge, N., and Guyot, J.P. (2000). Comparative characterization of complete and truncated forms of *Lactobacillus amylovorus* alpha-amylase and role of the C-terminal direct repeats in raw-starch binding. Appl. Environ. Microbiol. *66*, 3350–3356. https://doi.org/10.1128/AEM.66.8.3350-3356.2000.

Rubio, R., Jofré, A., Martín, B., Aymerich, T., and Garriga, M. (2014). Characterization of lactic acid bacteria isolated from infant faeces as potential probiotic starter cultures for fermented sausages. Food Microbiol. *38*, 303–311. https://doi.org/10.1016/j.fm.2013.07.015

Ruiz-Palacios, G.M., Cervantes, L.E., Ramos, P., Chavez-Munguia, B., and Newburg, D.S. (2003). *Campylobacter jejuni* binds intestinal H(O) antigen (Fucα1, 2Galβ1, 4GlcNAc), and fucosyloligosaccharides of human milk inhibit its binding and infection. J. Biol. Chem. *278*, 14112–14120. https://doi.org/10.1074/jbc.M207744200

Sangwan, V., Tomar, S.K., Singh, R.R., Singh, A.K., and Ali, B. (2011). Galactooligosaccharides: novel components of designer foods. J. Food Sci. *76*, R103–11. https://doi.org/10.1111/j.1750-3841.2011.02131.x

Saulnier, D.M.A., Molenaar, D., de Vos, W.M., Gibson, G.R., and Kolida, S. (2007). Identification of prebiotic fructooligosaccharide metabolism in *Lactobacillus plantarum* WCFS1 through microarrays. Appl. Environ. Microbiol. *73*, 1753–1765. https://doi.org/10.1128/aem.01151-06

Schnaar, R.L., Suzuki, A., and Stanley, P. (2009). Glycosphingolipids. In Essentials of Glycobiology, Varki, A., Cummings, R.D., Esko, J.D., Freeze, H.H., Stanley, P., Bertozzi, C.R., Hart, G.W., and Etzler, M.E. eds. (Cold Spring Harbor Laboratory Press, New York). https://doi.org/10.1101/glycobiology.3e.011

Sghir, A., Chow, J.M., and Mackie, R.I. (1998). Continuous culture selection of bifidobacteria and lactobacilli from human faecal samples using fructooligosaccharide as selective substrate. J. Appl. Microbiol. *85*, 769–777. https://doi.org/10.1111/j.1365-2672.1998.00590.x

Silvestroni, A., Connes, C., Sesma, F., De Giori, G.S., and Piard, J.C. (2002). Characterization of the *melA* locus for alpha-galactosidase in *Lactobacillus plantarum*. Appl. Environ. Microbiol. *68*, 5464–5471. https://doi.org/10.1128/AEM.68.11.5464-5471.2002

Sommer, F., Anderson, J.M., Bharti, R., Raes, J., and Rosenstiel, P. (2017). The resilience of the intestinal microbiota influences health and disease. Nat. Rev. Microbiol. *15*, 630–638. https://doi.org/10.1038/nrmicro.2017.58

Sonnenburg, J.L., and Bäckhed, F. (2016). Diet-microbiota interactions as moderators of human metabolism. Nature *535*, 56–64. https://doi.org/10.1038/nature18846

Stearns, J.C., Lynch, M.D., Senadheera, D.B., Tenenbaum, H.C., Goldberg, M.B., Cvitkovitch, D.G., Croitoru, K., Moreno-Hagelsieb, G., and Neufeld, J.D. (2011). Bacterial biogeography of the human digestive tract. Sci. Rep. *1*, 170. https://doi.org/10.1038/srep00170

Stefanovic, E., Fitzgerald, G., and McAuliffe, O. (2017). Advances in the genomics and metabolomics of dairy lactobacilli: A review. Food Microbiol. *61*, 33–49. https://doi.org/10.1016/j.fm.2016.08.009

Tannock, G.W. (2004). A special fondness for lactobacilli. Appl. Environ. Microbiol. *70*, 3189–3194. https://doi.org/10.1128/AEM.70.6.3189-3194.2004

Thongaram, T., Hoeflinger, J.L., Chow, J., and Miller, M.J. (2017). Prebiotic galactooligosaccharide metabolism by probiotic lactobacilli and bifidobacteria. J. Agric. Food Chem. *65*, 4184–4192. https://doi.org/10.1021/acs.jafc.7b00851

Thurl, S., Munzert, M., Henker, J., Boehm, G., Müller-Werner, B., Jelinek, J., and Stahl, B. (2010). Variation of human milk oligosaccharides in relation to milk groups and lactational periods. Br. J. Nutr. *104*, 1261–1271. https://doi.org/10.1017/S0007114510002072

Tieking, M., Korakli, M., Ehrmann, M.A., Gänzle, M.G., and Vogel, R.F. (2003). In situ production of exopolysaccharides during sourdough fermentation by cereal and intestinal isolates of lactic acid bacteria. Appl. Environ. Microbiol. *69*, 945–952. https://doi.org/10.1128/aem.69.2.945-952.2003

Tropini, C., Earle, K.A., Huang, K.C., and Sonnenburg, J.L. (2017). The gut microbiome: connecting spatial organization to function. Cell Host Microbe *21*, 433–442. https://doi.org/10.1016/j.chom.2017.03.010

Turnbaugh, P.J., Hamady, M., Yatsunenko, T., Cantarel, B.L., Duncan, A., Ley, R.E., Sogin, M.L., Jones, W.J., Roe, B.A., Affourtit, J.P., *et al.* (2009). A core gut microbiome in obese and lean twins. Nature *457*, 480–484. https://doi.org/10.1038/nature07540

Tzortzis, G., Jay, A.J., Baillon, M.L., Gibson, G.R., and Rastall, R.A. (2003). Synthesis of alpha-galactooligosaccharides with alpha-galactosidase from *Lactobacillus reuteri* of canine origin. Appl. Microbiol. Biotechnol. *63*, 286–292. https://doi.org/10.1007/s00253-003-1426-0

Urashima, T., Taufik, E., Fukuda, K., and Asakuma, S. (2013). Recent advances in studies on milk oligosaccharides of cows and other domestic farm animals. Biosci. Biotechnol. Biochem. *77*, 455–466. https://doi.org/10.1271/bbb.120810

Vaarala, O., Atkinson, M.A., and Neu, J. (2008). The 'perfect storm' for type 1 diabetes: the complex interplay between intestinal microbiota, gut permeability, and mucosal immunity. Diabetes *57*, 2555–2562. https://doi.org/10.2337/db08-0331

van Arkel, J., Sévenier, R., Hakkert, J.C., Bouwmeester, H.J., Koops, A.J., and van der Meer, I.M. (2013). Tailor-made fructan synthesis in plants: a review. Carbohydr. Polym. *93*, 48–56. https://doi.org/10.1016/j.carbpol.2012.02.001

Van Geel-Schutten, G.H., Faber, E.J., Smit, E., Bonting, K., Smith, M.R., Ten Brink, B., Kamerling, J.P., Vliegenthart, J.F.G., and Dijkhuizen, L. (1999). Biochemical and structural characterization of the glucan and fructan exopolysaccharides synthesized by the *Lactobacillus reuteri* wild-type strain and by mutant strains. Appl. Environ. Microbiol. *65*, 3008–3014.

van Hijum, S.A., van Geel-Schutten, G.H., Rahaoui, H., van der Maarel, M.J., and Dijkhuizen, L. (2002). Characterization of a novel fructosyltransferase from *Lactobacillus reuteri* that synthesizes high-molecular-weight inulin and inulin oligosaccharides. Appl. Environ. Microbiol. *68*, 4390–4398. https://doi.org/10.1128/aem.68.9.4390-4398.2002

Velikova, P., Petrov, K., and Petrova, P. (2017). The cell wall anchored β-fructosidases of *Lactobacillus paracasei*: overproduction, purification, and gene expression control. Process Biochem. *52*, 53–62. https://doi.org/10.1016/j.procbio.2016.10.010

Vera, C., Córdova, A., Aburto, C., Guerrero, C., Suárez, S., and Illanes, A. (2016). Synthesis and purification of galacto-oligosaccharides: state of the art. World J. Microbiol. Biotechnol. *32*, 197. https://doi.org/10.1007/s11274-016-2159-4

Wang, L., Yang, Y., Cai, B., Cao, P., Yang, M., and Chen, Y. (2014a). Coexpression and secretion of endoglucanase and phytase genes in *Lactobacillus reuteri*. Int. J. Mol. Sci. *15*, 12842–12860. https://doi.org/10.3390/ijms150712842

Wang, Y., Black, B.A., Curtis, J.M., and Gänzle, M.G. (2014b). Characterization of α-galacto-oligosaccharides formed via heterologous expression of α-galactosidases from *Lactobacillus reuteri* in *Lactococcus lactis*. Appl. Microbiol. Biotechnol. *98*, 2507–2517. https://doi.org/10.1007/s00253-013-5145-x

Weckx, S., Allemeersch, J., Van der Meulen, R., Vrancken, G., Huys, G., Vandamme, P., Van Hummelen, P., and De Vuyst, L. (2011). Metatranscriptome analysis for insight into whole-ecosystem gene expression during spontaneous wheat and spelt sourdough fermentations. Appl. Environ. Microbiol. *77*, 618–626. https://doi.org/10.1128/AEM.02028-10

Winters, Merry, Müller, Davies, Pahlow, and Müller (1998). Degradation of fructans by epiphytic and inoculant lactic acid bacteria during ensilage of grass. J. Appl.

Microbiol. *84*, 304–312. https://doi.org/10.1046/j.1365-2672.1998.00342.x.

Yazawa, K., and Tamura, Z. (1982). Search for sugar sources for selective increase of bifidobacteria. Bifidobact. Microfl. *1*, 39–44.

Yen, C.H., Tseng, Y.H., Kuo, Y.W., Lee, M.C., and Chen, H.L. (2011). Long-term supplementation of isomalto-oligosaccharides improved colonic microflora profile, bowel function, and blood cholesterol levels in constipated elderly people – a placebo-controlled, diet-controlled trial. Nutrition *27*, 445–450. https://doi.org/10.1016/j.nut.2010.05.012

Yu, Z.T., Chen, C., and Newburg, D.S. (2013). Utilization of major fucosylated and sialylated human milk oligosaccharides by isolated human gut microbes. Glycobiology *23*, 1281–1292. https://doi.org/10.1093/glycob/cwt065

Production of Lactate Using *Lactobacillus* 4

Mariana C. Allievi[1,2*], María Mercedes Palomino[1,2] and Sandra M. Ruzal[1,2]

[1]Universidad de Buenos Aires, Facultad de Ciencias Exactas y Naturales, Departamento de Química Biológica, Buenos Aires, Argentina.
[2]CONICET – Universidad de Buenos Aires, Instituto de Química Biológica de la Facultad de Ciencias Exactas y Naturales (IQUIBICEN), Buenos Aires, Argentina.

*Correspondence: mallievi@qb.fcen.uba.ar

https://doi.org/10.21775/9781910190890.04

Abstract

Lactic acid is used in the food, cosmetic, textile, pharmaceutical and chemical industries. In the last decades, the interest in lactate production by the fermentation route has increased due to the prospects of environmental friendliness and the use of renewable sources instead of petroleum products. Recently, the enlarged manufacture of the biodegradable polylactic acid polymer (PLA) has increased the global interest in the production of L-lactic acid and D-lactic acid. In this chapter we will focus on the importance of lactic acid, and the main difficulties and the solutions that were designed to increase the production of lactic acid will be addressed. It will focus on the genetic engineering approaches to improve the producing strains of the genus *Lactobacillus*.

Introduction

Lactic acid has applications in the pharmaceutical, chemical, textile and foods industries. Lactic acid is used as acidulant/flavouring/pH-buffering agent or inhibitor of bacterial spoilage in a wide variety of processed foods. It has received great attention as a precursor of the biodegradable polylactic acid (Narayanan *et al.*, 2004; John *et al.*, 2007). Great amounts of lactate are produced by lactic acid bacteria, such as *Lactobacillus*. The metabolism of sugars by *Lactobacillus* is characterized by the production of lactate through the action of lactate dehydrogenase enzyme, as the main product of its strictly fermentative metabolism. The enhancement of lactic acid production has been studied under various chemical, physical, biochemical and media components. There is consequently much research to find cheaper supplements, which could be used as alternatives. The desirable characteristics of industrial microorganisms are their ability to rapidly and completely ferment cheap raw materials.

Lactobacillus genus

The genus *Lactobacillus* is a heterogeneous group with a variable percentage of G + C in the DNA of 33 –55%. To date it contains 236 species, according to the LPSN site (http://www.bacterio.net/). It has a strictly fermentative metabolism with high nutritional requirements. They are associated with different habitats such as vegetables, dairy products, meats, gastrointestinal tract, genital and oral cavity.

The general utility of the *Lactobacillus* species is related to their GRAS (generally recognized as safe) status and will be dependent on the availability of cost-effective methods for production and delivery of viable cultures. They have high nutritional requirements, which need to be provided within the growth medium. Among them, carbohydrates, fatty acids or esters of fatty acids, salts, nucleic acid derivatives and vitamins should be supplied. They are also auxotrophic for many amino acids. For this reason, growing lactobacilli often becomes annoying. However, industrial wastes that provide all these elements can be used to lower the costs of producing desirable metabolites such as lactic acid, for which this chapter focuses.

Metabolism of sugar by *Lactobacillus*

Lactic acid is produced as a main product of the metabolism of sugars in *Lactobacillus*. The metabolism of carbohydrates is the main source of energy and involves phosphorylation at the substrate level. In addition to the ATP synthesized during glycolysis, ATP can be derived from acetyl phosphate, which is formed in reactions catalysed by pyruvate oxidase or pyruvate decarboxylase (Hammes and

Table 4.1 List of the species of the genus *Lactobacillus*, type of glucose fermentation of the species and lactic acid isomer produced

Name of species	Type of glucose fermentation	Lactic acid isomer (s)
L. acidophilus (Moro 1900) Hansen and Mocquot 1970	OHO	DL
L. amylophilus Nakamura and Crowell 1979	OHO	L
L. brevis (Orla-Jensen 1919) Bergey *et al.*, 1934; Doelle 1971	OHE	DL
L. buchneri (Henneberg 1903) Bergey *et al.;* 1923, Zou *et al.*, 2015	OHE	L
L. casei (Orla-Jensen 1916) Hansen and Lessel, 1971	FHE	L
L. coryniformis subsp. *coryniformis* Abo-Elnaga and Kandler, 1965	FHE	DL
L. crispatus (Brygoo and Aladame, 1953) Cato *et al.*, 1983	OHO	DL
L. curvatus subsp. *curvatus* (Troili-Petersson 1903) Abo-Elnaga and Kandler, 1965	FHE	DL
L. delbrueckii subsp. *delbrueckii* (Leichmann 1896) Beijerinck, 1901	OHO	D
L. fermentum Beijerinck, 1901	OHE	D
L. gasseri Lauer and Kandler, 1980	OHO	DL
L. helveticus (Orla-Jensen, 1919) Bergey *et al.*, 1925	OHO	DL
L. intestinalis (Hemme, 1974) Fujisawa *et al.*,1990	FHE	DL
L. jensenii Gasser *et al.*, 1970	FHE	D
L. johnsonii Fujisawa *et al.*,1992	OHO	DL
L. kefiri. Kandler and Kunath, 1983	OHE	DL
L. murinus Hemme *et al.*, 1980	FHE	L
L. paracasei subsp. *paracasei* Collins *et al.*, 1989	FHE	L
L. paraplantarum Curk *et al.*,1996	FHE	DL
L. pentosus (Fred *et al.*, 1921) Zanoni *et al.*, 1987	FHE	DL
L. plantarum (Orla-Jensen 1919) Bergey *et al.*, 1923	FHE	DL
L. reuteri Kandler *et al.*,1980, Ruiz-Moyano *et al.*, 2009	OHE	DL
L. rhamnosus (Hansen 1968) Collins *et al.*, 1989	FHE	L
L. sakei subsp. *sakei* corrig. Katagiri *et al.*, 1934	FHE	L (D<15%)
L. salivarius subsp. *salivarius* Rogosa *et al.*, 1953	OHO	L
L. vaginalis Embley *et al.*, 1989	OHE	DL
L. zeae (Kuznetsov 1959) Dicks *et al.*,1996	FHE	L

FHE, facultatively heterofermentative.
OHE, obligately heterofermentative.
OHO, obligately homofermentative.
In parentheses, the author and date of the first isolation.
Adapted and modified from Hammes and Hertel (2006).

Hertel, 2006). Lactobacilli can be grouped as both homofermenters and heterofermenters depending on the type of end product (Table 4.1). The physiological basis is the presence of the enzymes fructose 1-6 diphosphate aldolase and phosphoketolase. These are keys in the homo or heterofermentative metabolism of hexoses and pentoses, respectively (Kandler and Weiss, 1986) (Fig. 4.1). The division includes three groups:

- Group 1: Homofermentative group. They ferment glucose to lactic acid and do not ferment pentoses or gluconate. Homofermentative *Lactobacillus* can convert sugars exclusively into lactic acid through the Embden-Meyerhof pathway (EMP). They lack phosphoketolase enzyme and possess the enzyme fructose 1-6 diphosphate aldolase. Examples of the group are *L. helveticus*, *L. acidophilus*, and *L. delbrueckii*.

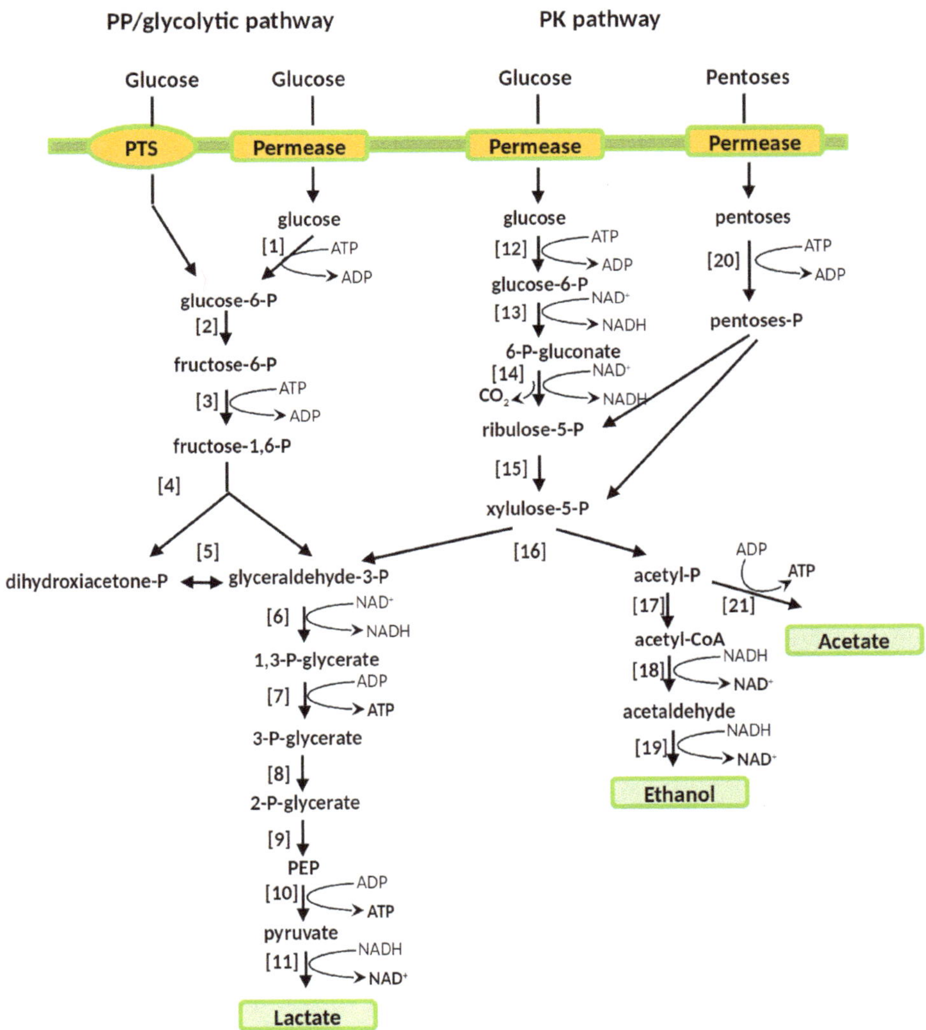

Figure 4.1 Metabolic pathways for lactic acid production by *Lactobacillus*. Enzymes: (1) hexokinase; (2) glucose 6-phosphate isomerase; (3) 6-phosphofructokinase; (4) fructose 1,6 bisphosphate aldolase; (5) triosephosphate isomerase; (6) glyceraldehyde 3-phosphate dehydrogenase; (7) phosphoglycerate kinase; (8) phosphoglycerate mutase; (9) enolase; (10) pyruvate kinase; (11) lactate dehydrogenase; (12) hexokinase; (13) glucose 6-phosphate dehydrogenase; (14) 6-phosphogluconate dehydrogenase; (15) ribulose 5-phosphate 3-epimerase; (16) phosphoketolase; (17) phosphotransacetylase; (18) aldehyde dehydrogenase; (19) alcohol dehydrogenase; (20) pentose kinase; (21) acetate kinase. PTS: phosphotransferase system. PP: pentose phosphate. PK: phosphoketolase. PEP: phosphoenolpyruvate. Modified from Fuente-Hernández *et al.*, 2013 and Abdel-Rahman *et al.*, 2013.

- Group 2: Facultative heterofermentative group. They ferment hexoses through EMP to lactate, use gluconate with gas production. They also ferment pentoses by a phosphoketolase (PK) inducible by pentoses (through the PK pathway) producing lactate and acetic acid. Examples are *L. casei*, *L. plantarum* and *L. paracasei*.
- Group 3: Heterofermentative group. They are obligate heterofermenters. They use hexose through the phosphoketolase (PK) pathway to produce lactic acid, acetic acid and/or ethanol and carbon dioxide. They use pentoses via 6-phosphogluconic acid with production of lactic and acetic acid. Examples of the group are *L. fermentarum* and *L. reuteri*.

The homolactic species use hexoses by the glycolytic route, while the heterofermentative species, by the 6-phosphogluconate route. This pathway requires phosphoketolase, an enzyme that is not present in obligate homofermenters, but is present in facultative heterofermentative species. The fermentable carbohydrates and the products formed depend on the environment. Glucose represses the activity of ketolases in the facultative heterofermenters, and the pH also exerts effects on the type of fermentable carbohydrates used (Carr *et al.*, 1987; Hammes and Hertel, 2006).

Glucose is transported by permease and is phosphorylated in the cytoplasm. Some homofermentative lactobacilli additionally contain a phosphoenol pyruvate (PEP): glucose phosphotransferase system that is

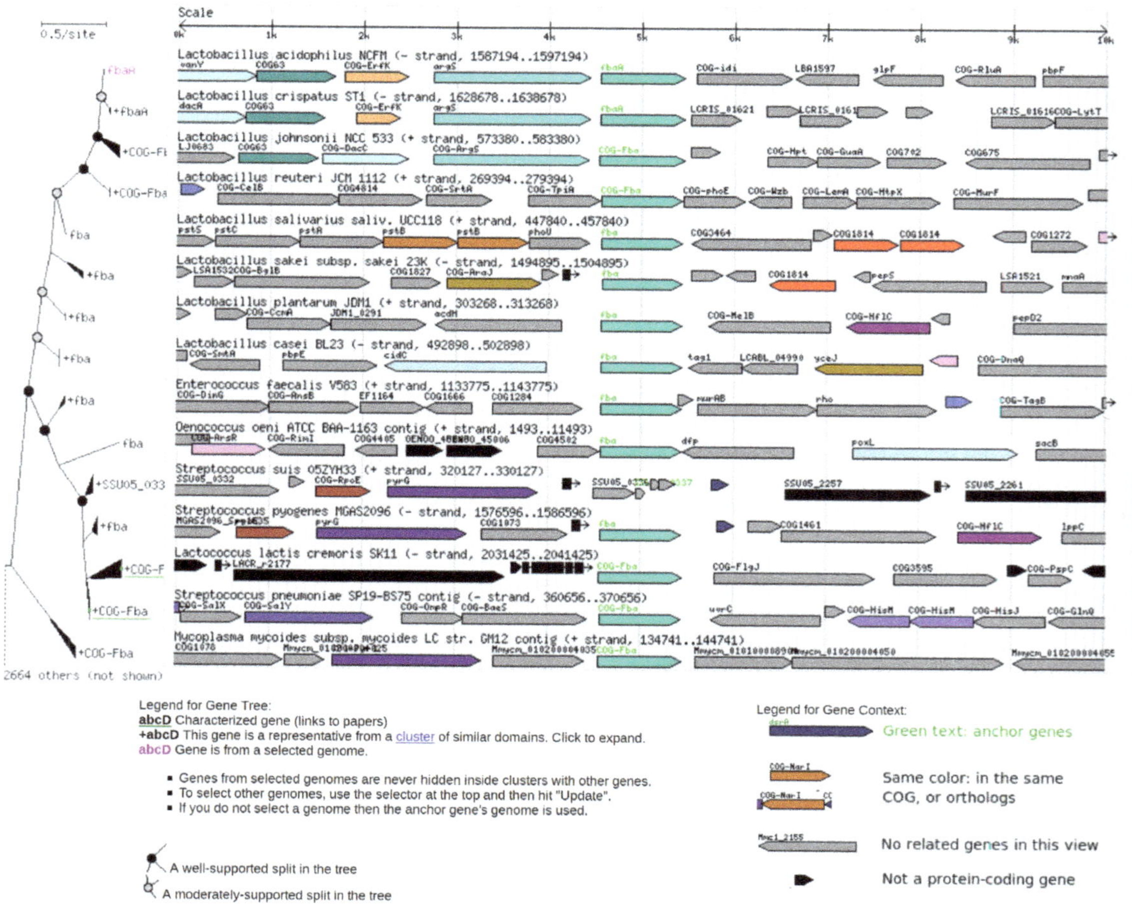

Figure 4.2 Gene tree for *fbaA*: fructose-bisphosphate aldolase (LBA1599), 303 amino acid from *Lactobacillus acidophilus* NCFM at http://www.microbesonline.org.

lacking in heterofermentative species. The presence of the specific PEP-dependent sugar: phosphotransferase system (PTS) in *L. casei* was detected (Romano *et al.*, 1979; London and Hausman, 1982; Chassy and Thompson, 1983; Viana *et al.*, 2005a). In lactose-fermenting lactobacilli, both permease and PTS appear to be present (Hammes and Hertel, 2006).

Fig. 4.2 shows the organization of the genes coding for aldolases, in a gene tree centred on the *fbaA* gene of *L. acidophilus* NCFM. The aldolase genes are orthologous in species that belong to the functional homofermentative group. Within the gene tree, aldolase genes have high sequence homology between species related by their fermentative metabolism, for example, when compared to the genetic sequence coding for fructose-bisphosphate aldolase from *L. acidophilus* NCFM: 91.75% *L. crispatus*, 81.64% *L. johnsonii*. These three strains have homofermentative metabolism. Then, within the same tree continue *L. reuteri* (75.35%) and *L. salivarius* (70.00%), both obligately heterofermenters; and *L. sakei* (62.72%) and *L. plantarum* (62.72%) facultative heterofermenters.

Figs. 4.3 and 4.4 illustrate the organization of the gene tree of the phosphoketolase enzymes of

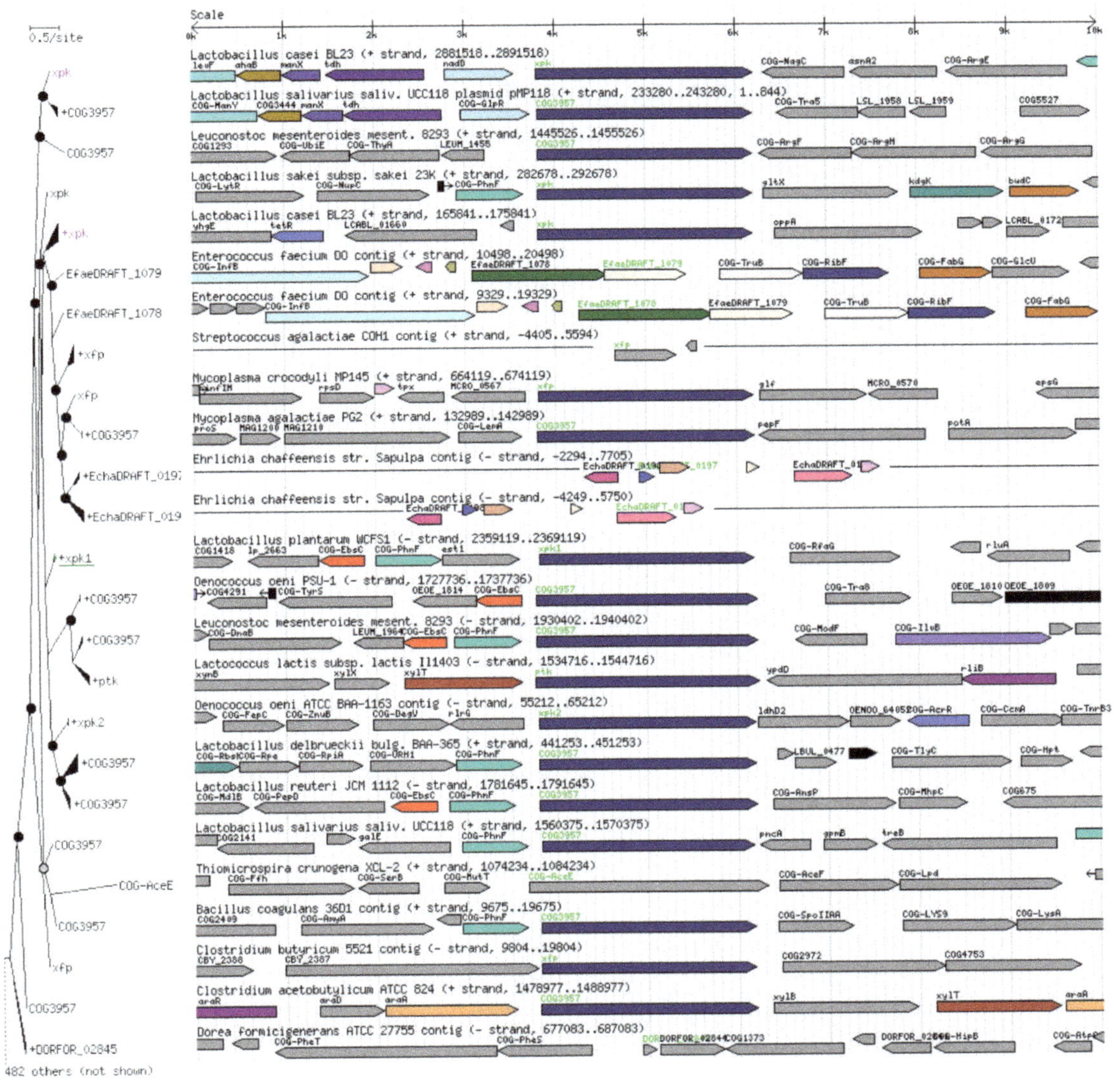

Figure 4.3 Gene tree for *xpk*: Xylulose-5-phosphate phosphoketolase (LCABL_29270), 795 amino acid from *Lactobacillus casei* BL23 at http://www.microbesonline.org.

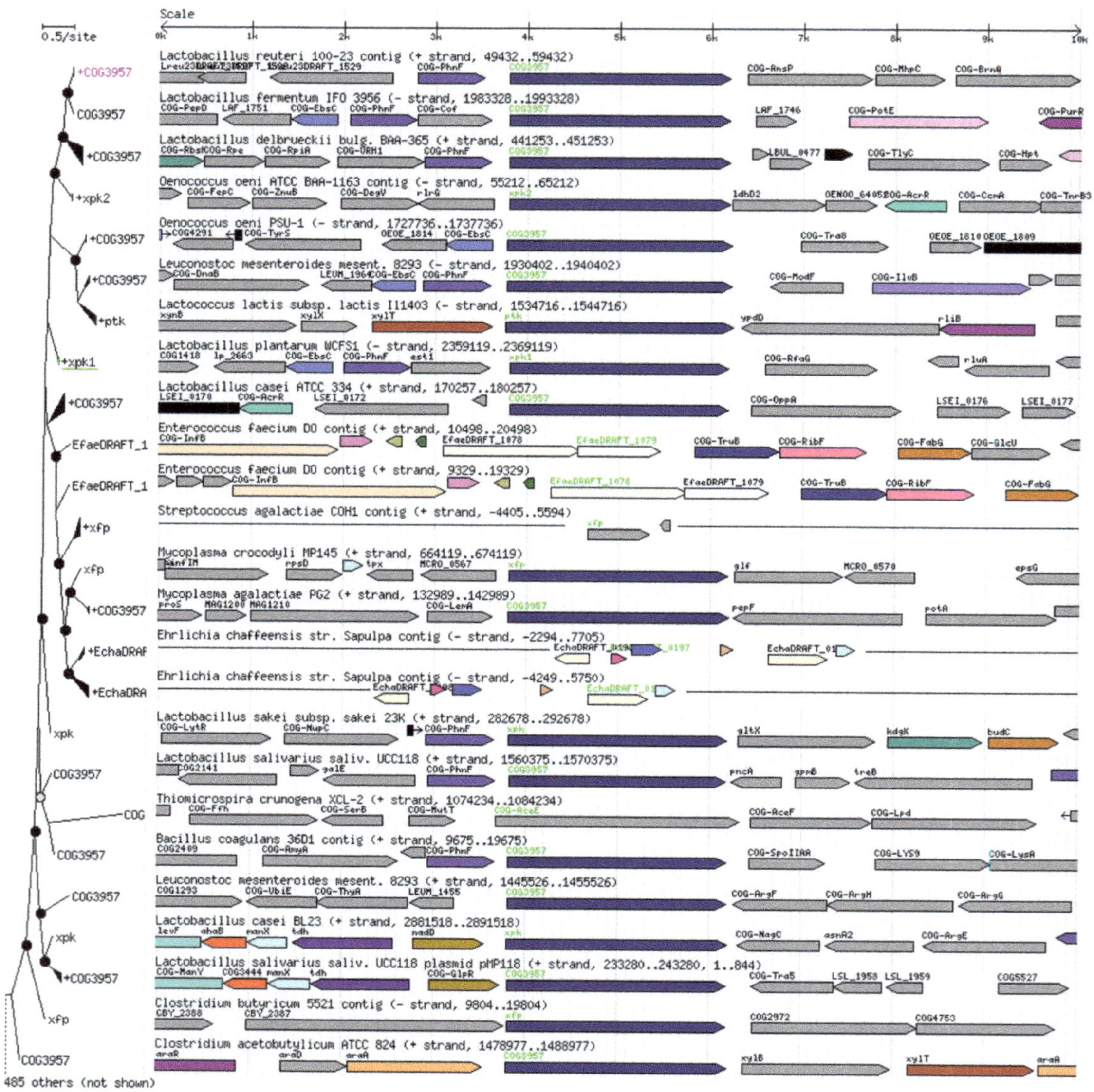

Figure 4.4 Gene tree for COG3957: Fructose-6-phosphate phosphoketolase (Lreu23DRAFT_1531), 803 amino acid from *Lactobacillus reuteri* at http://www.microbesonline.org.

some heterofermentative lactobacilli. Again, related species belonging to the group of facultative heterofermentative lactobacilli show greater sequence homology than obligate heterofermenters. On the other hand, when the tree centred on *L. casei* BL23 is analysed, the presence of paralogues genes encoded in the same strand is observed.

In the genus *Lactobacillus*, lactate is produced by the activity of the enzyme LDH. It would be expected that this function is encoded not only by one but by a group of genes. In *Lactobacillus casei* BL23 the presence of the *ldh* gene (main responsible for the synthesis of lactic acid) and three additional *ldh* genes were reported, which have a minor role in the reduction of pyruvic acid. On the other hand, the gene coding for the HicD enzyme synthesizes D-lactate, a product that increases when the *ldh* gene is mutated (Rico *et al.*, 2008). When *ldh* gene tree centred on *L. casei* BL23 is analysed, these other genes appear as paralogues (Fig. 4.5).

The L- and D-Ldh enzymes belong to different protein families and studies suggest that the processes of their evolution are also different. Very little analysis has been done regarding the molecular

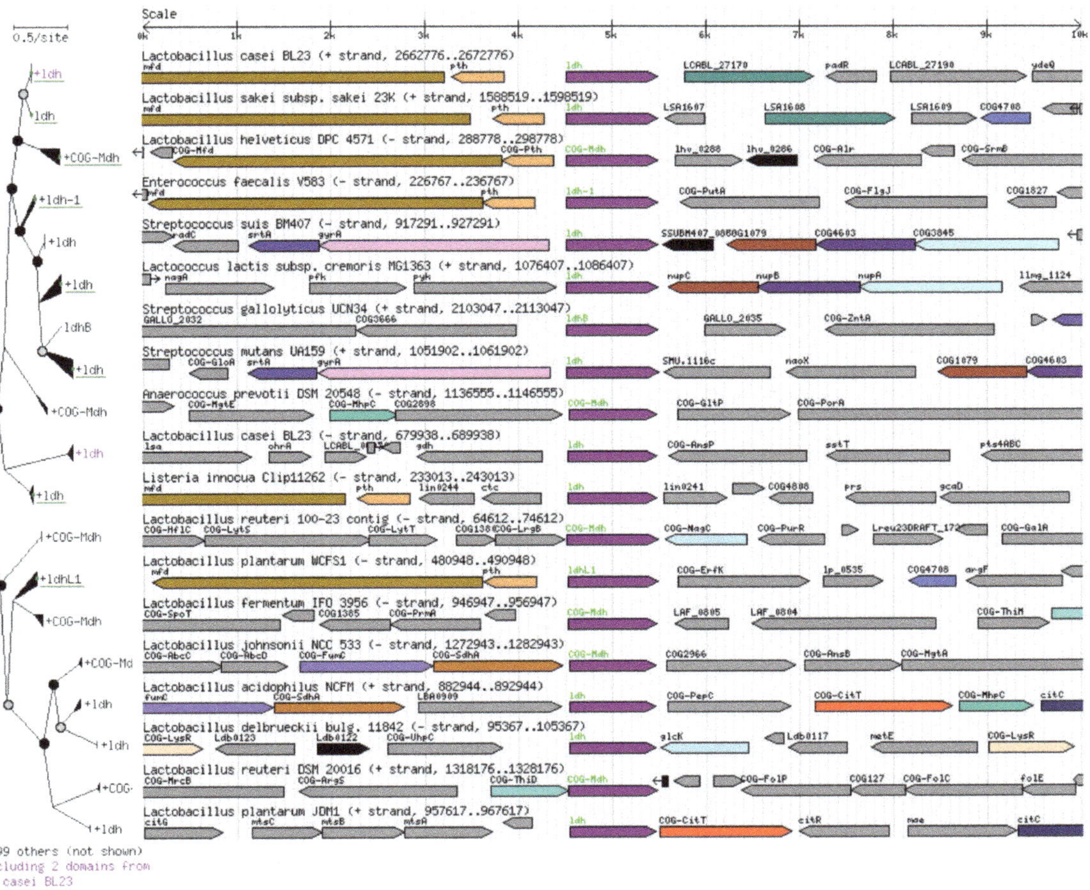

Figure 4.5 Gene tree for L-*ldh*: L-lactate dehydrogenase (L-LDH) (LCABL_27160), 326 amino acid from *Lactobacillus casei* BL23 at http://www.microbesonline.org.

evolution of these two enzymes, mainly due to the presence of multiple genetic copies. Zhang and colleagues (2015) evaluated both the phylogenetic tree of the *Lactobacillus* species based on the *ldh* genes and 16S rRNA showed that the tree generated by the 16S rRNA gene is not the best marker, but can be complemented with the *d-ldh* gene tree. This tree better reconstructed the relationship phylogenetically. Unlike what happens with the *l-ldh* gene, *d-ldh* is a molecular marker for phylogenetic studies. Thus, the tree made with the *d-ldh* gene (Fig. 4.6) shows that there is evolutionary correspondence.

Homofermentative lactobacilli are of great industrial interest because the unique product is lactate. Heterofermentative lactobacilli produce by-products besides lactic acid, and metabolic or genetic engineering is performed to improve the yield.

Lactic acid

Lactic acid was discovered in 1780 by the Swedish chemist Carl Wilhelm Scheele, who isolated it from sour milk. In 1857, Pasteur in his study of the production of lactic acid by fermentation showed that the fermentation was carried out by living organisms. Subsequently, Charles Avery (in Littleton, 1881) performed the first lactic fermentation on an industrial scale.

The main use of lactic acid is in the food related industry. It has a mild acidic taste, non-volatile, odourless, good preservative and GRAS grade. Lactate, 2-hydroxypropionic acid, can be achieved by chemical synthesis or fermentation. Chemical synthesis provides only the racemic mixture, while the fermentative production route allows obtaining optically pure D- and L-lactic acid. The main species used for lactate production by fermentative pathway belong to the genus

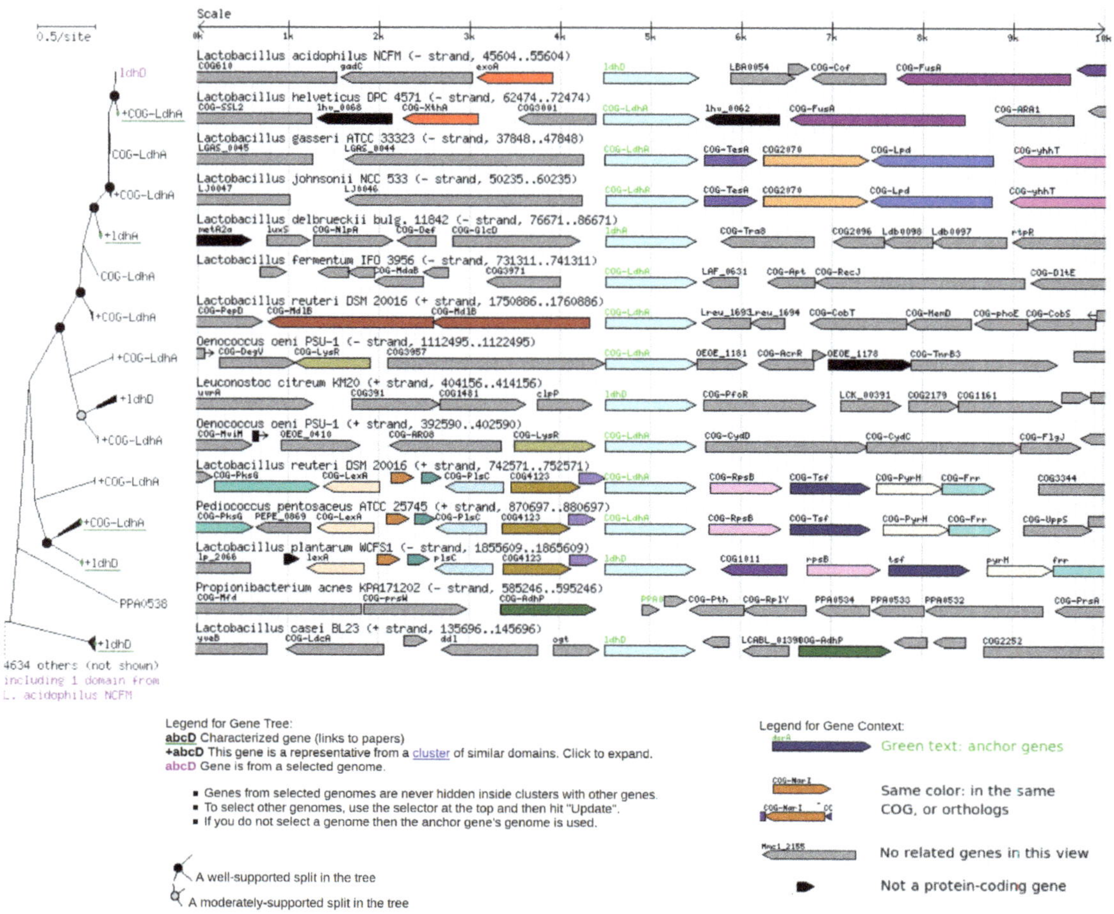

Figure 4.6 Gene tree for *ldhD*: D-lactate dehydrogenase (LBA0055), 349 amino acid from *Lactobacillus acidophilus* NCFM at http://www.microbesonline.org.

Lactobacillus and *Lactococcus*. Other advantages of the use of microorganisms for the production of lactate are that it can be produced from industrial waste, cheap substrates, and allows high product specificity (Pandey *et al.*, 2000).

The United States is the largest market for lactic acid. However, Europe has been projected as a growing market in the last five years. Food additives represent the largest application, but biodegradable plastics are the fastest end-use application for this acid. The annual price of lactic acid in the market is between $1500 and 1700 per ton (ICIS site). The production of lactic acid in 2009 was 258,000 tons, and it is estimated that by 2017 will be close to 370,000 tons (Global Industry Analysts Inc., 2011). With the growing interest in PLA production, annual demand has grown significantly.

Polylactic acid: PLA

Lactic acid is a precursor of polylactic acid (PLA), which is a biodegradable plastic with renewed importance. One of the first applications of PLA was in food packaging (Yoshito *et al.*, 1987). A number of characteristics make PLA important in the biomedical industry: biocompatibility, stability, thermophysical and mechanical properties, strength, low toxicity and plasticity. Within degradable plastics, PLA is a key polymer. The advantages of biodegradable plastics in the growing market are based on the increase in the price of plastics from petroleum products and their increasingly stringent regulations. There is also an interest on the part of users to use environmentally friendly plastics.

This plastic has different characteristics depending on the composition of the D- and L-isomers. The stereocomplex PLA composed of both poly-D- and

L-lactic acid has high thermostability. When poly-D-lactic and poly-L-lactic acid are mixed in an equimolar ratio the thermostability increased up to 200°C, which is 20–25°C above the temperature of the homopolymers (Sawai *et al.*, 2007).

Due to the importance in the ratio between L-lactic acid and D-lactic acid in the properties of PLA, there was an increase demand to obtain the pure isomers (Okano *et al.*, 2010; Tashiro *et al.*, 2011). The organisms that predominantly yield the L-isomer are *Lactobacillus amylophilus*, *Lactobacillus sakei*, *Lactobacillus casei*, *Lactobacillus maltaromicus*, and *Lactobacillus salivarius*. *Lactobacillus delbrueckii*, *Lactobacillus jensenii*, and *Lactobacillus acidophilus* yields the D-isomer or mixtures of both (Madhavan Nampoothiri *et al.*, 2010). Research has historically been focused on obtaining L-lactate. However, with increasing demand for PLA on the market, many groups began using *Lactobacillus* to produce D-lactate. In general, the PLA fulfils many requirements as a thermoplastic. When plasticized with its monomers, the PLA is flexible, so it can serve the same as PVC, PP and PS products. Because it is biodegradable, it can also be used to produce loose fill packaging, compost bags, food containers and disposable tableware. Because of the resistance to UV light discoloration, it can be used in the form of fibres, textiles and awnings.

In 2005, Wal-Mart announced that it would use PLA as containers for products of vegetable origin such as herbs, which implied a growth in the demand for bioplastic. The PLA has applications in the biomedical field, and it may be one of the most promising areas. It has been used as suture material, bone fixation, in the form of microspheres for drug administration and tissue engineering (Zhao *et al.*, 2004), producing an adjuvant effect, being able to release antigens after intraperitoneal injection. These PLA vaccine vehicles have been shown to induce a sustained and targeted release of different antigens and bacterial, viral and tumour-associated adjuvants *in vivo*, triggering distinct immune responses. PLA-based delivery systems have also been shown to be protective of plasma enzyme activity (Peres *et al.*, 2017). These systems showed protective responses after nasal, oral and pulmonary administration. This application in mucosa is very important since the response is not only humoral but local, so as to prevent infections of agents whose infective mechanism is by mucosa (Berquist *et al.*, 1997; Liang *et al.*, 2001).

Problems and solutions in the formation of lactate by fermentation process with *Lactobacillus* species

The main limitations of the fermentation process involve (a) production of both L- and D-lactic acid via L-lactate dehydrogenase (L-Ldh) and D-lactate dehydrogenase (D-Ldh), respectively; (b) low yield due to by-product formation or inhibition of production by acidity; (c) use of nutritionally rich medium; and (d) high risk of bacteriophage infection that results in cell lysis and subsequent cessation of lactic acid production (Abdel-Rahman *et al.*, 2013). Different strategies are described below to deal with those limitations.

Production of optically pure stereoisomer

Lactic acid can form part of the cell wall, which confers resistance to the vancomycin antibiotic (Goffin *et al.*, 2005). The microorganisms can metabolize both isomers through stereospecific lactate dehydrogenases. One goal of current research is to obtain a desired stereoisomer. A low optical purity has a negative effect on the industry due to the high cost in purification. In *Lactobacillus*, the final step in lactic fermentation is the reduction of pyruvate to lactate. Each lactate isomers is synthesized by a specific enzyme: L-lactate dehydrogenases (L-Ldh) and D-lactate dehydrogenases (D-Ldh), with no structural relationship. A second enzyme, D-hydroxyisocaproate dehydrogenase (HicDH) can generate D-lactate with lower affinity (Viana *et al.*, 2005). When only one stereospecific lactate dehydrogenase is present, the production or utilization of the other isomer can proceed by lactate isomerization involving a specific racemase. This activity has been identified in several species of *Lactobacillus* (Desguin *et al.*, 2014).

In order to produce the pure L- or D-stereoisomer much work has been done to obtain knockout strains for the *d-ldh* or *l-ldh* genes. Because D-lactate cannot be metabolized by mammals, it has not been considered historically of commercial interest. Thus, efforts were initially directed to the improvement of strains to produce L-lactate (Kylä-Nikkilä *et al.*, 2000).

In order to decrease the levels of D-lactate produced by *Lactobacillus casei* BL23, Viana *et al.* (2005b) inactivated the *hicD* gene, one of those responsible for the synthesis of the stereoisomer. The mutant showed no detectable D-lactate production, with similar levels of L-lactate final concentration to the total lactate in the wild type strain (35 mM, in MRS medium supplemented with 0.5% glucose). Kuo *et al.* (2015) were able to inactivate the *d-ldh* gene in *Lactobacillus paracasei* 7B. This strain was isolated from rotten fruit and the *ldhD*-deficient strain 7BL was able to ferment lignocellulosic hydrolysates achieved high productivity rates and yields. The optical purity of L-lactic acid produced by *L. paracasei* was enhanced from 95% to 100%.

On the other hand, in *L. helveticus* CNRZ32 two negative mutants with different genetic strategies were constructed. The promoter region of the *d-ldh* gene was deleted, constructing a strain with 53% greater L-Ldh activity than the wild-type strain. In the second strain, the structural gene of the enzyme L-Ldh was incorporated under control of the *d-ldh* promoter. In the latter case, the enzymatic activity was 93% higher, and finally 20% more lactic acid was reached than with the wild type strain (Kylä-Nikkilä *et al.*, 2000).

The production of D-lactic acid has become more important due to the need to have the D-isomer to synthesize PLA with equimolar amounts of both stereoisomers. Recently, in 2017, Okano and colleagues achieved the production of optically pure D-lactic acid (99.6%) from brown rice without any nutrient supplementation. They used an engineered *Lactobacillus plantarum* NCIMB 8826 strain, in which the L-lactate dehydrogenase gene was disrupted, produced 97.7 g/l D-lactic acid from 20% (w/v) brown rice without any nutrient supplementation. In addition, the production was improved by an acid adaptation technique with an increased glucose consumption and lactic acid production. Thus, with an acid pre-adaptation, 127.1 g/l of lactic acid was produced from 20% (w/v) of brown rice.

A good possibility to improve the production of lactate with lower cost of culture is to use the structural genes of lactate synthesis in other strains with lower nutritional requirements. In this regard, genetic modifications were made in *Corynebacterium glutamicum* by which the gene coding for the endogenous L-Ldh was deleted and the coding sequence for the D-lactate dehydrogenase of *L. bulgaricus* was cloned. A modified strain was obtained which produced high-purity D-lactic acid (optical purity higher than 99.9%). After 16 hours of fermentation, the production was 17.92 g/l which was 32.25% higher than the production of lactic acid than the wild type strain (Jia *et al.*, 2011).

Most homofermentative lactobacilli are not able to use pentoses, which are the main sugars of hemicellulose, so there is a decrease in productivity when homofermentative group is used. In turn, heterofermentative lactobacilli, which can use pentose sugars, generate by-products. Oriented to achieve a high production of D-lactic acid from lignocellulosic biomass a plasmid for xylose assimilation was constructed. It was introduced in *L. plantarum*, mutant for the enzyme L-Ldh. The production of D-lactic acid from maize stubble and soybean meal extract as substrates in fed batch fermentation was demonstrated with the resulting recombinant strain. Recombinant *L. plantarum* was able to simultaneously use all the main sugars derived from maize stubble to produce D-lactic acid with high optical purity (>99%). The production reached was 61.4 g/l of lactic acid with an overall yield of 0.77 g/g (Zhang *et al.*, 2016).

One possible way to increase the production of D-lactic acid might be to choose the correct strain and increase the culture temperature to accelerate the fermentation. *Lactobacillus delbrueckii*, *L. bulgaricus*, *L. jensenii* and *L. coryniformis* are good producers of the D-isomer. However, by choosing these bacterial species and increasing the temperature, an increase in the production of the L-isomer is observed. As each isomer has a specific Ldh enzyme, it is assumed that the greater enzymatic thermostability is a determining factor by which the optical purity of D-lactic acid decreases. Gu *et al.* (2014) tested this hypothesis and found that the L-Ldh of *L. coryniformis* was much more thermostable than the D-Ldh. A genetically improved strain, with a more thermostable D-Ldh enzyme, would be a good candidate for large-scale production.

Sugarcane molasses, sugar cane juice and sugar beet juice do not require heating or enzymatic treatment since the content is almost all in the form of sucrose. *Lactobacillus delbrueckii*, which produces mostly D-lactic acid, was grown in those complex growing media. The culture condition was at 40 degrees in batch fermentation. Optical purities of D-lactic acid of raw materials ranged from 97.2% to 98.3% (Calabia *et al.*, 2007).

In a recent work (Singhvi *et al.*, 2017) the role of lactate dehydrogenase enzymes and the use of nitrogen compound for synthesis of lactate in the presence of waste was related. Strains of mutant *Lactobacillus* by UV radiation were generated and subjected to fermentation studies in modified media to evaluate the production of both isomers of lactic acid. A mutant strain showed an increase in the synthesis of D-lactic acid; however it was accompanied by the appearance of the L-isomer. Both dehydrogenase enzymes were stimulated by the diammonium hydrogen phosphate. The mutant strain showed an approximately 17-fold increase in D-Ldh and a 1.38-fold increase in L-Ldh, whereas the original strain showed a slight change in L-Ldh expression in the case of modified media. These results indicated that the parental and mutant strains already have the L-Ldh gene that is expressed in the presence of the ammoniated compound. That is, a production of D-lactic acid is seen but the undesired synthesis of the L-isomer appears. This appearance may be due to the enzyme lactate racemase which in most species acts on the substrate D-lactate and produces L-lactate. These results suggest that *l-ldh* genes can be expressed, decrypted in the presence of inducers by the producer organism of pure D-lactic acid. Therefore, this information shows that genetic approaches should be evaluated since there are functional limits for the strains.

Acid tolerance and decrease in by-product formation

As the culture progresses, there is a decrease in the pH of the media causing growth arrest and even death. One of the strategies used to develop resistant or tolerant strains of acid stress is to use adaptive evolution to obtain them. In *L. casei* acid tolerant strain was generated by serially exposing mid-exponential phase strains to low pH conditions. Evolution culture was started by transferring the culture into fresh MRS medium at pH 5.5 when the culture was at mid-exponential phase. The culture was then sequentially transferred to MRS at decreasing pH concentrations. The entire protocol was carried out for 70 days. The 'evolved mutant' exhibited an increase of more than 60% and 13% in biomass and lactate concentration, respectively (Zhang *et al.*, 2012).

Genome shuffling is a relatively new genetic manipulation technique that uses complete genome transfer primarily to improve cellular phenotypes. This approach uses recursive recombination of protoplasts throughout the genome without the need for genome sequence data or network information (Patnaik *et al.*, 2002; Stephanopoulos *et al.*, 2002). This strategy was used to improve the acid tolerance of *L. plantarum*, while simultaneously enhancing lactic acid production. First, mutant libraries were generated by gradual adaptation to low pH levels and ultraviolet radiation. Subsequently, protoplast fusion was carried out. Then, the process was repeated with shuffled mutants for three rounds in order to obtain microorganisms that grow best at pH below 4.0 and consumes glucose faster than the wild type. Although the production of lactate was 12 g/l, it meant an increase of 64% compared to the initial strain, which supports this approach to develop improved strains to tolerate conditions of high acidity.

The conventional batch fermentation process has some disadvantages such as inhibition by the substrate or the end product. One strategy to improve the cell density increase and on the other hand to avoid the inhibition by end product is to use the cellular immobilization technique. *Lactobacillus rhamnosus* ATCC 7469 was immobilized onto zeolite, and lactate production was compared in free and immobilized cells. A maximal process productivity of 1.69 g/l, maximal lactic acid concentration of 42.19 g/l (18% higher than in the free-cell system) and average yield coefficient of 0.96 g/g were achieved (Djukić-Vuković *et al.*, 2013).

Lactobacillus casei-immobilized cells were used to produce L-lactic acid from the whey medium. Calcium pectate gel was used to obtain the stable system during fermentation. A high conversion of lactose (94.37%) in lactic acid (32.95 g/l) was achieved. The long-term viability of bacterial cells trapped in pectate was tested by reusing the immobilized bacterial biomass and it was observed that there was no change in the conversion of lactose to lactic acid up to 16 batches, which demonstrated its high stability and potential for commercial application (Panesar *et al.*, 2007).

Using the technique of immobilization with *Lactobacillus delbrueckii* in different matrices, the behaviour of lactate production compared to by-products was evaluated. The succinylated alginate beads provided better stability and durability in an acidic environment than alginate beads. In addition, there was a decrease in the formation of by-products (acetic and propionic acid) compared to free-cell fermentation. The immobilized beads showed increased cell-mass entrapment and production of L-lactic acid with higher yields (0.93 Yp/s in grams) and enantioselectivity (99%) (Rao *et al.*, 2008).

Cell immobilization by gel entrapment has some disadvantages that Chantawongvuti *et al.* (2011) tried to solve. Some of them are the possibility of rupture due to the pressure generated during cell division, the speed of lactate diffusion and the gradual solubilization of the beads when trying to take an industrial step. Immobilization on a solid surface support through physical adsorption can overcome all these disadvantages, and has no limitations when it is adapted for mass production. The fermentation of lactic acid by *Lactobacillus salivarius* immobilized in a sponge was evaluated. To increase the surface area of the sponge for cellular immobilization, H_2O_2 and chitosan were introduced as surface modifying reagents. All experiments were performed in stirred flask mode. The immobilized cell system produced a concentration 1.5 times higher than the free cells during 24 h of fermentation. In addition, the immobilized cells were able to shorten the fermentation time by two times compared to the free cells at the same level of lactate concentration.

Another strategy to avoid the problems associated with inhibition by substrate or by end product focused on different feeding strategies using whey. This method has the advantage of not requiring hydroxide for pH control. This strategy was used to produce L-lactate with *Lactobacillus rhamnosus*. The pH-stat methodology showed the maximum production of lactic acid of 143.7 g/l. The concentration of L-lactic acid in the discontinuous fermentation was 57.00 g/l at the end of the fermentation (Bernardo *et al.*, 2016).

Lactobacillus can be grouped as homofermenters or heterofermenters depending on the species, so that depending on the type of producer during lactate synthesis some unwanted by-products can originate. This is very evident in lactobacilli that metabolize pentoses through the use of phosphoketolases (which generate as by-products CO_2, ethanol and acetic acid). Genetic engineering has been performed on these species to redirect the carbon flux. The goal is to avoid the phosphoketolase pathway and create the pentose-phosphate pathway (Fig. 4.1). *L. pentosus, L. brevis, L. plantarum* ferment arabinose to xylulose-5-phosphate by arabinose isomerase, and then by another series of enzymes so that xylulose-5-phosphate is converted to equimolar amounts of lactic acid and acetic acid. From lignocellulose, *L. brevis* can use glucose and xylose without the phenomenon of catabolic repression. This fact makes it a species with interest in the production of lactate. However, due to its fermentative metabolism, there is a high production of by-products. Expressing the genes encoding fructose-6-phosphate kinase and fructose-1,6-biphosphate aldolase, which are key enzymes in the EMP glycolytic pathway of *Lactobacillus rhamnosus*, Guo *et al.* (2014) constructed a strain that produces lactate through a constructed homofermentative pathway. The yield of lactic acid obtained was higher than native strain.

Okano *et al.* (2009), succeeded in modifying a strain of *Lactobacillus plantarum* by removing the genes of the L-Ldh enzyme and substituting the phosphoketolase genes for transketolase genes from *Lactococcus lactis*. Thus they obtained D-lactic acid from arabinose (38.6 g/l from 50 g/l of arabinose).

Use of substrates from industrial waste

The substrate is one of the highest production costs for lactate synthesis. The use of refined sugar (glucose for example) makes the process more expensive. Food-waste materials are a good alternative as a substrate because of their high content of carbohydrates. Many efforts have been made to use waste material by *Lactobacillus* to produce lactate.

Using residual yoghurt whey, 41.5 g/l lactate was produced by *Lactobacillus casei* ATCC 393 under pH-controlled batch fermentation conditions. In the same work, Alonso and colleagues (2014) observed the existence of population heterogeneity, which suggests that a bacterial subpopulation was adapted to that regulation, allowing the overproduction. Using a lactose-enriched dairy waste effluent, casein whey permeate (CWP), two species from *Lactobacillus* were grown to produce D-lactic acid. Casein whey permeate is a by-product generated by subjecting cheese whey to ultra-filtration to concentrate whey protein. Prasad *et al.* (2014), cultivated *Lactobacillus delbrueckii* subsp. *lactis* ATCC 4797 in a synthetic casein whey permeate medium, again with or without pH control, and produced 24.3 g/l lactic acid with an optical purity >98%.

Using two genomes of Gram-positive bacteria, with genome shuffling technique, an exfusant that can use liquefied cassava bagasse to produce lactate was developed. *Lactobacillus* is a fastidious genus, with high nutritional requirements. John *et al.* (2008) generated by mutagenesis using nitrous acid and genome shuffling an exfusant of *Lactobacillus delbrueckii* and *Bacillus amyloliquefaciens* carrying an amylase activity. Focused on developing a non-fastidious lactate-producing strain having better growth rate, low pH tolerance and good productivity, after the third cycle of protoplast fusion, lactic acid production by exfusants was monitored and selected. *Lactobacillus* exfusant was able to use the liquefied cassava bagasse starch directly to produce 40 g/l of lactic acid with a productivity of 0.42 g/l/h and 96% conversion of starch to lactic acid.

During the production of biodiesel, algae-free biomass (algae cake) is generated as a residual by-product. This substrate is fermentable. Recent works on waste exploitation of the refinery industry were carried out (Overbeck *et al.*, 2016), exploring the ability of *Lactobacillus casei* to ferment algae cake for the co-production of lactic acid. As the main limitation is the bioavailability of carbohydrates and amino acids, to optimize the production of lactic acid, an enzymatic hydrolysis with a non-specific protease, α-amylase and cellulase was performed and lactate production was successfully achieved. New substrates have also been explored as a carbon source.

In 2016, inulin was used as a carbohydrate source for lactic fermentation. Inulin can be hydrolysed into fructose and glucose by enzymatic or acid catalysis. Xu *et al.* (2016) used inulin derived from chicory for the biosynthesis of D-lactic acid by *Lactobacillus delbrueckii* subsp. *bulgaricus*. Regarding the process, the best production yield was obtained by saccharification and simultaneous fermentation, with a concentration of D-lactic acid of 123 g/l. The use of *L. delbrueckii* subsp. *bulgaricus* allows high optical purity of D-lactic acid (99.9%). The main advantage of using inulin is obtaining a higher value product from non-food biomass.

By-products of bioethanol production can be used as substrates in the fermentation media. This is the case of thin stillage (TS) and condensed soluble distillers (CDS) that were used as nitrogen source. A strain of *L. paracasei* was used. Values of 90 g/l of lactic acid were obtained. Therefore, both wastes could be used for the economic production of lactic acid (Moon *et al.*, 2015).

Vegetable straw has been sought as a substrate for production. The fermentation was investigated, after the treatment with ammonia, of soy straw, wheat, maize and rice, composed mainly of cellulose, hemicellulose and lignin. Soybean straw was chosen as the substrate for the production of lactic acid by *L. casei* because of its high protein content (Wang *et al.*, 2015). The experimental results demonstrated the feasibility of using an enzymatic soybean straw hydrolysate as a substrate for the production of lactic acid. Although the results are relatively low compared to other studies that used *L. casei* to produce lactic acid from simple sugars and other substrates from waste, it is a good substrate to continue with heat and enzymatic treatments and then fermentation.

Comparative studies of different types of waste substrates in the industry have been carried out in order to produce lactate by fermentation. In particular, Djukić-Vuković *et al.* (2016) analysed as wasted bread substrates and potato waste from the production of bioethanol and the hydrolysate of spent grains from beer production. As a producing microorganism they used *Lactobacillus rhamnosus*. The kinetics of sugar consumption in the wasted bread and wasted potato substrates was similar, while the conversion of sugar into the spent grain hydrolysate of the brewers was slower and less efficient. The concentration of L-lactic acid after 32 hours of batch culture was 49.20, 42.97, 16.01 g/l for the wasted bread stillage, wasted potato stillage, brewers spent grains hydrolysate media, respectively. The lignocellulosic hydrolysate of the beer production required to be supplemented with nitrogen. These results show the limitations in the choice of the substrates that the *Lactobacillus* species have in relation to the nutritional requirements of nitrogen.

Another technique that has been very efficient to expand the substrates was the use of bacterial co-cultures. In this way it is possible to obtain a better use of the waste substrates with respect to the single culture. The group of Nancib (2009) investigated the production of lactic acid from date juice in single and mixed cultures of *Lactobacillus casei* and *Lactococcus lactis*. The results suggest that the mixed culture system of lactic acid bacteria gave better results than the individual cultures with respect to the concentration of lactic acid and the use of sugar (60.3 g/l vs 53 and 46 g/l of pure cultures). Using a mixture of xylose and glucose, which simulates a ligncellulose hydrolysate, it was possible to produce L-lactic acid of high yield and purity.

A combination of co-culture with immobilization was experimented with strains of *Lactobacillus paracasei* subsp. *paracasei* and a fast growing *L. delbrueckii* subsp. *delbrueckii* mutant, with the aim of increasing lactate productivity using cassava bagasse starch hydrolysate. The culture was imbibed in polyurethane foam cubes as a biofilm and used for fermentation. The cells were trapped in calcium alginate. Although the productivity was lower when coated with alginate, the cell loss was decreased. Thus, cubes were reused repeatedly and production did not decrease (John and Madhavan Nampoothiri, 2011).

The methodology of co-cultivation with the lactic acid bacterium *Enterococcus casseliflavus* and *Lactobacillus casei* was used. In addition, the products were compared in the individual species. When a mixture of 50 g/l of xylose and 100 g/l of glucose was used as a carbon source in a culture of *E. casseliflavus* alone, the xylose was hardly consumed. In the co-culture, there were two stages: the first was 40 h with inoculum of *L. casei*, and then *E. casseliflavus* was added to the fermentation system. A complete consumption of both sugars was achieved: 95 g/l of lactic acid with a high optical purity of 96%. This is an example of an inoculation design for an efficient process based on the mixture of lignocellulosic sugars (Taniguchi *et al.*, 2004).

The co-culture methodology has also been used to reuse by-products of dairy production. In particular, by-products of ricotta, which have a high content of lactose and protein, have been used in fermentative processes for the production of L-lactic acid. Mixed and pure cultures of *Lactobacillus casei*, *Lactobacillus helveticus* and *Streptococcus thermophilus* species were analysed in ricotta by-product based media with or without added supplements. It was observed that the use of mixed cultures reduced the need for nutritional supplements. In addition, the process could be directed to the production of the L-isomer by choosing the combination *L. casei* and *S. thermophilus* (Secchi *et al.*, 2012). The results showed a synergistic interaction in the fermentations with co-cultures that improves the production of lactic acid a better utilization of the substrate.

Prevention of phage infection

Bacteriophages (phage) are viruses that attack bacteria, and cannot replicate outside their host cell. The phages attach themselves to the outside of the cell and inject their DNA through their tail into the cell, where it takes over the biosynthetic machinery of the host cell to make more phage particles. The possibility of being infected by phage is a problem for lactic bacteria producing lactic acid. Phages can ultimately result in complete failure of acid production by the producer (Fuquay *et al.*, 2011). The dairy industry has been at the forefront of developments and strategies to avoid phage contamination, since the costs of economic losses are significant. Traditional strategies include phage inhibitory media, aseptic processing conditions, strain rotation schemes and improvement of phage-resistant starter cultures. In this case, for the production of lactate, the rotation of a good lactic acid producer culture would be undesirable. In addition, damage by bacteriophages to a strain designed especially for this activity would be nonsense. A technique used for the use of probiotic strains (where the strain cannot be rotated either) is a secondary culture technique. With this methodology, mutants resistant to phages were found and these strains would replace phage-sensitive probiotic strains. Another approach is the use of strains with industrial interest that are transformed with genes that give phage resistance. A new strain refractory to bacteriophages is generated (Daly *et al.*, 1996; Capra *et al.*, 2011). Regarding the selection of sanitizing agents in bioprocess plants, work has been carried out with temperate bacteriophages from species of interest in the dairy industry. In phages isolated from *Lactobacillus paracasei*, physical treatments were tested (cold storage at different temperatures, survival after heat treatment, high pressure homogenization) and classical chemical treatments and novel commercial disinfectants. The results that inactivate the phages would be candidates as sanitizers (Mercanti *et al.*, 2012). Despite the constant search for strains resistant to viruses, due to the multiplication rate and the frequency of mutation, the phages overcome the resistance and the fermentative processes are again unprotected. Constant vigilance is required. For further information, see Chapter 7.

An example of lactate production with *Lactobacillus casei* in cheese whey supplement with different alternatives nitrogen sources

The desirable characteristics of industrial microorganisms are their ability to rapidly and completely ferment cheap raw materials. Cheese whey contains significant amounts of lactose. In addition, it contains other essential nutrients necessary for the growth of lactic acid bacteria, so it can serve as a substrate for fermentations, being a good destination for effluents. Production from industrial waste often starts from mixed sugars. The efficient use of nutrients is associated with carbon catabolite repression (CCR), which allows microbes to select their preferred source from multiple offerings of carbon catabolites. The catabolite control protein A (CcpA) is a mediator of CCR in low-G+C Gram-positive bacteria. It can therefore be hindered by the process of catabolic repression. Thus, the catabolic repression can be an important limiting factor in the lactic acid synthesis, the regulatory protein CcpA has been shown to play a major part in the CCR process (see Chapter 3). In this context, *Lactobacillus casei* BL23 strain and its *ccpA* mutant (BL71), were evaluated to produce lactic acid from cheese whey based medium with different nitrogen sources as supplement (yeast extract, YE; or maize steep liquor, CSL) to evaluate the effect of *ccpA* mutation in the lactic acid production. Lactate concentration was determined for both isomers. Increase lactic concentration for the *ccpA* mutant was obtained. L-lactate is represented at the same proportion in wild-type and *ccpA* mutant in either media. Higher acidification rate was observed for the *ccpA* mutant in all conditions. When yeast extract was the supplement of choice it was possible to verify almost a 2-fold increase in lactic acid production early in the growth compare to the wild-type. The values obtaining show that the mutation of global regulator CcpA is a good approach to use culture medium with different carbon and nitrogen sources. Considering the preliminary results obtained, *ccpA* mutant could be an excellent candidate for future technological developments, enabling the production of lactic acid in an effective and inexpensive way, using cheese whey as carbon sources from dairy industrial wastes.

In another microorganism, *L. plantarum*, growth was also studied in the context of a *ccpA* deletion and, unlike what was observed in *L. casei*, growth and lactate production under aerobic conditions were markedly reduced. However, in the stationary phase of growth, the cells were more tolerant to stress (especially oxidative stress) than in the exponential phase (Zotta *et al.*, 2012). This finding could be important especially for the preservation of starters.

Concluding remarks

This chapter described the importance of lactic acid in different industrial areas, and its growing interest due to the appearance on the market of PLA biodegradable plastic. Due to the qualitative differences of the physical characteristics of this plastic, the production of optically pure isomers becomes important. Therefore, the fermentation synthesis through lactic acid bacteria is the most promising. Particularly, the genus *Lactobacillus* possesses strains that are natural producers in high concentration. The most common problems in fermentation for the production of lactic acid were pointed out. To avoid racemic mixtures and obtain a product of high optical purity, strategies of classical genetics (genetic inactivation, for example), novel techniques (genome shuffling) or changes in culture conditions (changes in temperature or pH) have been reported. Acid tolerance was improved by generation of mutants by adaptive evolution, or genome shuffling. Most reports to avoid inhibition by end product are associated with immobilizing cells or feeding cultures. To take advantage of industrial wastes and use them as substrates, the techniques of co-cultures or strains with new metabolic pathways have been developed. However, some aspects still remain unsolved, such as the generation of phage-resistant strains or the development of strains that inhibit replication. The generation of phage-resistant strains or strains that inhibit steps in the life cycle need to be developed.

A high quality product, pure optically, that comes from industrial waste, generated through a high-productivity process is desired. However, it is necessary to take into account the needs and limitations of the producer, in this case, *Lactobacillus*. To modify metabolic pathways, it is necessary to pay attention to the adaptive equilibrium of these processes. A case that illustrates the concept is the work of Singhvi *et al.* (2017), in which *Lactobacillus delbrueckii* subsp. *lactis*, producer of D-lactic acid, was modified to increase lactic acid production. However, as a result of the increase of total lactate the L-isomer was obtained as an undesired product. In *L. casei* the inactivation of the enzyme L-ldh did not suppress the production of L-lactate, although it decreased it. In addition, part of the pyruvate flow was diverted. Other effects were surprising as the appearance of other products such as acetate, acetoin, pyruvate, ethanol, diacetyl, mannitol and CO_2. Also, a lack of carbon catabolite repression of lactose metabolism and N-acetyl-glucosaminidase activity was observed (Viana *et al.*, 2005b). Trying to obtain favourable conditions to produce lactic acid respecting the philosophy of keeping the bacteria alive and in balance with its environment, is an interesting challenge.

Web resources

Global Industry Analysts Inc. 2011. Global market opening for lactic acid [WWW document]. URL http://www.prweb.com/releases/2011/1/prweb8043649.htm/

ICIS. https://www.icis.com/chemicals/channel-info-chemicals-a-z/

List of Prokaryotic names with Standing in Nomenclature (LPSN). http://www.bacterio.net/

Microbes on line: http://www.microbesonline.org

References

Abdel-Rahman, M.A., Tashiro, Y., and Sonomoto, K. (2013). Recent *advances* in lactic acid production by microbial fermentation processes. Biotechnol. Adv. *31*, 877–902. https://doi.org/10.1016/j.biotechadv.2013.04.002

Abo-Elnaga, I.G., and Kandler, O. (1965). Zur Taxonomie der Gattung *Lactobacillus* Beijerinck. I: Das Subgenus Streptobacterium Orla Jensen. Zentralbl. Bakteriol. Parasitenkde. Infektionskrankh. Hyg. Abt. 2 *119*, 1–36.

Alonso, S., Herrero, M., Rendueles, M., and Díaz, M. (2014). Physiological heterogeneity in *Lactobacillus casei* fermentations on residual yoghurt whey. Process Biochemistry, *49*, 732–739. https://doi.org/10.1016/j.procbio.2014.01.033

Beijerinck, M.W. (1901). Uber oligonitrophile mikroben. *Zentralbl. Bakterol. Parasitenkd. Infektionskr. Hyg. Abt. II.*, 7, 561–582.

Bergey, D.H., Harrison, F.C., Breed, R.S., Hammer, B.W., and Huntoon, F.M. (1923). Bergey's Manual of Determinative Bacteriology, 1st edn. (Williams and Wilkins, Baltimore).

Bergey, D.H., Harrison, F.C., Breed, R.S., Hammer, B.W., and Huntoon, F.M. (1925). Bergey's Manual of Determinative Bacteriology, 2nd edn. (Williams and Wilkins, Baltimore).

Bergey, D.H., Breed, R.S., Hammer, B.W., Huntoon, F.M., Murray, E.G.D., and Harrison, F.C. (1934). Bergey's Manual of Determinative Bacteriology, 4th edn. (Williams and Wilkins, Baltimore).

Bergquist, C., Johansson, E.L., Lagergård, T., Holmgren, J., and Rudin, A. (1997). Intranasal vaccination of humans with recombinant cholera toxin B subunit induces systemic *and* local antibody responses in the upper respiratory tract *and* the vagina. Infect. Immun. *65*, 2676–2684.

Bernardo, M.P., Coelho, L.F., Sass, D.C., and Contiero, J. (2016). l-(+)-Lactic acid production by *Lactobacillus rhamnosus* B103 from dairy industry waste. Braz. J. Microbiol. *47*, 640–646. https://doi.org/10.1016/j.bjm.2015.12.001

Brygoo, E.R., and Aladame, N. (1953). Etude d'une espèce nouvelle anaerobic stricte du genre Eubacterium: E. crispatum n. sp. Ann. Inst. Pasteur 84:640–651.

Calabia, B.P., and Tokiwa, Y. (2007). Production of D-lactic acid from sugarcane molasses, sugarcane juice and sugar beet juice by *Lactobacillus delbrueckii*. Biotechnol. Lett. *29*, 1329–1332. https://doi.org/10.1007/s10529-007-9408-4

Capra, M.L., Mercanti, D.J., Rossetti, L.C., Reinheimer, J.A., and Quiberoni, A. (2011). Isolation and phenotypic characterization of *Lactobacillus casei* and *Lactobacillus paracasei* bacteriophage-resistant mutants. J. Appl. Microbiol. *111*, 371–381. https://doi.org/10.1111/j.1365-2672.2011.05056.x

Carr, J.G. (1987). The microbiology of ciders and wines. Prevention (USA).

Cato, E.P., Moore, W.E.C., and Johnson, J.L. (1983). Synonymy of strains of '*Lactobacillus acidophilus*' group A2 (Johnson *et al.* 1980) with the type strain of *Lactobacillus crispatus* (Brygoo and Aladame 1953) Moore and Holdeman 1970. Int. J. Syst. Evol. Microbiol. 33, 426–428.

Champagne, C.P., Moineau, S., Lafleur, S., and Savard, T. (2017). The effect of bacteriophages on the acidification of a vegetable juice medium by microencapsulated *Lactobacillus plantarum*. Food Microbiol. 63, 28–34.

Chantawongvuti, R., Veerajetbodithat, J., Jaturapiree, P., and Muangnapoh, C. (2010). Immobilization *of Lactobacillus salivarius* ATCC 11741 on loofa sponge coated with chitosan for lactic acid fermentation. J. Microbiol. Biotechnol. 20, 110–116.

Chassy, B.M., and Thompson, J. (1983). Regulation *of* lactos e-phosphoenolpyruvate-dependent phosphotransferase system and beta-D-phosphogalactoside galactohydrolase activities in *Lactobacillus casei*. J. Bacteriol. *154*, 1195–1203.

Chen, L., Bai, Y., Fan, T.P., Zheng, X., and Cai, Y. (2017). Characterization *of* a D-lactate dehydrogenase from *Lactobacillus fermentum* JN248 with high phenylpyruvate reductive activity. J. Food Sci. *82*, 2269–2275. https://doi.org/10.1111/1750-3841.13863

Collins, M.D., Phillips, B.A., and Zanoni, P. (1989). Deoxyribonucleic acid homology studies of *Lactobacillus casei, Lactobacillus paracasei* sp. nov., subsp. paracasei and subsp. tolerans, and *Lactobacillus rhamnosus* sp. nov., comb. nov. Int. J. Syst. Evol. Microbiol. *39*, 105–108.

Curk, M.C., Hubert, J.C., and Bringel, F. (1996). *Lactobacillus paraplantarum* sp. nov., a new species related *to Lactobacillus plantarum*. Int. J. Syst. Evol. Microbiol. *46*, 595–598. https://doi.org/10.1099/00207713-46-2-595

Daly, C., Fitzgerald, G.F., and Davis, R. (1996). Biotechnology of lactic acid bacteria with special reference to bacteriophage resistance. Antonie Van Leeuwenhoek *70*, 99–110.

Desguin, B., Goffin, P., Viaene, E., Kleerebezem, M., Martin-Diaconescu, V., Maroney, M.J., Declercq, J.P., Soumillion, P., and Hols, P. (2014). Lactate racemase is a nickel-dependent enzyme activated by a widespread maturation system. Nat. Commun. *5*, 3615. https://doi.org/10.1038/ncomms4615

Dicks, L.M., Du Plessis, E.M., Dellaglio, F., and Lauer, E. (1996). Reclassification of *Lactobacillus casei* subsp. *casei* ATCC 393 and *Lactobacillus rhamnosus* ATCC 15820 as *Lactobacillus zeae* nom. rev., designation of ATCC 334 as the neotype of *L. casei* subsp. *casei*, and rejection of the name *Lactobacillus paracasei*. Int. J. Syst. Evol. Microbiol. *46*, 337–340. https://doi.org/10.1099/00207713-46-1-337

Djukić-Vuković, A.P., Mojović, L.V., Jokić, B.M., Nikolić, S.B., and Pejin, J.D. (2013). Lactic acid production on liquid distillery stillage by *Lactobacillus rhamnosus* immobilized onto zeolite. Bioresour. Technol. *135*, 454–458. https://doi.org/10.1016/j.biortech.2012.10.066

Djukić-Vuković, A., Mladenović, D., Radosavljević, M., Kocić-Tanackov, S., Pejin, J., and Mojović, L. (2016). Wastes from bioethanol and beer productions as substrates for l(+) lactic acid production - A comparative study. Waste Manag. *48*, 478–482 https://doi.org/10.1016/j.wasman.2015.11.031

Doelle, H.W. (1971). Nicotinamide adenine dinucleotide-dependent and nicotinamide adenine dinucleotide-independent lactate dehydrogenases in homofermentative and heterofermentative lactic acid bacteria. J. Bacteriol. *108*, 1284–1289.

Embley, T.M., Faquir, N., Bossart, W., and Collins, M.D. (1989). *Lactobacillus vaginalis* sp. nov. from the human vagina. Int. J. System. Evolut. Microbiol. *39*, 368–370.

Fred, E.B., Peterson, W.H., and Anderson, J.A. (1921). The characteristics of certain pentose-destroying bacteria, especially as concerns their action on arabinose and xylose. J. Biol. Chem. *48*, 385–412.

Fuente-Hernández, A., Corcos, P.O., Beauchet, R., and Lavoie, J.M. (2013). Biofuels and co-products out of hemicelluloses. In Liquid, Gaseous and Solid Biofuels-Conversion Techniques (Intech Open). http://doi.org/10.5772/52645

Fujisawa, T., Benno, Y., Yaeshima, T., and Mitsuoka, T. (1992). Taxonomic study of the *Lactobacillus acidophilus* group, with recognition of *Lactobacillus gallinarum* sp. nov. and *Lactobacillus johnsonii* sp. nov. and synonymy of *Lactobacillus acidophilus* group A3 (Johnson *et al.* 1980) with the type strain of *Lactobacillus amylovorus* (Nakamura 1981). Int. J. Syst. Evol. Microbiol. *42*, 487–491. https://doi.org/10.1099/00207713-42-3-487

Fujisawa, T., Itoh, K., Benno, Y., and Mitsuoka, T. (1990). *Lactobacillus intestinalis* (ex Hemme 1974) sp. nov., nom. rev., isolated from the intestines *of* mice *and* rats. Int. J. Syst. Evol. Microbiol. *40*, 302–304. https://doi.org/10.1099/00207713-40-3-302

Fuquay, J.W., Fox, P.F., and McSweeney, P.L. (2011). Encyclopedia of Dairy Sciences (Academic Press).

Gasser, F., Mandel, M., and Rogosa, M. (1970). *Lactobacillus jensenii* sp.nov., a new representative of the subgenus *Thermobacterium*. J. Gen. Microbiol. *62*, 219–222. https://doi.org/10.1099/00221287-62-2-219

Goffin, P., Deghorain, M., Mainardi, J.L., Tytgat, I., Champomier-Vergès, M.C., Kleerebezem, M., and Hols, P. (2005). Lactate racemization as a rescue pathway for supplying D-lactate to the cell wall biosynthesis machinery in *Lactobacillus plantarum*. J. Bacteriol. *187*, 6750–6761. https://doi.org/10.1128/JB.187.19.6750-6761.2005

Gu, S.A., Jun, C., Joo, J.C., Kim, S., Lee, S.H., and Kim, Y.H. (2014). Higher thermostability of l-lactate dehydrogenases is a key factor in decreasing the optical purity of d-lactic acid produced from *Lactobacillus coryniformis*. Enzyme Microb. Technol. *58–59*, 29–35. https://doi.org/10.1016/j.enzmictec.2014.02.008

Guglielmotti, D.M., Mercanti, D.J., Reinheimer, J.A., and Quiberoni, A.delL. (2011). Review: efficiency of physical and chemical treatments on the inactivation of dairy bacteriophages. Front. Microbiol. *2*, 282. https://doi.org/10.3389/fmicb.2011.00282

Guo, W., He, R., Ma, L., Jia, W., Li, D., and Chen, S. (2014). Construction of a constitutively expressed homo-fermentative pathway in *Lactobacillus brevis*. Appl. Microbiol. Biotechnol. *98*, 6641–6650. https://doi.org/10.1007/s00253-014-5703-x

Hammes, W.P., and Hertel, C. (2006). The genera *Lactobacillus* and *Carnobacterium*. In The Prokaryotes, Dworkin, M., Falkow, S., Rosenberg, E., Schleifer, KH., and Stackebrandt, E., eds. (Springer, New York), pp. 320–403.

Hansen, P.A., and Lessel, E.F. (1971). *Lactobacillus casei* (Orla-Jensen) comb. nov. *Int. J. Syst. Evol. Microbiol. 21*, 69–71.

Hemme, D., Raibaud,P., Ducluzeau, R., Galpin, J.-V., Sicard, P., and Van Heijenoort, J.. (1980). *Lactobacillus murinus* n. sp., une nouvelle espèce de la flore dominante autochtone du tube digestif du rat et de la souris. Ann. Microbiol. Inst. Pasteur *131*, 297–308.

Henneberg, W. (1903). Zur Kenntnis der Milchsäurebakterien der Brennereimaische, der Milch, des Bieres, der Presshefe, der Melasse, des Sauerkohls, der sauren Gurken und des Sauerteigs, sowie einige Bemerkungen über die Milchsäurebakterien des menschlichen Magens: Aus dem botanischen Laboratorium des Instituts für Gärungsgewerbe zu Berlin. (P. Parey, Germany).

Jia, X., Liu, P., Li, S., Li, S., and Wen, J. (2011). D-lactic acid production by a genetically engineered strain *Corynebacterium glutamicum*. World J. Microbiol. Biotechnol. *27*, 2117–2124. https://doi.org/10.1007/s11274-011-0675-9

John, R.P., and Madhavan Nampoothiri, K. (2011). Co-culturing of *Lactobacillus paracasei* subsp. paracasei with a *Lactobacillus delbrueckii* subsp. delbrueckii mutant to make high cell density for increased lactate productivity from cassava bagasse hydrolysate. Curr. Microbiol. *62*, 790–794. https://doi.org/10.1007/s00284-010-9793-0

John, R.P., Gangadharan, D., and Madhavan Nampoothiri, K. (2008). Genome shuffling of *Lactobacillus delbrueckii* mutant and *Bacillus amyloliquefaciens* through protoplasmic fusion for L-lactic acid production from starchy wastes. Bioresour. Technol. *99*, 8008–8015. https://doi.org/10.1016/j.biortech.2008.03.058

Kandler, O., and Kunath, P. (1983). *Lactobacillus kefir* sp.nov., a component of the microflora of Kefir. Syst. Appl. Microbiol. *4*, 286–294. https://doi.org/10.1016/S0723-2020(83)80057-5

Kandler, O., and Weiss, N. (1986). Genus *Lactobacillus* Beijer-inck 1901, 212^{AL} In Bergey's Manual of Systematic Bacteriology, Sneath, P.H.A., Mair, N.S., Sharpe, M.E., and Holt, J.G., eds. (Williams and Wilkins, Baltimore) vol. 2, pp. 1209–1234.

Kandler, O., Stetter, K.O., and Köhl, R. (1980). *Lactobacillus reuteri* sp. nov., a new species of heterofermentative lactobacilli. Zentralblatt für Bakteriologie: I. Abt. Originale C: Allgemeine, angewandte und ökologische Mikrobiologie, *1*, 264–269.

Katagiri, H., Kitahara, K., Fukami, K., and Sugase, M. (1934). The characteristics of the lactic acid bacteria isolated from moto, yeast mashes for saké manufacture. *Bulletin of the Agricultural Chemical Society of Japan, 10*, 153–157.

Kuo, Y.C., Yuan, S.F., Wang, C.A., Huang, Y.J., Guo, G.L., and Hwang, W.S. (2015). Production of optically pure L-lactic acid from lignocellulosic hydrolysate by using a newly isolated and D-lactate dehydrogenase gene-deficient *Lactobacillus paracasei* strain. Bioresour. Technol. *198*, 651–657. https://doi.org/10.1016/j.biortech.2015.09.071

Kuznetsov, V.D. 1959. A new species of lactic acid bacteria. Mikrobiologiya *28*:248–351.

Kylä-Nikkilä, K., Hujanen, M., Leisola, M., and Palva, A. (2000). Metabolic engineering of *Lactobacillus helveticus* CNRZ32 for production of pure L-(+)-lactic acid. Appl. Environ. Microbiol. *66*, 3835–3841.

Lauer, E., and Kandler, O. (1980). *Lactobacillus gasseri* sp. nov., a new species of the subgenus Thermobacterium. Zentralblatt für Bakteriologie: I. Abt. Originale C: Allgemeine, angewandte und ökologische Mikrobiologie, *1*, 75–78.

Liang, B., Hyland, L., and Hou, S. (2001). Nasal-associated lymphoid tissue is a site *of* long-term virus-specific antibody production following respiratory virus infection *of* mice. J. Virol. *75*, 5416–5420. https://doi.org/10.1128/JVI.75.11.5416-5420.2001

London, J., and Hausman, S. (1982). Xylitol-mediated transient inhibition *of* ribitol utilization by *Lactobacillus casei*. J. Bacteriol. *150*, 657–661.

Moon, S.K., Wee, Y.J., and Choi, G.W. (2014). Utilization *of* by-products derived from bioethanol production process for cost-effective production *of* lactic acid. J. Ind. Microbiol. Biotechnol. *41*, 1525–1531. https://doi.org/10.1007/s10295-014-1495-3

Moro, E. (1900). Uber den *Bacillus acidophilus* n. sp. *Jahrb. Kinderheilk*, *52*, 38–55.

Nakamura, L.K., and Crowell, C.D. (1979). *Lactobacillus amylophilus*, a new starchhydrolyzing species from swine waste-corn fermentation. *Dev. Ind. Microbiol, 20*, 531-540.

Madhavan Nampoothiri, K., Nair, N.R., and John, R.P. (2010). An overview of the recent developments in polylactide (PLA) research. Bioresour. Technol. *101*, 8493–8501. https://doi.org/10.1016/j.biortech.2010.05.092

Nancib, A., Nancib, N., and Boudrant, J. (2009). Production of lactic acid from date juice extract with free cells of single and mixed cultures of *Lactobacillus casei* and *Lactococcus lactis*. World J. Microbiol. Biotechnol. 25, 1423–1429.

Okano, K., Hama, S., Kihara, M., Noda, H., Tanaka, T., and Kondo, A. (2017). Production of optically pure D-lactic acid from brown rice using metabolically engineered *Lactobacillus plantarum*. Appl. Microbiol. Biotechnol. *101*, 1869–1875. https://doi.org/10.1007/s00253-016-7976-8

Okano, K., Yoshida, S., Tanaka, T., Ogino, C., Fukuda, H., and Kondo, A. (2009). Homo-D-lactic acid fermentation from arabinose by redirection of the phosphoketolase pathway to the pentose phosphate pathway in L-lactate dehydrogenase gene-deficient *Lactobacillus plantarum*. Appl. Environ. Microbiol. 75, 5175–5178. https://doi.org/10.1128/AEM.00573-09

Orla-Jensen, S. (1916). Maelkeri-Bakteriologi (Schonberske Forlag, Denmark).

Orla-Jensen, S. (1919). The lactic acid bacteria. Andr. Fred Høst and Son. Mém. Acad. Roy. Sci., Copenhagen, Danemark, Sect. Sci., 8 Sér. 5:81–197.

Overbeck, T., Steele, J.L., and Broadbent, J.R. (2016). Fermentation of de-oiled algal biomass by *Lactobacillus casei* for production of lactic acid. Bioprocess Biosyst. Eng. *39*, 1817–1823. https://doi.org/10.1007/s00449-016-1656-z

Pandey, A., Soccol, C.R. (2000). Economic utilization of crop residues for value addition: a future approach. J. Sci. Ind. Res. 59:12–22.

Panesar, P.S., Kennedy, J.F., Knill, C.J., and Kosseva, M.R. (2007). Applicability of pectate-entrapped *Lactobacillus casei* cells for L(+) lactic acid production from whey. Appl. Microbiol. Biotechnol. *74*, 35–42. https://doi.org/10.1007/s00253-006-0633-x

Patnaik, R., Louie, S., Gavrilovic, V., Perry, K., Stemmer, W.P., Ryan, C.M., and del Cardayré, S. (2002). Genome shuffling of *Lactobacillus* for improved acid tolerance. Nat. Biotechnol. *20*, 707–712. https://doi.org/10.1038/nbt0702-707

Peres, C., Matos, A.I., Conniot, J., Sainz, V., Zupančič, E., Silva, J.M., Graça, L., Sá Gaspar, R., Préat, V., and Florindo, H.F. (2017). Poly(lactic acid)-based particulate systems are promising tools for immune modulation. Acta Biomater. *48*, 41–57. https://doi.org/10.1016/j.actbio.2016.11.012

Prasad, S., Srikanth, K., Limaye, A.M., and Sivaprakasam, S. (2014). Homo-fermentative production of D-lactic acid by *Lactobacillus* sp. employing casein whey permeate as a raw feed-stock. Biotechnol. Lett. *36*, 1303–1307. https://doi.org/10.1007/s10529-014-1482-9

Rao, C.h.S., Prakasham, R.S., Rao, A.B., and Yadav, J.S. (2008). Functionalized alginate as immobilization matrix in enantioselective L (+) lactic acid production by *Lactobacillus delbrucekii*. Appl. Biochem. Biotechnol. *149*, 219–228. https://doi.org/10.1007/s12010-007-8052-8

Rogosa, M., Wiseman, R.F., Mitchell, J.A., Disraely, M.N., and Beaman, A.J. (1953). Species differentiation of oral lactobacilli from man including descriptions of *Lactobacillus salivarius* nov spec and *Lactobacillus cellobiosus* nov spec. Journal of Bacteriology, *65*, 681.

Romano, A.H., Trifone, J.D., and Brustolon, M. (1979). Distribution *of* the phosphoenolpyruvate:glucose phosphotransferase system in fermentative bacteria. J. Bacteriol. *139*, 93–97.

Ruiz-Moyano, S., Martín, A., Benito, M.J., Casquete, R., Serradilla, M.J., and Córdoba, M.de G. (2009). Safety and functional aspects of pre-selected lactobacilli for probiotic use in Iberian dry-fermented sausages. Meat Sci. *83*, 460–467. https://doi.org/10.1016/j.meatsci.2009.06.027

Sawai, D., Tamada, M., and Kanamoto, T. (2007). Development of oriented morphology and mechanical properties upon drawing of stereo-complex of poly (L-lactic acid) and poly (D-lactic acid) by solid-state coextrusion. Polymer Journal, 39, 953–960. https://doi.org/10.1295/polymj.PJ2007038

Secchi, N., Giunta, D., Pretti, L., García, M.R., Roggio, T., Mannazzu, I., and Catzeddu, P. (2012). Bioconversion of ovine scotta into lactic acid with pure and mixed cultures of lactic acid bacteria. J. Ind. Microbiol. Biotechnol. 39, 175–181. https://doi.org/10.1007/s10295-011-1013-9

Singhvi, M., Zendo, T., Iida, H., Gokhale, D., and Sonomoto, K. (2017). Stimulation *of* D- *and* L-lactate dehydrogenases transcriptional levels in presence *of* diammonium hydrogen phosphate resulting to enhanced lactic acid production by *Lactobacillus* strain. J. Biosci. Bioeng. *124*, 674–679. https://doi.org/10.1016/j.jbiosc.2017.07.009

Södergård, A., and Stolt, M. (2002). Properties of lactic acid based polymers and their correlation with composition. Prog. Pol. Sci. 27, 1123–1163. https://doi.org/10.1016/S0079-6700(02)00012-6

Stephanopoulos, G. (2002). Metabolic engineering by genome shuffling. Nat. Biotechnol. *20*, 666–668. https://doi.org/10.1038/nbt0702-666

Taniguchi, M., Tokunaga, T., Horiuchi, K., Hoshino, K., Sakai, K., and Tanaka, T. (2004). Production of L-lactic acid from a mixture of xylose and glucose by co-cultivation of lactic acid bacteria. Appl. Microbiol. Biotechnol. *66*, 160–165. https://doi.org/10.1007/s00253-004-1671-x

Tashiro, Y., Kaneko, W., Sun, Y., Shibata, K., Inokuma, K., Zendo, T., and Sonomoto, K. (2011). Continuous D-lactic acid production by a novel thermotolerant *Lactobacillus delbrueckii* subsp. *lactis* QU 41. Appl. Microbiol. Biotechnol. *89*, 1741–1750. https://doi.org/10.1007/s00253-010-3011-7

Triratna, L., Saksono, B., Sukmarini, L., and Suparman, A. (2011). Genome-shuffling-improved acid tolerance and lactic acid production in *Lactobacillus plantarum* for commercialization. Microbiology Indonesia, 5, 4. http://doi.org/10.5454/mi.5.1.4

Troili-Petersson, G. (1903). Studien über die Mikroorganismen des schwedischen Güterkäses. Zentralbl. Bakteriol. Parasitenkd. Infektionskr. Hyg. II Abt. *11*, 120–143.

Viana, R., Pérez-Martínez, G., Deutscher, J., and Monedero, V. (2005a). The glycolytic genes pfk and pyk from *Lactobacillus casei* are induced by sugars transported by the phosphoenolpyruvate:sugar phosphotransferase system and repressed by CcpA. Arch. Microbiol. *183*, 385–393. https://doi.org/10.1007/s00203-005-0003-6

Viana, R., Yebra, M.J., Galán, J.L., Monedero, V., and Pérez-Martínez, G. (2005b). Pleiotropic effects of lactate dehydrogenase inactivation in *Lactobacillus casei*. Res. Microbiol. *156*, 641–649. https://doi.org/10.1016/j.resmic.2005.02.011

Wang, J., Wang, Q., Xu, Z., Zhang, W., and Xiang, J. (2015). Effect of fermentation conditions on L-lactic acid production from soybean straw hydrolysate. J. Microbiol. Biotechnol. 25, 26–32.

Xu, Q., Zang, Y., Zhou, J., Liu, P., Li, X., Yong, Q., and Ouyang, J. (2016). Highly efficient production of D-lactic acid from chicory-derived inulin by *Lactobacillus bulgaricus*. Bioprocess Biosyst. Eng. *39*, 1749–1757. https://doi.org/10.1007/s00449-016-1650-5

Yoshito, I., Jamshidi, K., Tsuji, H., Hyon, S.H. (1987). Stereocomplex formation between enantiomeric poly(lactides). Macromolecule 20(4):904–906

Zanoni, P.J., Farrow, A.E., Phillips, B.A., and Collins, M.D.. (1987). *Lactobacillus pentosus* (Fred, Peterson, and Anderson) sp. nov., nom. rev. Int. J. Syst. Bacteriol. 37, 339–341.

Zhang, J., Gong, G., Wang, X., Zhang, H., and Tian, W. (2015). Positive selection on D-lactate dehydrogenases of *Lactobacillus delbrueckii* subspecies *bulgaricus*. IET Syst. Biol. *9*, 172–179. https://doi.org/10.1049/iet-syb.2014.0056

Zhang, J., Wu, C., Du, G., and Chen, J. (2012). Enhanced acid tolerance in *Lactobacillus casei* by adaptive evolution and compared stress response during acid stress. Biotechnology and Bioprocess Engineering, *17*, 283. https://doi.org/10.1007/s12257-011-0346-6

Zhang, Y., Vadlani, P.V., Kumar, A., Hardwidge, P.R., Govind, R., Tanaka, T., and Kondo, A. (2016). Enhanced D-lactic acid production from renewable resources using engineered *Lactobacillus plantarum*. Appl. Microbiol. Biotechnol. *100*, 279–288. https://doi.org/10.1007/s00253-015-7016-0

Zhao, Y., Wang, Z., Wang, J., Mai, H., Yan, B., and Yang, F. (2004). Direct synthesis of poly (D, L-lactic acid) by melt polycondensation and its application in drug delivery. J. Appl. Pol. Sci. *91*, 2143–2150.

Zotta, T., Ricciardi, A., Guidone, A., Sacco, M., Muscariello, L., Mazzeo, M.F., Cacace, G., and Parente,

E. (2012). Inactivation of ccpA and aeration affect growth, metabolite production and stress tolerance in *Lactobacillus plantarum* WCFS1. Int. J. Food Microbiol. *155*, 51–59. https://doi.org/10.1016/j.ijfoodmicro.2012.01.017

Zou, H., Wu, Z., Xian, M., Liu, H., Cheng, T., and Cao, Y. (2013). Not only osmoprotectant: betaine increased lactate dehydrogenase activity and L-lactate production in lactobacilli. Bioresour. Technol. *148*, 591–595. https://doi.org/10.1016/j.biortech.2013.08.105

Modifications of *Lactobacillus* Surface Under Environmental Stress Conditions

5

Mariana C. Allievi[1,2], Sandra M. Ruzal[1,2] and María Mercedes Palomino[1,2*]

[1]Universidad de Buenos Aires, Facultad de Ciencias Exactas y Naturales, Departamento de Química Biológica, Buenos Aires, Argentina.
[2]CONICET – Universidad de Buenos Aires, Instituto de Química Biológica de la Facultad de Ciencias Exactas y Naturales (IQUIBICEN), Buenos Aires, Argentina.

*Correspondence: mpalomino@qb.fcen.uba.ar

https://doi.org/10.21775/9781910190890.05

Abstract

Lactobacilli are a diverse group of food-grade microorganisms widely applied in the fermented food industry due to their technological and health-promoting properties; these bacteria have been extensively used as starter cultures and as probiotics. In this way, *Lactobacillus* is exposed to different stress factors during food fermentation. In this chapter we will discuss the ability of *Lactobacillus* to respond to environmental conditions, focusing on osmotic stress, by altering the nature of their cell wall for adaptation. We will describe their modification, in particular those related to the structure of peptidoglycan, secondary cell wall polymers, namely teichoic acids and surface-layer proteins.

Introduction

Lactobacilli are Gram-positive, non-sporulated, anaerobic bacteria. They are normal inhabitants of the oral cavity and the digestive tract in humans. *Lactobacillus* are members of the lactic acid bacteria (LAB). Some strains are used in food fermentation, are found in the dairy industry for the production of cheese, yoghurt, cultured milks and other fermented milk products. They are also frequently used as probiotics in foods and various pharmaceutical preparations (Sanders 1999; Remus *et al.*, 2011; Ahmed *et al.*, 2010). Fermented foods using LAB are among the oldest forms of processed foods and nowadays they are present in our diet. The process is based on the lactic-acid fermentation performed by a relatively wide range of LAB.

Since LAB are used in food production and probiotics food/formulas to enhance the health of the host, it is worth noting that they are one of the best-studied microorganism.

Over the past four decades, there has been a growing interest in the physiology, biochemistry, genetics, and evolution of LAB and therefore these microorganisms have become the centre of attention for research in many laboratories around the world.

With the advent of genome sequencing and meta-omics technologies our studies about LAB have been revolutionized and the new approaches allow us to increase our knowledge on their behaviour in food production and the microbiome ecosystem.

Like all other microorganisms, LAB are exposed to stressful conditions, but understanding the stress responses of the different categories of LAB is important for different reasons.

A fundamental aspect of LAB used as starters is their robustness during the production and storage of fermented foods. Vulnerability of starters when faced with the conditions during the technological processes may influence the fermentation process *per se*, which can produce some unwanted modification of food quality and safety.

In the same way, LAB need to be able to resist technological stresses during the preparation of probiotic formulas to maintain high viable counts. After consumption, probiotic LAB must survive the hostile conditions in the gastrointestinal tract (GIT), since a probiotic strain must reach a high viable number in the intestine to exert a beneficial effect on the host.

The maintenance of a healthy intestinal microbial ecosystem is carried out by lactobacilli and these bacteria are very important for providing protection against pathogen infection (Jacobsen *et al.*,1999; Hirano *et al.*, 2003; Henderson *et al.*, 2012). Lactobacilli are part of the intestinal microbial ecosystem and are present throughout the GIT in varying proportions. They are dominant in the proximal small intestine (Bongaerts *et al.*, 2001), a nutrient rich environment, whereas in the faecal microbiota they are present at most 0.01–0.6% and this proportion varies significantly between individuals (Maukonen *et al.*, 2008; Neville *et al.*, 2012). Lactobacilli are able to adhere to the mucosal layers and have the ability of surviving the hostile conditions of the luminal environment and the competing microbiota (Buck *et al.*, 2009).

The cell-surface architecture found in lactobacilli presents a great diversity and it is known that *Lactobacillus* have the ability to changing their surface properties in response to environmental changes (Taranto *et al.*, 2003; Fozo *et al.*, 2004).

Different macromolecules constituting the cell wall of lactobacilli have been shown to contribute to maintaining bacterial cell integrity during environmental stress (Guerzoni *et al.*, 2001).

In this chapter we present the state of the art for *Lactobacillus* stress behaviour and their modification in the cell envelope compare to other LAB. In a first section we will describe how the cell wall of *Lactobacillus* is composed. In a second section we will discuss the experimental evidence obtained to study the stress responses focusing on osmotic stress and its influence on the cell wall architecture and the adaptation in the adverse microenvironment.

Lactic acid bacteria

For a long time we have used the ability of lactic acid bacteria (LAB) to produce lactic acid from fermentable substrates as a food preservation method. The consequent decrease in pH prevents the growth of the competing microflora that could produce some modification on the quality and safety in the final products or become pathogenic for the host.

From the beginning of the microbiology, fermentation processes with LAB were used in the food manufactured. In 1873, Lister trying to determine which microorganisms were involved in milk fermentations, isolated for the first time, the first pure bacterial culture from milk which was called *Bacterium lactis*, today known as *Lactococcus lactis*.

Although for a long time the food production processes were carried out in an empirical way, nowadays they have acquired a high technological development in order to maintain the quality and regularity of the final product. This technological development was possible in part by the accumulation of knowledge of LAB biology.

The common characteristics of the genera are: Gram-positive, non-spore-forming, immobile, aerotolerant anaerobic, non-catalase microorganisms. Its metabolism is strictly fermentative and presents high nutritional requirements. Some genera only produce lactic acid as the main product of the fermentation of carbohydrates (homofermentative) or a mixture of lactic acid, carbon dioxide, ethanol and/or acetic acid (heterofermentative) (Kandler, 1983).

This broad LAB description includes genera of cocci (*Lactococcus, Streptococcus, Vagococcus, Pediococcus, Aerococcus, Leuconostoc, Oenococcus, Tetragenococcus*), as well as bacilli (*Lactobacillus* and *Carnobacterium*).

There is genetic evidence, mainly on the basis of the comparison of 16S rRNA sequences (Stackebrand *et al.*, 1988) that the LAB classification, defined under the physiological characteristics mentioned above, is consistent with the actual phylogenetic level.

The most used species in fermentation processes belong to the genera *Lactobacillus, Lactococcus, Leuconostoc, Pediococcus* and *Streptococcus*, which can easily be found in meat products, vegetables, dairy products, fruits and beverages. But these microorganisms are also inhabitants of the intestinal and genital tracts of human being and other animals.

The genus *Lactobacillus* is very heterogeneous and presents a great variability in the percentage of G + C of DNA ranging from 32% to 54%.

Kandler and Weiss (1986) divided the genus into three groups, regarding the presence or absence of the enzymes fructose 1-6 diphosphate aldolase and phosphoketolase, key enzymes in the homo- or heterofermentative metabolism of hexoses and pentoses, respectively.

- Group I: obligately homofermentative lactobacilli, ferment the hexoses by the Embden Meyerhof-Parnas pathway, producing two moles of lactic acid per mole of glucose. The microorganisms possess the enzyme fructose 1-6 diphosphate aldolase but lack phosphoketolase and therefore neither gluconate nor pentoses are fermented.. They do not use pentose or gluconate. Examples: *L. helveticus*, *L. acidophilus*, and *L. delbrueckii*.
- Group II: facultatively heterofermentative lactobacilli: ferment hexose, using gluconate and pentoses with lactic and acetic acid production. 1-6 diphosphate aldolase and phosphoketolase activity are present but the latter is inducible by pentose. Examples: *L. casei* and *L. plantarum*.
- Group III: obligately heterofermentative lactobacilli: fermenting hexose with formation of lactic acid, acetic acid (or ethanol) and carbon dioxide in equimolar amounts. They use pentoses via 6-phosphogluconic acid with the production of lactic and acetic acid. Examples: *L. reuteri* and *L. fermentum*. See Chapter 4 for more details on classification.

Starter cultures

Starter cultures containing one or more microorganisms are used in the industry to assure not only the organoleptic qualities but also the quality and safety of the final products. The starter cultures are commercially produced by numerous industries and laboratories to ensure the reproducibility and standardization of processes.

Advances in recombinant DNA techniques have allowed the construction of strains in order to improve quality, variety and nutritional value to the products, as well as amelioration in the industrial production processes.

Lactic acid bacteria and among them *Lactobacillus*, are excellent candidates for metabolic engineering since their use for human consumption in fermented foods is very ancient. *Lactobacillus* genus is widely used as starter to manufacture a broad variety of fermented products, such as cheese, yoghurt, vegetables, etc. The extended use of *Lactobacillus* species is mainly related to their GRAS (generally recognized as safe) status, a qualification granted by the Food and Drug Administration (USA).

Currently LAB are employed as starter cultures in the fermentation of a wide range of foods: vegetables and fruits in brine, meat products such as the production of sausages and in the production of dairy foods.

Several metabolic properties of LAB have a direct or indirect impact on processes such as the development of flavour, aroma and maturation of dairy products (such as cheeses and yoghurts). The degradation of the caseins present in milk plays a crucial role in the development of these processes: while certain peptides contribute to the formation of flavour, other peptides are undesirable because of producing a bitter and unwanted taste.

However, when focusing the study on LAB metabolism in a biotechnological manner in order to improve their characteristics, the microenvironment in which the fermented food is produced is not taken into account.

The addition of salt, to preserve food, is a common technique used since ancient times. In many foods, fermentations processes occur in an environment with a high concentration of solutes. In the same way, the ripening process in hard-paste cheeses occurs after the salted moment in which LAB are fundamental to determine the flavour and aroma of the final product.

Facing this problem, the main objective of this chapter has been the study of the physiological modifications of *Lactobacillus* during growing in high-salt media and mainly related to the modifications that happen at their cell envelopes.

Cell envelopes

The cell wall of Gram-positive bacteria is a complex arrangement of macromolecules. In *Lactobacillus* the cell envelope is composed of the following structures from the intracellular to the outer medium: the cytoplasmic membrane that is covered by pentapeptide stem-connected layers of peptidoglycan (PG), a polymer composed of alternating residues of N-acetylglucosamine (GlcNAc) and β-1-4-linked N-acetyl muramic acid (MurNAc).

In addition, all lactobacilli produce lipoteichoic acid (LTA), a glycosyl-substituted and/or D-ala-substituted glycerol phosphate (GroP) polymer which is anchored in the cytoplasmic membrane through a (diacylated and/or tri-acylated) glycolipid anchor.

Most lactobacilli also synthesize wall teichoic acid (WTA), which typically consists of GroP or ribitolphosphate (RboP) backbone polymers that are covalently anchored to the Mur- NAC residue of PG via a phosphodiester bond.

In some species there are additional envelopes such as S-layer, capsules and exopolysaccharides (EPS) (Fig. 5.1).

The cell envelope represents the first line of defence against environmental disturbances so it is a physical barrier to separate the cell from its environment and, as such, acts as the first sensor to monitor such modifications. Changes in chemical composition of both the cell wall and the cell membrane triggered by stress have been shown to aid in cell survival (Machado *et al.*, 2004, Piuri *et al.*, 2005). Maintaining the integrity of the cell wall under stress conditions is a matter of life or death for bacteria. The cell envelope is also a major cellular organelle with several physiological functions. It has multiple functions during bacterial growth, including maintaining bacterial cell integrity and shape as well as resisting internal turgor pressure. Furthermore, it must remain flexible to accommodate the remodelling that is required for cell division and growth. Since it acts as the interface between the bacterial cell and its environment, the cell wall also mediates bacterial interactions with abiotic surfaces, bacteriophages, or eukaryotic host cells. Also an important fraction of the probiotic effector molecules resides in the bacterial cell envelope, as this part of the microbial cell is the first to interact with intestinal host cells.

It is known that a plethora of antimicrobials exert their activity against the cell wall (Bush, 2012). Over the past years, it has been demonstrated that LAB, like other bacteria, have the ability of closely monitoring the integrity of the cell envelope and to induce specialized repair mechanisms in case of damage (Jordan *et al.*, 2008).

Cytoplasmic membrane

The first envelope surrounding the bacterium is the cytoplasmic membrane. It consists of a double layer of phospholipids, among which there are associated proteins that play structural roles, sensors or metabolic functions. Several enzymatic activities are associated with cytoplasmic membrane including transport systems and various polymer synthesis systems. Up to 90% of the ribosomes can be isolated as an aggregate DNA-polyribosome-membrane.

Membranes are responsible for about 30% or more of the cell weight. They contain between 60 and 70% of proteins, about 40% of lipids and small amounts of carbohydrates. Among lipids, a high proportion of the polar lipids present in the

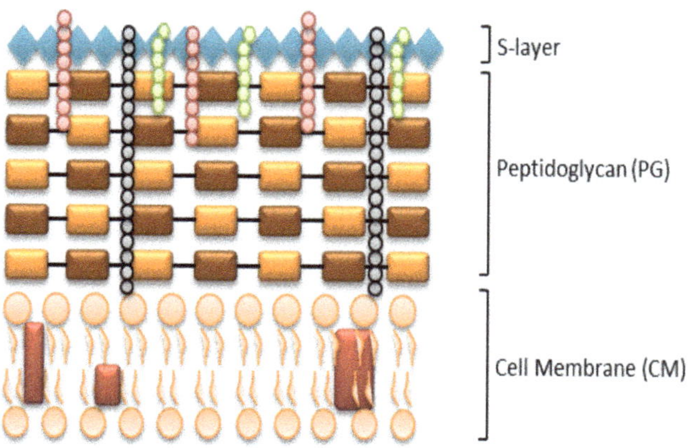

Figure 5.1 Schematic representation of the cell envelope of lactobacilli. The bilipidic cell membrane (CM) with associated proteins is covered by a multilayered peptidoglycan (PG) decorated with lipoteichoic acids, LTA (grey circles) wall teichoic acids, WTA (red circles), and exopolysaccharides (green hexagons). Blue diamonds represent the outer envelope of S-layer proteins.

membrane are glycolipids and in a lesser amount phospholipids (Machado 2004). Reports on polar lipids of lactobacilli showed that most of them were mainly phosphatidylglycerol (PG), with small amounts of phosphatidic acid, diphosphatidylglycerol or cardiolipin (CL), lysylphosphatidylglycerol, phosphoglycolipid and diglycosyldiglycerides (DGDG).

Also diacylphosphatidylglycerol, triglycosyldiglycerides and neutral lipids are found (Fernandez Murga, 2000).The presence of polyunsaturated fatty acid (PUFA) in the growth medium was shown to induce changes in bacterial fatty acids in relation to the regulation of the degree of fatty acid unsaturation, cyclization, and proportions of CLA and PUFA containing 20–22 carbons (Kankaanpaa et.al., 2004).

Peptidoglycan: chemical composition

Peptidoglycan (PG) is the main component of the Gram-positive cell wall. It consists of glycan chains made of alternating N-acetylglucosamine (GlcNAc) and N-acetylmuramic acid (MurNAc) that are linked via β-1,4 bonds (Fig. 5.2). Peptidic chains are linked covalently through their N-terminus to the lactyl group of MurNAc. These peptidic chains vary in composition across species and can be cross-linked directly or indirectly, through short chains of one or more amino acids that generate a three-dimensional structure around the cell, which ensures bacterial integrity. In LAB, the amino acid sequence of the stem peptide is L-Ala-γ-D-Glu-X-D-Ala, while the third amino acid (X) is a di-amino acid. It is most often L-Lys (e.g. in *L. lactis* and most lactobacilli) but can also be mesodiaminopimelic acid (mDAP) (e.g. in *L. plantarum*) or L-ornithine (e.g. in *L. fermentum*) (Schleifer and Kandler, 1972). Among LAB, D-Ala predominates at position five in newly synthesized PG; however, D-lactate residues are found in naturally vancomycin-resistant lactobacilli such as *L. casei* and *L. plantarum*.

Cross-linking between adjacent stem peptides takes place between the D-Ala in position four of one peptide chain and the diamino acid in position three (4–3 cross-link) of another chain.

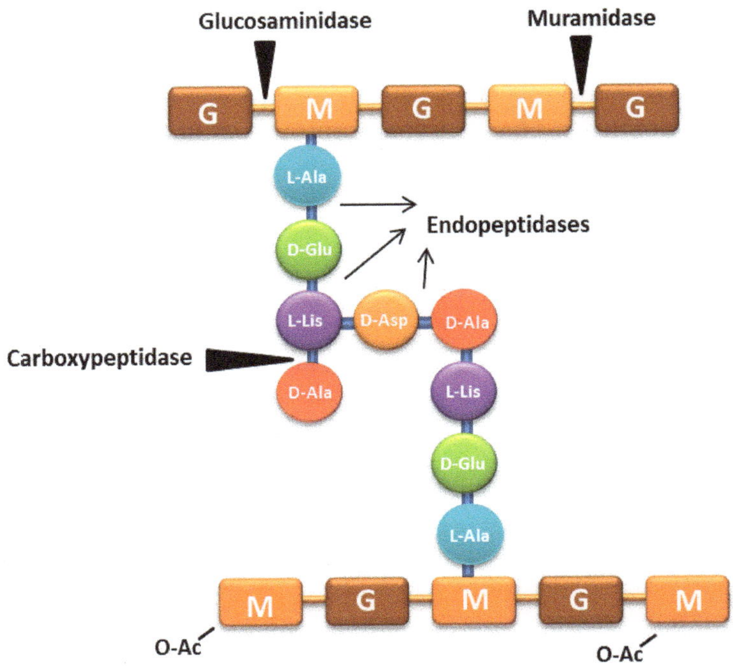

Figure 5.2 Schematic representation of the structure of peptidoglycan. This is the type of structure found in *L. casei* and numerous lactobacilli. G: N-acetylglucosamine, M: N-acetylmuramic acid. The latter may be O-acetylated (O-Ac). The cleavage sites of the different classes of PG hydrolases are indicated with arrows.

Schleifer and Kandler (1972) have meticulously reviewed and classified the different types of peptidoglycan and their taxonomic implication.

The primary structure of peptidoglycan in most lactobacilli belongs to the A4α-type peptidoglycan, with a monomer primary structure (GlcNAc -MurNAc-L-Ala-α-D-Glu-L-Lys-D-Ala) and a D-Asp in the interpeptide bridge, attached to the ε-amino group of Lys as *Lactococcus lactis*. This differentiated from the peptidoglycans found in streptococci belong to variation A and 3 with L-Lys in position 3 of the peptide subunit and interpeptide bridges consisting of monocarboxylic L-amino acids or glycine, or both. The presence of mesodiaminopimelic acid (A2pm) in the pentapeptide stem of *L. plantarum* has been verified (Schleifer and Kandler 1972).

Diverse technology such as HPLC and mass spectrometry has allowed analysing the structural PG in a detail manner. In several *Lactobacillus* species, detailed information is known about the nature of peptide cross bridges and the degree of cross-linking of PG. Detailed PG structure has been studied for several *Lactobacillus*, including *L. buchneri* (Anzengruber *et al.*, 2014), *L. casei* (Regulski et al., 2012), *L. rhamnosus* (Claes *et al.*,2012), and *L. plantarum* (Bernard *et al.*, 2011). The first two species were found to have D-Ala4-D-Asp/Asn-L-Lys3 cross-bridges, while the latter has a direct D-Ala4–mDAP3 cross-bridge. The PG layer remains in a dynamic state throughout a bacterium's life, and its structure is the result of complex biosynthetic, maturation, and degradation reactions. It has also revealed the existence of covalent PG modifications, such as O-acetylation, N-deacetylation, or amidation. In *L. casei*, 59% of the peptidoglycan is O-acetylated (in N-acetyl muramic), similar to the 60% found in *L. acidophilus* (Billot-Klein *et al.*, 1997). Although the physiological role of O-acetylation is unclear, it has been suggested that it would serve to protect peptidoglycan chains from the hydrolytic activity of muramidase autolysins. These modifications may play essential roles in bacterial physiology. Additional modifications of PG have been described in *Lactobacillus* species, including N-deacetylation of GlcNAc and/or MurNAc. Notably, the amino acids D-Glu and D-Asp are presented in their amidated forms (D-iso-Gln and D-iso-Asn, respectively) to exemplify one of the several post-translational modifications that has been observed in PG. (Bron *et. al.*, 2013).

Penicillin-binding proteins (PBP)

Polymerisation and cross-linking of PG are mediated by penicillin-binding proteins or PBP, so named because they have high affinity and are inhibited by penicillin and other β-lactam antibiotics (Ghuysen 1991). PBP are cytoplasmic membrane proteins, and are classified according to sequence similarity in three groups:

Class A: PBP of high molecular weight, with bifunctional characteristics, containing both transglycosylation and transpeptidation domains located at the N- and C-terminal of the protein, respectively.

Class B: are high molecular weight PBP, with a C-terminal domain of transpeptidation whereas its N-terminal domain has yet unknown function. These ones play an essential role in septation and maintenance of the cellular form.

Finally, Class C: the low molecular weight PBP with D-, D- or D-, L-carboxypeptidase or endopeptidase activity, whose most important function would be to regulate the number of cross-links between adjacent glycans chains.

Most species of Gram-positive bacteria possess a variable number of PBP, which implies their functional redundancy and essentiality. *Bacillus subtilis* genomic sequence analysis revealed 16 genes encoding PBP, four class A, six class B and six low molecular weight PBP. An extended study has been carried out on *Bacillus subtilis* PBP to reveal their specific functions in peptidoglycan biosynthesis, maintenance of cell morphology, cell elongation and septation, sporulation and germination. Analysis of the genome of *L. lactis*, an ovococcus species, has revealed the presence of six PBPs: five high molecular weight (HMW) PBP (PBP1a, PBP1b, PBP2a, PBP2b, and PBPx) and one low molecular weight (LMW) PBP (D-Ala-D-Ala-carboxypeptidase DacA) (David 2013). Through a genomic analysis, it was possible to find different PBP in *Lactobacillus* species belonging to the three classes regarding their fermentative type. Fig. 5.3 shows the gene tree of *pbp1A*, a bifunctional glycosyltransferase transpeptidase penicillin-binding, *pbpF*, with a carboxypeptidase domain, and *dacC* a, serine-type D-Ala-D-Ala carboxypeptidase.

Although PBP enzymology has been extensively studied in many bacteria (Ghuysen 1991; Popham and Young 2003), there is limited information on the *in vivo* functions of these proteins during

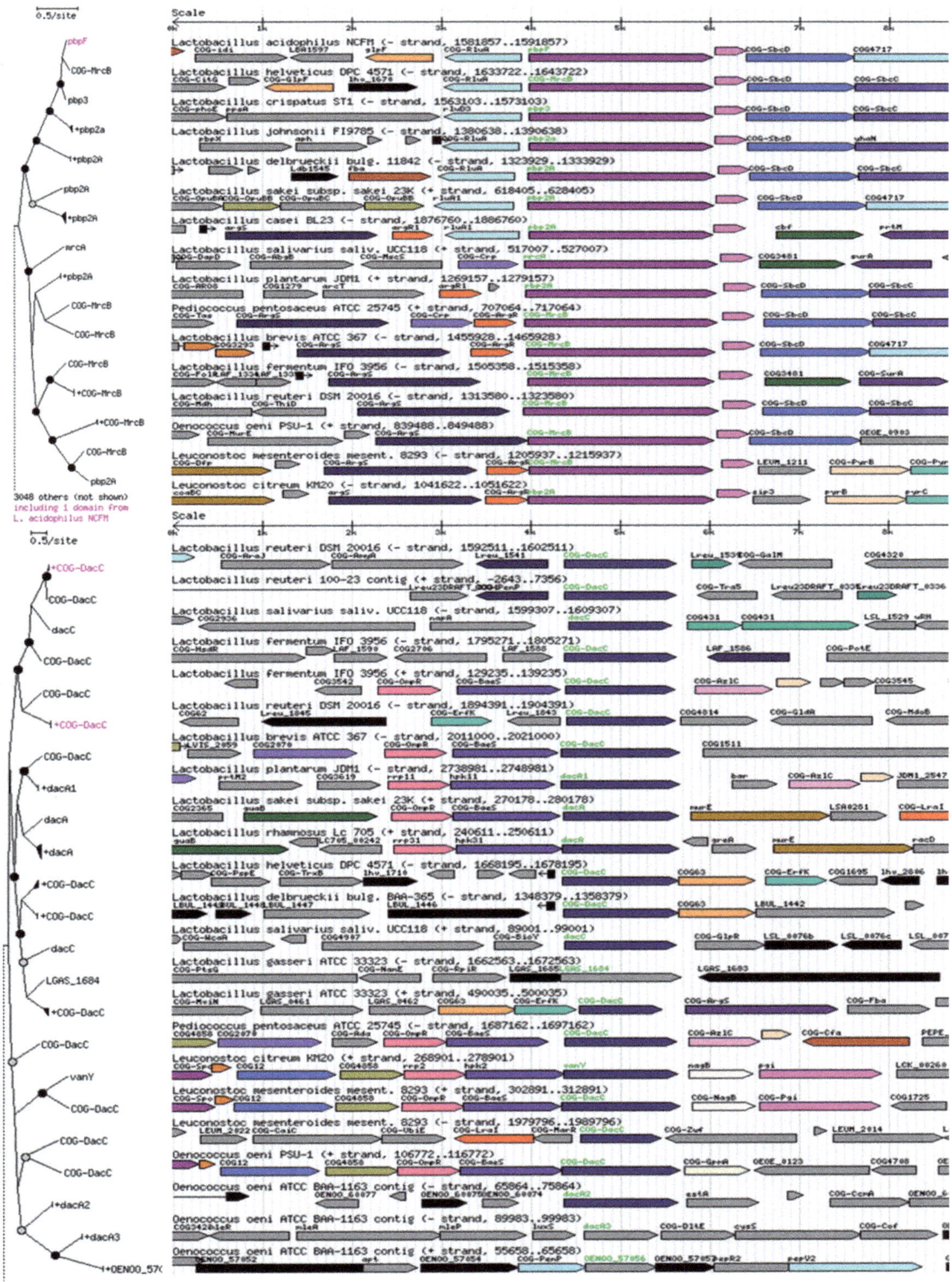

Figure 5.3 Gene tree for *pbp1A*: Bifunctional glycosyltransferase transpeptidase penicillin-binding protein 1A (LCABL_17030), 762aa from *Lactobacillus casei* BL23.
Gene tree for *pbpF*: penicillin-binding protein (LBA1593), 685aa from *Lactobacillus acidophilus* NCFM.
Gene tree for *dacC*: Serine-type D-Ala-D-Ala carboxypeptidase (Lreu_1540), 414aa from *Lactobacillus reuteri* DSM-20016.
http://www.microbesonline.org

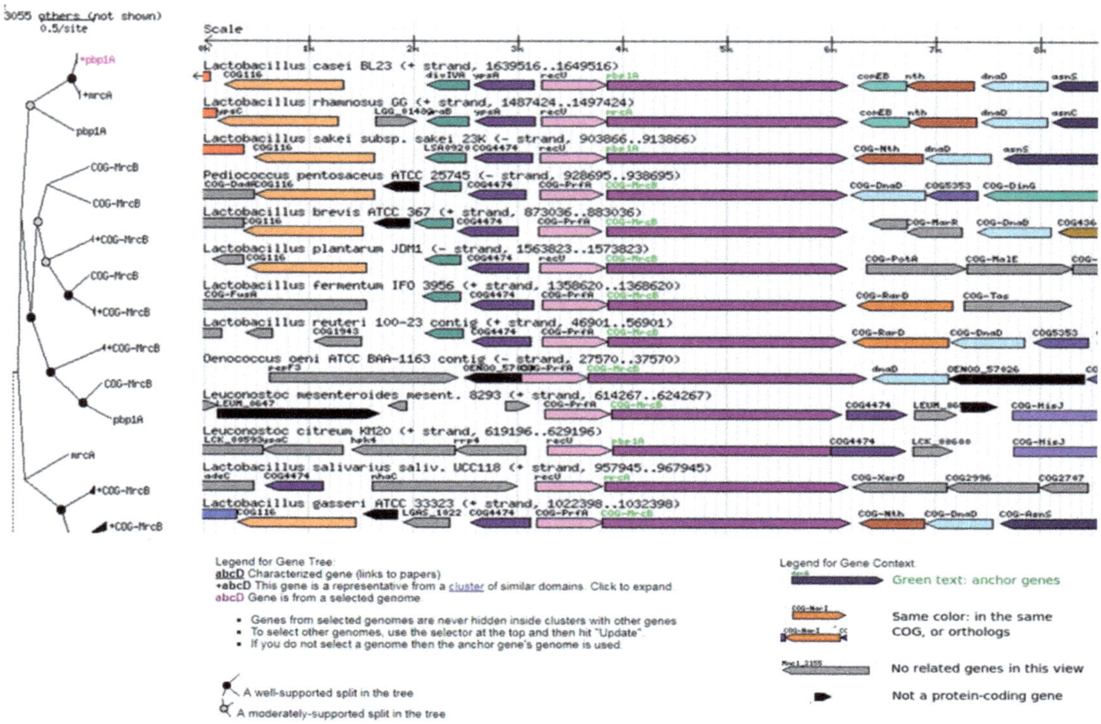

Figure 5.3 Continued.

Lactobacillus stress adaptation. We will discuss in this chapter the modifications on PBP activity when lactobacilli are subjected under high-salt condition and its consequences on cell envelopes. Piuri and colleagues (2005) observed a difference in the concentration and affinity of *Lactobacillus casei* PBP when the cells were pre-cultured in a hypersaline medium. This fact was correlated with a decrease in PG cross-linking in that microorganism.

Teichoic acids

Besides peptidoglycan, the cell wall of lactobacilli comprises TA (teichoic acids), which are essential anionic polymers of Gram-positive bacteria and represent up to 50% of the cell wall dry weight. TA are involved in various aspects of the functionality of the cell wall. Together with peptidoglycan, they form a polyanionic matrix or gel contributing to the porosity, elasticity and electrostatic steering of the cell envelope. They are also involved in cation homeostasis, in particular of Mg^{2+} and protons, the latter are important in the maintenance of a proton gradient in the cell wall. Among a large range of identified biological functions, TA participate in the modulation of the activity of PGHs, the binding of surface proteins, phage adsorption, cell adhesion and interaction with the immune system (counterpart of Gram-negative lipopolysaccharide) (Delcour *et al.*, 1999; Neuhaus and Baddiley, 2003).

Lactobacilli deserve specific mention in terms of TA since they were initially discovered by Baddiley and colleagues in *L. plantarum* (Baddiley, 1989). Most lactobacilli investigated for their TA content possess two types of anionic polymers: WTA that are covalently bound to MurNAc of peptidoglycan glycan strands via a linkage unit and LTA that are anchored on the cytoplasmic membrane through a glycolipid, but that are also found to be loosely bound to peptidoglycan and even released into the extracellular medium. Both TA types are decorated by D-alanyl esters associated or not with glycosyl (mainly glucose) substitutions (Neuhaus and Baddiley, 2003). Notably, specific *Lactobacillus* species; for example, *L. rhamnosus*, *L. casei* and *L. reuteri*, do not have the genetic capacity to produce WTA

Three fundamental functions of the D-alanyl-esters of TA have been proposed:

1. to maintain cationic homeostasis;
2. to modulate the activity of the autolysins; and
3. to define the electromechanical properties of the cell wall.

Furthermore, additional functions have been described in terms of adhesion, biofilm formation, acid tolerance, coaggregation, protein folding, antibiotic resistance, UV sensitivity and virulence.

S-layer

Surface-layer (S-layer) proteins are ubiquitous in both Eubacteria and Archaea. S-layers are arrays of a single protein that constitutes the outermost cell envelope. They are composed of numerous identical (glyco) protein subunits, 40–200 kDa in molecular weight, which form a two-dimensional and regular symmetry. The subunits are held together and attached to the underlying cell surface by non-covalent interactions and have an intrinsic, entropy-driven tendency to form regular structures either in solution or on a solid support *in vitro* (Sára and Sleytr 2000). It is worth considering that S-layer proteins from *Lactobacillus* are biochemically unique: they do not possess SLH domains (found in *Bacillus* S-layer), they are smaller (25–71 kDa) than those of other Gram-positive bacteria, and they are non-glycosylated and highly basic with pI values ranging from 9.35 to 10.4 (Åvall-Jääskelläinen and Palva, 2005).

S-layers proteins are involved in important cell functionalities such as acting as a protective barrier against environmental hazards, controlling the transfer of nutrients and metabolites, maintaining the cell shape and envelope rigidity, and promoting cell adhesion and surface recognition, among others (Vidgren *et al.*, 1992; Buck *et al.*, 2005). In some *Lactobacillus* species, S-layer proteins mediate bacterial adherence to host cells or to the extracellular matrix (Hynönen *et al.*, 2014). In *L. acidophilus* NCFM, the SlpA protein, one of the three major S layer proteins of this microorganism, plays a role in adhesion to Caco-2 intestinal epithelial cells *in vitro* and modulates dendritic cell and T cell functionalities with murine dendritic cells (Johnson *et al.*, 2013). Another S-layer property is to contribute to the pathogen exclusion in the GIT niche. The S-layer of *Lactobacillus acidophilus* ATCC 4356 provokes the inhibition of infection by Junin Virus (JUNV) that is mediated by DC-sign lectin (Martinez *et al.*, 2012). The same protein has been described as having endopeptidase activity against the cell wall of *Salmonella enterica* serovar Newport, with the hydrolase activity predicted to be located at the C-terminus (Prado Acosta *et al.*, 2008). Both antiviral and antibacterial activities support the probiotic status described in this lactobacilli strain. See Chapter 6 for more details about the S-layer.

Wall polysaccharides

Bacterial polysaccharides (PS), also named neutral polysaccharides, to distinguish them from the anionic polysaccharide teichoic acid, exhibit great diversity in the nature of the sugar monomers (rhamnose is usually found as a constituent in LAB) but also in linkage, branching, and substitution (Delcour *et al.*, 1999)

In a formal manner we can classify PS (polysaccharides) into three groups: exopolysaccharides (EPS), which are loosely coupled to the bacterial envelope and are released into the surrounding environment; Capsular polysaccharides (CPS), which are permanently, attached to the cell as an armoured layer; and cell wall polysaccharides (CWPS), which are or are not covalently attached to the cell wall but that do not form a capsule. Sometimes it is difficult to discriminate into the three groups at an experimental level and nomenclature cannot be strictly formal; for instance, EPS is also used to talk about extracellular polysaccharide in general.

Recently, Vinogradov and colleagues (2016) have described the structure of CWPS in *Lactobacillus casei* BL 23 and found that all constituent polysaccharides were rich in rhamnose. They established its chemical structure by periodate oxidation, methylation analysis and 2D NMR spectroscopy. Three variants of CWPS were revealed. Variant 1 was shown to contain α-Rha, α-Glc, β-GlcNAc and β-GalNAc forming a branched heptasaccharide repeating unit with an additional partial substitution with α-Glc. Variant 2 was a modified non-reducing end octasaccharide, corresponding to a terminal unit of the CWPS and variant 3, was also identified and allowed to define the biological repeating unit of the CWPS.

A number of roles have already been assigned to CWPSs in LAB in bacterial physiology as well as in interactions with bacteriophages or eukaryotic hosts. Knowing the *L. casei* CWPS structure is useful to understand the mechanisms involved in the immunomodulatory properties of this strain as a probiotic microorganism. A cluster of genes from *L. casei* Shirota involved in the biosynthesis of CWPS have been identified and the immunomodulation activities of the mutants defective in these genes on mouse macrophages and spleen cells *in vitro* have been determined (Yasuda *et al.*,2008). Furthermore, the chemical structure of CWPS provides valuable information to understand the molecular interactions between phages and the target bacterial surface. There are recent studies indicating that certain bacteriophages infecting *L. casei* strains recognize saccharide-containing receptors at the bacterial surface during the adsorption step of phage infection (Dieterle *et. al.*, 2014, 2017).

In *Lactobacillus helveticus* differential architecture of CWPS structure was observed. Strains of *L. helveticus* are commonly used in the dairy industry as a starter or an adjunct culture for manufacture of cheese and some types of fermented milk. Its autolysis releases intracellular enzymes which is a prerequisite for optimum cheese maturation, and is known to be strain dependent. It has been hypothesized that these differences in CWPS may partially explain the variation in autolytic properties among the strains studied (Vinogradov *et al.*, 2013).

Exopolysaccharides (EPS) have a weak union to the cell wall, so they tend to be excised from the envelope and are usually found in the surrounding medium. Conformation of EPS varies widely between strains and their genes are usually associated with mobile elements. (Forde *et al.*, 1999). EPS of variable composition, branching and size can be found between different strains of the same genus (de Vuyst *et al.*, 1999).

In the food industry these polysaccharides are of great importance because they give the fermented product organoleptic characteristics. They are used as densifiers, stabilizers and emulsifiers (de Vuyst *et al.*, 1999). Furthermore, they have been proposed as probiotics and there are reports of their antitumor importance and as immunomodulators (Nikolic *et al.*, 2012; Salazar *et al.*, 2015).

Osmotic stress

Lactobacilli are often exposed to changes in the solute concentrations of their natural habitats. Nevertheless, their cytoplasmic solute concentration needs to be relatively constant. A sudden increase in the osmolarity of the environment (hyperosmotic stress) results in the movement of water from the cell to the outside, which causes a detrimental loss of cell turgor pressure, changes the intracellular solute concentration and changes the cell volume, causing plasmolysis (i.e. the cytoplasmic membrane may retract from the cell wall) in extreme cases. On the other hand, a hypotonic shock causes the entrance of water into the cell, an increase in cell volume and turgidity pressure, and eventually cellular lysis. The two types of osmotic variations are deleterious to bacteria. To maintain turgor pressures within specific cell viability ranges and to avoid the effects of plasmolysis and cell lysis, the microorganisms must adjust the concentration of intracellular solutes. To counteract the outflow of water, microorganisms increase their intracellular solute pool by accumulating large amounts of organic osmolytes, the so-called compatible solutes (Fig. 5.4).

The accumulation of the compatible solutes may be due to two mechanisms: an active transport system and by *de novo* synthesis (Fig. 5.4). Some compatible solutes can alleviate the inhibitory effect of high osmolarity when they are present in the culture medium, suggesting that their accumulation is due to transport and not to *de novo* synthesis and are called osmoprotectants (Poolman and Glaasker 1998).To survive osmotic stress, the cells need to adapt by accumulating specific solutes under hyperosmotic condition and releasing them under hypoosmotic condition. The major solutes described are: ions as K^+; amino acids (e.g. glutamate, proline); amino acids derivatives (peptides, N-acetylated amino acids); quaternary amines (e.g. glycine betaine, carnitine); sugars (e. g. sucrose, trehalose); and tetrahydropyrimidines (ectoines) (Csonka and Hanson 1991). In Table 5.1 the main functions involved in osmostress protection found in Lactobacilli are described. The Opu family of transporters, the synthesis route for glycine betaine from exogenously provided choline, the K^+ importers, and mechanosensitive channels serving as safety valves for the rapid release of ions and organic solutes upon sudden osmotic down-shocks.

In nature and in industrial processes LAB are frequently exposed to adverse environmental

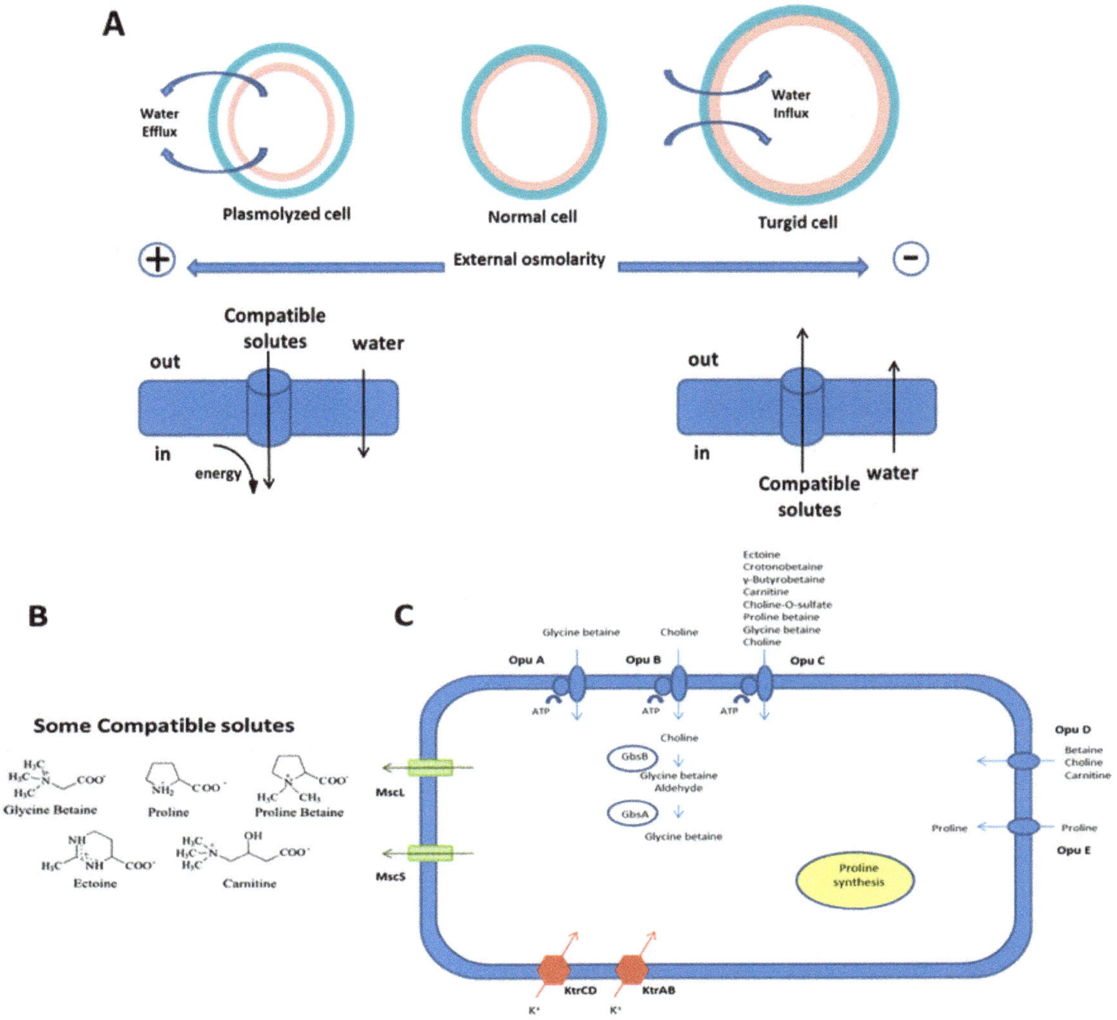

Figure 5.4 A. Changes in external osmolarity caused by hyperosmotic stress (water efflux) or hypoosmotic stress (water influx). The cell reacts by accumulating or releasing compatible solutes, respectively. In all cases, an attempt is made for water fluxes to be redirected. B. Chemical structures of some compatible solutes. C. The accumulation of the compatible solutes may be due to two mechanisms: an active transport system and by de novo synthesis.

conditions. Osmotic stress is one of the most dangerous environmental perturbations that can cause a decrease in growth rate or survival and affect metabolic activities. In the industrial environment, salt is a major agent used during cheese production and ripening, meat fermentation, and the yoghurt-making process. Salt is an important tool for the cheesemaker to influence the ripening process towards the desired variety and quality. Therefore, different cheese varieties need different salt contents, from 0.4% for Emmentaler up to 4% or even 5% for blue cheeses.

LAB employed as probiotics have to survive and proliferate within the gastrointestinal tract, so they must tolerate several environmental hurdles, including the elevated osmolarity in the upper small intestine.

Variations in membrane compositions under osmotic stress

The stability of the cellular membranes plays a very important role on the adaptation to different kinds of stresses and they might be closely related to the lipid and fatty acid composition. Most organisms

Table 5.1 Orthologous genes involved in osmostress protection found in lactobacilli

Gene (*B. subtilis*)	Description	*L. casei*	*L. rhamnosus*	*L. plantarum*	*L. delbrueckiii*	*L. farraginis*	*L. brevis*	*L. rennini*	*L. kefiri*	.. *buchneri*
Opu AC	ABC compatible solutes transporter. Glycine betaine binding protein	+	+	–	–	+	+	–	+	+
Opu BC	ABC compatible solutes transporter. Choline binding protein	+	+	–	+	–	+	+	–	–
OpuCC	ABC compatible solutes transporter. Glycine betaine, carnitine, choline binding protein	+	+	+	+	–	+	–	–	–
OpuD	BCCT (betaine, choline carnitine) transporter	–	–	+	–	+	+	–	+	–
OpuE	Sodium solute symporter. L-Proline transporter	–	–	+	–	–	–	+	–	–
GbsB	Glycine betaine synthesis from exogenously provided choline. Betaine aldehyde dehydrogenase	+	–	–	–	+	+	–	+	+
GbsA	Glycine betaine synthesis. Alcohol dehydrogenase	+	–	–	–	+	+	–	+	+
MscL	Large conductance mechanosensitive channel. It opens in response to sudden osmotic down-shocks.	–	–	+	–	+	+	–	+	+
MscS	Small conductance mechanosensitive channel. It opens in response to sudden osmotic down-shocks.	–	+	–	–	–	–	–	–	+
Ktr	K^+ importers	–	+	+	–	–	+	+	–	–

alter their membrane lipid composition in response to changes in the environmental.

Lactobacillus species are used as a probiotic in humans so it is a fundamental they have to survive into the intestine micro environment. The resistance of these bacteria to the action of the bile salts appears to be dependent on the physico-chemical properties of the cellular envelopes.

Under osmotic stress conditions *Lactobacillus* produce an increased hydrophobicity in their envelope to tolerate the stress imposed by the hypertonicity of the medium. The nature of the molecules conferring hydrophobicity in Gram-positive bacteria is related to external polymers as lipoteichoic acids and envelopes proteins. The glycolipid composition of *L. casei* BL23 membranes from cultures grown under hyper-osmotic medium showed a significant increment of tetrahexosyldiacylglycerol and a small increase of trihexosyldiacylglycerol and the acyltrihexosyldiacylglycerol was only present when cells were grown in this medium.

These glycolipids would be related to the increased cell surface hydrophobicity observed during osmotic stress (Machado *et al.*, 2004). This high content of glycolipids was reported previously for other *L. casei* strain as well as for *Lactobacillus acidophilus* (Fernandez Murga *et al.*, 1999).

It was shown that the presence of NaCl in the medium resulted in an increase of the saturated/unsaturated ratio and promotes the formation of lactobacillic acid (cyc C19:0) from the corresponding monounsaturated precursor 18:1 n7. The formation of cyclopropane-containing fatty acids is induced when certain microorganisms have to confront adverse environmental conditions such as osmotic stress or when the temperature is modified (Perly *et al.*, 1985). The increase in the glycolipid: phospholipid ratio has been related to a greater stability of the cell membrane to environmental stress due to its ability to undergo interlipid hydrogen bonding via the glycosyl headgroups (Curatolo, 1987).

Lactobacillus reuteri, a species with beneficial properties on the health of the gastrointestinal tract for its probiotic characteristics has shown changes in its lipid profile when it was subjected to bile salt stress. It was reported that changes induced in the lipid profiles in the cell membrane probably constitute one of the main physiological responses of the cells for survival in the gastrointestinal tract.

It was described that the bile salts in the culture medium affected both the glycolipids and phospholipids fractions of the cell membrane of *L. reuteri* CRL 1098 (Taranto *et al.*, 2003). The analysis of the fatty acids profile revealed five major fatty acids corresponding to C16:0, C18:1, C18:2, C18:0 and C19-cyc acids. The presence of the unusual 10-oxo- and 10-hydroxy-octadecanoic is probably one of the most prominent characteristics of the lipids of bile salts-growing cultures. These modifications induced in the lipid profiles in the cell membrane, would probably constitute one of the main physiological responses of the cells for survival in the gastrointestinal tract, enabling the cells to maintain constant membrane fluidity in harmful environments, which is fundamental for cellular functions.

Freezing was represented as a combination of cold and osmotic stress. Meneghel *et al.* (2017) investigated the relative incidence of increasing sucrose concentrations coupled or not with subzero temperatures. Exposing *L. bulgaricus* cells to osmotic stress (i.e. to high concentrations of sucrose, a non-permeating molecule) induces water transport from the intracellular medium of the cell to the extracellular medium, which results in cell dehydration and cell volume contraction. Depending on the flexibility of the membrane, cell volume reduction could easily occur or be associated with considerable mechanical constraints possibly leading to membrane leakage. The membrane of a freezing-resistant strain was characterized by a higher content of unsaturated fatty acid or UFA (49%), a lower lipid phase transition following cooling ($T_s = 6.9°C$), and higher membrane fluidity at the centre and external surface at 0°C than the membrane of a sensitive one with UFA = 39%; $T_s = 17.5°C$. The higher the UFA/SFA ratio, the more fluid the membrane and the higher the freeze resistance are. Membrane fluidity could therefore be correlated with a combination of different fatty-acid structures, including unsaturations and chain length.

Osmotic stress induced by a high-salt concentration has been reported to increase the autolytic activity and survival following lyophilization in *Lactobacillus delbrueckii* subsp. *lactis* (Koch *et. al.*, 2007).

Variations in the cell wall composition under osmotic stress

LAB have the ability to respond to environmental conditions, one way is by altering the nature of their cell wall. In this part of the book we describe their modifications when cells of *Lactobacillus* are grown in high-salt condition; in particular those related to the structure of PG, considering the degree of cross-linking, and the presence and behaviour of PBP. Afterwards we consider the modification of cell wall polymers.

Piuri and colleagues (2005) described the cell wall modifications taking place in *L. casei* during growth in high-salt medium. Cell size was also affected when cells were growth in 1 M NaCl. Growth in 1 M NaCl resulted in a significant increase in cell size compared with cells grown in medium without NaCl (Fig. 5.4). A significant difference was observed in the mean cell diameter between control cells (415 ± 20 nm) and NaCl-grown cells (660 ± 60 nm). This significant enlargement of cells under high ionic stress has been correlated with slow growing in *L. casei* where growth of *L. casei* in high-salt condition was drastically retarded when compared with control growth (Piuri *et al.*, 2005).

Modifications in cell size were accompanied by changes in the cell envelope structure that were visualized by transmission electron microscopy (TEM). The modified cell wall structure observed by TEM could be correlated with a decrease in PG crosslinking. Fig. 5.5 shows the cell wall structural composition of a non-S-layer-forming strain such as *L. casei* and containing a species such as *Lactobacillus acidophilus* which contains an S-layer.

As explained before, the polymerization of PG involves the action of penicillin-binding proteins (PBP). These proteins carry out the glycosyltransferase and transpeptidase reactions required for PG synthesis. A decrease in PG cross-linking was attributed by a lower expression of PBP genes and/ or differences in their correct assembling in the envelope.

Sensitivity of whole cells and wall preparations to hydrolysis was also studied. Whole cells and wall preparations were assayed with mutanolysin, an enzyme that hydrolyses the N-acetylmuramyl-1,4–acetylglucosamine bonds in the PG structure. When cells were grown in salt medium in the presence of mutanolysin they lysed faster than control cells. The same pattern of lysis was observed with the cell wall preparation. These results clearly showed that the differences in sensitivity to lysis of whole cells would be due to differences in their cell wall structure. Modifications of PG can affect sensitivity to PG hydrolases.

Another modification found in *L. casei* BL 23 when it was growth under high-salt conditions is concerned with secondary cell wall polymers such teichoic acids. It was shown (Palomino *et al.*, 2013) that when growing in high-salt condition, *L. casei* produces less of the cell wall polymer lipoteichoic acid (LTA), and that this anionic polymer has a shorter mean chain length and a lower level of D-alanyl-substitution. Analysis of the transcript levels of the *dltABCD* operon, encoding the enzymes required for the incorporation of D-alanine into anionic polymers, showed a 16-fold reduction in mRNA levels, which is consistent with a decrease in D-alanine substitutions on LTA. Furthermore, a 13-fold reduction in the transcript levels was observed for the gene LCABL_09330 coding for a putative LTA synthase (Palomino *et al.*, 2013). These modifications of surface properties lead to an increased ability of this bacterium to form biofilms and to bind cations in high-salt conditions.

Table 5.2 summarizes the differences found in the LTA between bacteria grown in high-salt and control conditions. A lower recovery of LTA can be observed from strains grown in high-salt conditions. NMR analysis verified that LTA is shorter and less D-alanylated when *L. casei* BL23 is grown in high-salt medium.

It has been reported that growth in high-salt medium, leads to modifications in the cell envelope of this bacterium. Those modifications produce

Table 5.2 Modifications in LTA

	C	N
LTA yield (mg/g culture)	0.59	0.35
LTA chain length	42 ± 3	31 ± 4
D-Ala substitution	64 ± 4	27 ± 9

LTA was purified from *L. casei* BL23 cultures grown in MRS control medium (C) or medium containing 0.8 M NaCl (N), quantified and further analysed. Based on the NMR analysis of three independently isolated LTA samples, mean glycerolphosphate chain lengths was calculated and alanine substitution was evaluated as the ratio of integral values for D-alanine to Gro-P.

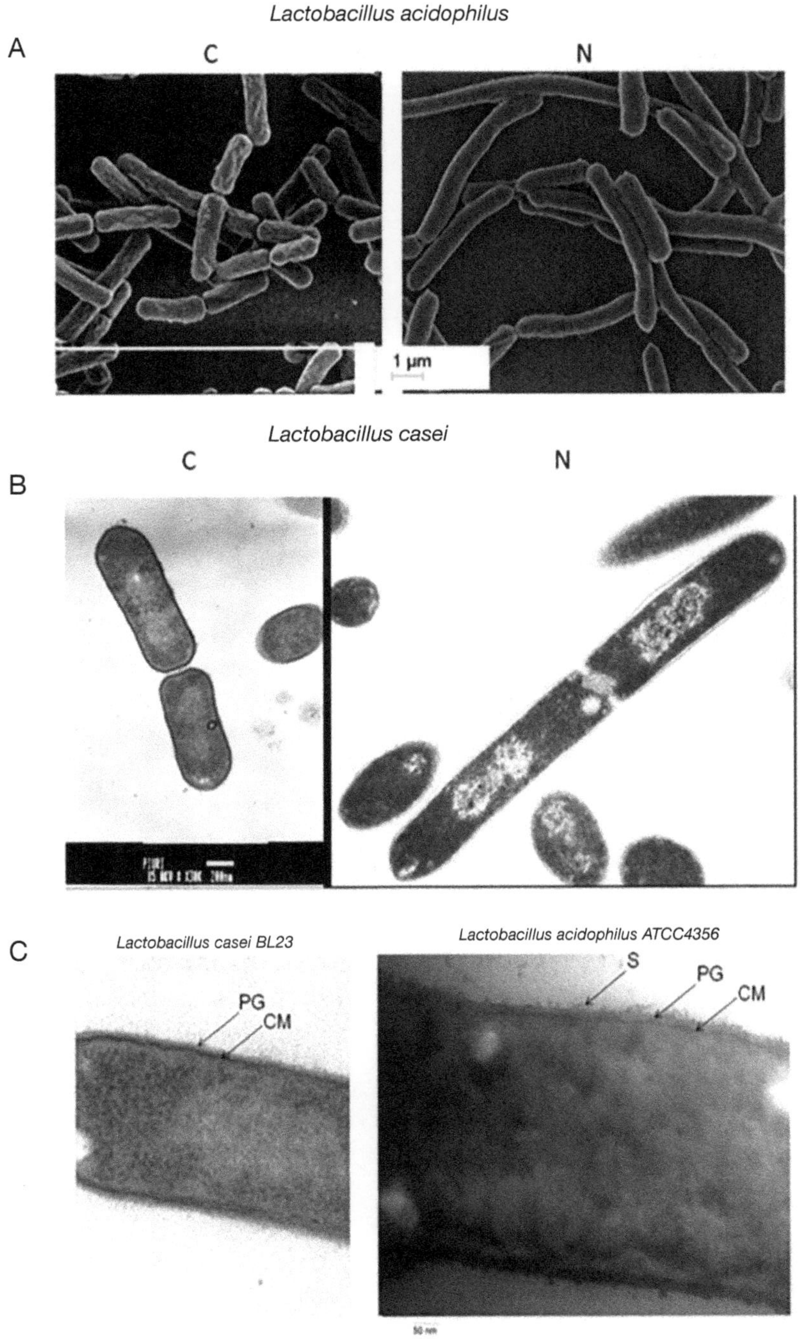

Figure 5.5 A: Scanning Electron Microscopy (SEM) images. Cultures of *Lactobacillus acidophilus* ATCC 4356 grown in MRS medium and MRS medium containing 1 M NaCl (N). B: Transmission electron micrographs (TEM). TEM of thin sections of exponential phase cultures of *Lactobacillus casei* BL23 grown in C and N conditions Magnification of micrographs is 30000-fold. C: Scanning Electron Microscopy (SEM) images. Abbreviations: Peptidoglycan containing layer (PG), S-layer (S) and cytoplasmic membrane (CM).

several pleiotropic effects including a difference in susceptibility to enzymic lysis, increased sensitivity to cationic antimicrobials, such as nisin, and an increased capacity to form biofilms on artificial surfaces. Evidence reported here shows that the consequent behaviour not only results from changes in the PG structure but also to changes to the zwitterionic character of the LTA molecule. A decrease in LTA D-alanylation will result in increased negative charge of the cell wall, and this might help to expel toxic Na^+, which is present in excess in the high-salt conditions, from the cell, preventing it from reaching the cytoplasm and interfering with metabolic processes. In the last section of this chapter we postulate a model to explain the cell-envelope adaptation under high-salt stress conditions.

Variation in S-layer protein under environmental stress conditions

Several species of *Lactobacillus*, including mucosal-associated species (e.g. *Lactobacillus acidophilus*, *Lactobacillus crispatus*, *Lactobacillus amylovorus*, and *Lactobacillus gallinarum*) and dairy-fermentation-associated species (e.g. *Lactobacillus helveticus* and *Lactobacillus kefiranofaciens*) are able to form an S-layer envelope which constitute the outermost structure of the cell envelope composed of a crystalline array of single non-covalently bound proteins. The presence of multiple S-layer genes seems to be quite common for lactobacilli for example in *Lactobacillus acidophilus* ATCC 4356 three S-layer-protein-coding genes are found: *slpA*, *slpB*, and *slpX*.

Several studies have reported the multiple functions in which the S-layers are described or assumed to have a role. Examples of these functions include providing a protective envelope against hostile environmental agents and playing an important role in the establishment of LAB probiotic strains in the GIT niche.

There is also increasing evidence that S-layer carrying bacteria may modify S-layer protein expression depending on the culture conditions. In *Lactobacillus acidophilus* ATCC 4356, a variation in the pattern of S-layer expression under osmotic stress due to a high-salt condition was found. The amounts of the predominant and the auxiliary S-layer proteins SlpA and SlpX were strongly influenced by the growth phase and high-salt conditions (0.6 M NaCl). This up-regulated expression of *slpA* and *slpX* in high-salt conditions agrees with an increase in S-layer protein synthesis, modifying the proportion between SlpX and SlpA in the S-layer profile when both conditions are compared (Palomino *et al.*, 2016). On the other hand, *slpB* remained silent in concordance with previous reports where attempts to demonstrate expression of the *slpB* gene were unsuccessful (Boot *et al.*, 1996; Pouwels *et al.*, 1997).

The increase in the S-layer proteins in high-salt medium, involving active transcription of S-layer genes, could be due to a direct stress or to the response of the genes to the modification of the cell wall needing more S-layer to counteract its fragility. Table 5.3 compares the differences found in the cell wall of *L. acidophilus* and a non-S-layer-forming strain such *L. casei*.

It was shown that both PG and also lipoteichoic acids were significantly lower for *L. acidophilus* (Palomino *et al.*, 2016)

Others authors have also shown a modification in the level of SlpA of *Lactobacillus acidophilus* ATCC 4356 when the strain is confronted with detrimental culture conditions. High concentrations of S-layer proteins were extracted when the strain grew at 45°C but also when 0.05% bile salts or 2% NaCl were added to the growth medium. (Khaleghi *et al.*, 2010, 2012).

Evidence showed that S-layer proteins were present during all growth phases of *Lactobacillus acidophilus* M92 under heat stress (Frece *et al.*, 2005). This suggested that the S-layer protein is preferentially expressed under conditions which

Table 5.3 Cell envelope structure of *Lactobacillus acidophilus* ATCC 4356 in the presence or absence of 0.6 M NaCl and *L. casei* BL23

Sample	CW	PG	LTA
L. acidophilus	15 ± 1.0 mg	7.5 ± 0.8 mg	0.23 ± 0.02 mg
L acidophilus NaCl 0.6 M	10.5 ± 1.0 mg	4.5 ± 0.5 mg	0.08 ± 0.01 mg
L. casei	50.0 ± 3.0 mg	35.0 ± 2.0 mg	0.56 ± 0.03 mg

CW, cell wall; PG, peptidoglycan; LTA, lipoteichoic acid. Adapted from Palomino *et al.* (2016).

are not optimal for bacterial growth. Furthermore, it was shown that *L. acidophilus* M92, usually grows at 50°C but did not grow at this temperature when S-layer protein is removed. This can be a proof for the protective role of S-layer for survival of *L. acidophilus* M92 during preparation of culture for probiotic products (usually utilized method is freeze-drying). Viability at high populations, preferred at 10^6–10^8 cells/g, is required in probiotic products.

In *Lactobacillus acidophilus* IBB 801, a potentially probiotic strain, production of S-layer under different stress culture conditions including high incubation temperatures, presence of bile salts or NaCl, and acidic pH was assayed. It was found that an increased S-layer synthesis by the strain was detected when it was grown at 42°C and in the presence of 0.05% bile salts or 2.0% NaCl. The increased expression of S-layer protein under environmental stress conditions may probably help the strain to maintain cell viability when it is confronted against hostile culture agents. (Grosu-Tudor *et al.*, 2016).

Lactobacillus as probiotics microorganisms and stress response mechanisms

In the last decades there has been a preponderant interest in using LAB as probiotics microorganism within functional foods. A functional food is considered to be one that 'is sufficiently proven to act beneficially on one or more functions of the body, beyond its nutritional effect, improving health or reducing the risk of disease' (Diplock *et al.*, 1999). In recent years, a renewed interest in the use of microorganisms in food has emerged due to its contribution to flavour and aroma, but mainly due to its beneficial aspects in restoring health and treating diseases. Probiotics, as defined by FAO/WHO (2002), are 'those living microorganisms that when administered in adequate amounts confer some beneficial effect on the health of the host'. Various species of bacteria and yeast have been used as probiotics. The most commonly used are strains of species of *Lactobacillus* and *Bifidobacterium*. Microorganisms for probiotic use are faced with stressful conditions at various stages from processing to storage and gut transit, which could affect viability (Lacroix and Yildirim 2007). Stresses could affect the physiological activity of the probiotic microorganisms, and as a consequence, affect their functionality.

LAB used in fermented food processes are exposed to several stress factors, during culture preparation, storage of the products and gastrointestinal passage. These include acid, bile, osmotic, oxidative, heat and cold stresses, which can affect viability and functionality. It is worth considering that a probiotic strain must reach high viable numbers in the intestine to exert a beneficial effect on the human host. For beneficial effects to be achieved, it has been recommended that minimum levels of probiotics of around 10^6–10^7 CFU/ml should be present in the product by the time of consumption/end of shelf-life (Silva *et al.*, 2015).

Table 5.4 summarizes the types of stress factors that LAB are confronted with and the response mechanisms that allow bacteria to survive during exposure to hostile conditions.

Concluding remarks

LAB have cell envelopes that are typical of Gram-positive bacteria. As seen in this chapter, they consist of a cytoplasmic membrane and a cell wall decorated with teichoic acids, polysaccharides and proteins. Some species have an external proteinous S-layer.

The cell envelope represents the physical barrier to separate the cell from its surrounding media and, as such, is the first line of defence against environmental perturbations.

Under stress conditions, LAB have the ability to produce changes in chemical composition and structural modifications in both the cell membrane and the cell wall, allowing them to adapt and survive to the new and hostile conditions.

Many of the environmental or technological stressors that LAB can encounter have an impact on the cell envelope. As seen earlier, when *Lactobacillus casei* is subjected to osmotic stress, changes on its cell wall conformation happen. High-salt concentrations lead to an increase in the size of the cells and to a sensitivity to antimicrobial peptides targeting PG, such as nisin. These effects are associated with a reduction of the cross-linking of PG molecules due to altered expression and activity of PBPs (Piuri *et al.*,2005, Palomino *et al.*, 2013). A model of the proposed overall modifications taking place in the cell envelope as a consequence of the

Table 5.4 Types of stress factors and response mechanisms

Stress type	Stage	Stress response mechanism	References
Acid	Organic acids in fermented dairy products and beverages used for probiotic delivery Low pH in the stomach	Metabolizable sugars, to enable proton exclusion by providing adenosine triphosphate (ATP) to F-ATPase Amino acid decarboxylation combined with an amino acid antiporter leads to the biochemical consumption of protons Increased expression of the *clpP*, *clpE*, *clpL* and *clpX* genes. Clp chaperones actively promote refolding or degradation of damaged proteins Changes in the membrane fatty acid composition increases the level of saturated fatty acids	Corcoran *et al.* (2005), Trip *et al.*, (2012), Linares *et al.* (2012), Ferreira *et al.* (2013), Muller *et al.* (2011)
Bile	Bile in small intestine	Express a range of bile salt hydrolases that confer protection against bile via bile salt deconjugation Degradation of misfolded proteins by proteases, e.g. ClpC Bile salt or acid detoxification by multidrug transporters and bile efflux pumps (BetA, Ctr) Alteration of cell surface by production of extracellular exopolysaccharides; changes in fatty acid composition; changes in surface-associated proteins (enolase, oligopeptide binding proteins)	Jones *et al.* (2008), Ferreira *et al.* (2013), Alcántara and Zúñiga (2012), Duary *et al.* (2012)
Thermal	Extremes of temperature during spray drying and freeze drying Storage	Induction of the chaperones GroEL, GroES, GrpE, DnaJ, DnaK, ClpB, Hsp20 associated to pre-treatment to heat shock Induction of CSP (cold shock proteins) associated with the stabilization of mRNA	du Toit *et.al.* (2013), Spano *et al.* (2004), Derzelle *et al.* (2000)
Osmotic	High osmotic pressure and low water activity during cell dehydration	Adaptation by accumulating compatible solutes, such as betaine, carnitine and trehalose GroES/GroEL chaperone was positively regulated Modification of cell envelope	Wasko *et al.* (2013), Prasad *et al.* (2003), Palomino *et al.* (2013, 2016)
Oxidative	Oxygen exposure during fermentation, drying and storage Presence of oxygen in carrier product Oxygen in mouth during consumption	Reduce or eliminate the deleterious effects of oxygen radicals by using reactive oxygen species (ROS)-scavenging enzymes, e.g. NADH oxidase, NADH peroxidase, Mn-superoxide dismutase Changes in the fatty acid composition of the cell membrane	van de Guchte *et al.* (2002), De Angelis and Gobbetti (2004), Guerzoni *et al.* (2001)

high-salt adaptation is presented in Fig. 5.6. The model includes multiple factors affecting different surface properties:

1. an increase in cyclic fatty acids is observed under high-salt conditions (Machado *et al.*, 2004);
2. the PG shows a lower number of layers and a lower level of cross linking. (Piuri *et al.*, 2005; Palomino *et al.*, 2009);
3. modifications in LTA involving shorter LTA chains and with a reduced amount of D-alanine substitutions. All these modifications may lead to an increase in the negative charge of the outermost cell wall layers that would ensure the extrusion of toxic Na^+ ions and the survival of the bacteria.

Others experimental evidences have shown the ability of LAB to modify the cell wall properties to cope with stress conditions. Osmotic stress induced by a high-salt concentration in *L. delbrueckii* subsp. *lactis* led to an increase of autolytic activity and survival following lyophilization (Koch *et al.*, 2007). *L. rhamnosus* also showed adaptations to technological stress conditions by decreasing the amounts

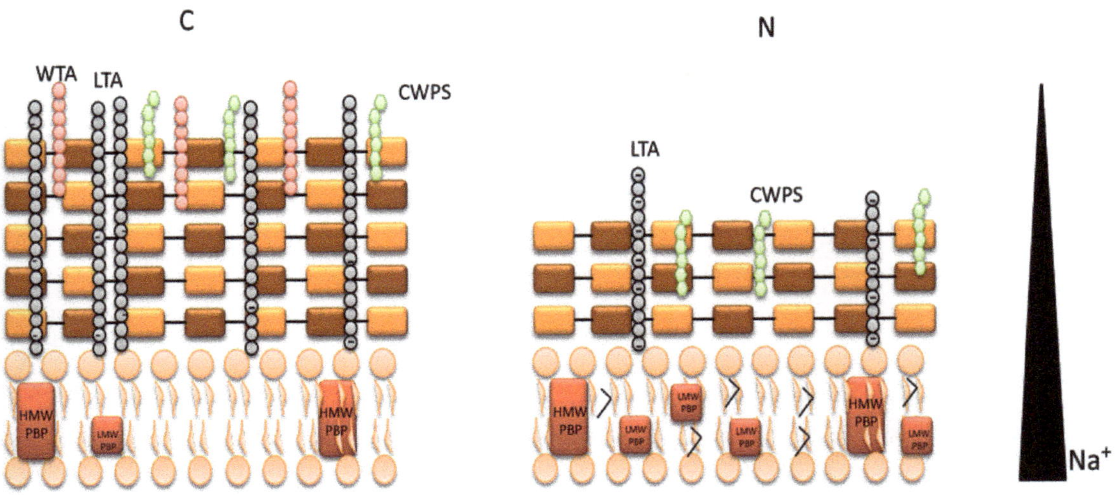

Figure 5.6 Proposed model of the cell envelope adaptation under high-salt stress conditions. The Gram-positive cell envelope comprises a thick PG layer (brown blocks), stabilized by wall teichoic acids (WTA, red circles), the membrane-linked lipoteichoic acid (LTA, grey circles)and cell wall polysaccharides (green hexagons). D-Alanine substitutions are represented as filled Gro-P monomer circles (grey) and the absence of esterification is shown with a minus symbol within the circles. Membrane phospholipid bilayer (pink) is modified in high-salt conditions by an increase in cyclic fatty acids (grey triangles). Increased amounts of low molecular mass PBPs (LMW PBP) and decreased amounts of high molecular mass PBPs (HMW PBP) are found in high-salt conditions (Piuri *et al.*, 2005). The secondary wall polymers decrease; LTA chains are shorter and less substituted. All these modifications lead to an increase of the negative charges that would ensure the extrusion of toxic Na+ ions (black gradient represents Na+ concentration from the inside to the outside). MRS medium alone, low-salt conditions, versus MRS with 0.8 M NaCl, high-salt conditions. (Adapted from Palomino 2013).

of proteins responsible for cell wall biosynthesis (MurD, MurC, and UDP-GlcNAc-2-epimerase) (Zuljan *et al.*, 2014).

As we have described earlier, in the strains carrying an S-layer such as *L. acidophilus*, the amounts of the predominant and the auxiliary S-layer proteins SlpA and SlpX were strongly influenced by growing in high-salt conditions.

As shown in Table 5.3 when comparing the differences at a structural level in the cell wall between *L. casei* and a non -S-layer producer, it is worth considering the need of this additional envelope.

The requirement of the S-layer for growth even in normal conditions in the strains carrying an S-layer could be explained by the 3-fold lower amount of peptidoglycan found when compared to a non-S-layer producer such as *L. casei*. Under high-salt conditions, a 2-fold decrease in peptidoglycan and increased fragility determined the need of the external highly compact S-layer component. The fragility of the cell walls of *Lactobacillus* strains carrying S-layers may be due to the presence of the S-layer or, otherwise, seems to force these strains to be dependent on the presence of the S-layer for growth. Further studies on other S-layer-containing and non-containing species is the aim of future work.

Several strategies have been exploited to achieve more robust bacterial strains; one of them is the preadaptation. The pre-exposure to sublethal stress is a manner of enhancing the stability of LAB in the dairy industry or those used as probiotic strains. The effects of stress pretreatments have been widely documented (Corcoran *et. al.*, 2008, Mills *et al.*, 2011, Sanchez *et al.*, 2013), and results have shown that general stress responses, by inducing shock proteins, DNA repair and energy metabolism enhance survival and viability.

Therefore it is worth thinking that pre-growth of lactobacilli in high-salt conditions would result in an advantage for the probiotic nature of cells: the increased production and release of the S-layer might be appropriate for their antimicrobial capacity (Prado-Acosta *et al.*, 2008, 2010; Martínez *et al.*, 2012; Meng *et al.*, 2015). Also a reduction in the LTA content, as before observed, would be welcome in probiotic use since several lactobacilli strains present inflammatory reactions as a consequence of their LTA (Lebeer *et al.*, 2012; Mohamadzadeh *et al.*, 2011).

References

Ahmed, Z., Wang, Y., Cheng, Q., and Imran, M. (2010).'*Lactobacillus acidophilus* bacteriocin, from production to their application: an overview,' African J. Biotechnol. 9, 2843–2850

Alcántara, C., Zúñiga, M. (2012). Proteomic and transcriptomic analysis of the response to bile stress of *Lactobacillus casei* BL23. Microbiology *158*, 1206–1218. https://doi.org/10.1099/mic.0.055657-0

Åvall-Jääskeläinen, S., and Palva, A. (2005). *Lactobacillus* surface layers and their applications. FEMS Microbiol. Rev. *29*, 511–529. https://doi.org/10.1007/s00253-013-4962-2

Anzengruber, J., Courtin, P., Claes, I.J.J., Debreczeny, M., Hofbauer, S., Obinger, C., Chapot-Chartier M.-P., Vanderleyden, J., Messner, P., and Schäffer, C. (2014). Biochemical characterization of the major N-acetylmuramidase *from Lactobacillus buchneri*. Microbiology *160*, 1807–1819. https://doi.org/10.1099/mic.0.078162-0

Baddiley, J. (1989). Bacterial cell walls and membranes. Discovery of the teichoic acids. Bioessays *10*, 207–210. https://doi.org/10.1002/bies.950100607

Bernard, E., Rolain, T., Courtin, P., Guillot, A., Langella, P., Hols, P., and Chapot-Chartier, M.P. (2011). Characterization of O-acetylation of N-acetylglucosamine: a novel structural variation of bacterial peptidoglycan. J. Biol. Chem. *286*, 23950–23958. https://doi.org/10.1074/jbc.M111.241414

Billot-Klein, D., Legrand, R., Schoot, B., van Heijenoort, J., and Gutmann, L. (1997). Peptidoglycan structure of *Lactobacillus casei*, a species highly resistant to glycopeptide antibiotics. J. Bacteriol. *179*, 6208–6212.

Bongaerts, G.P., and Severijnen, R.S. (2001). The beneficial, antimicrobial effect of probiotics. Med. Hypotheses *56*, 174–177. https://doi.org/10.1054/mehy.2000.1135

Boot, H.J., Kolen, C.P., and Pouwels, P.H. (1996). Interchange of the active and silent S-layer protein genes of *Lactobacillus acidophilus* by inversion of the chromosomal slp segment. Mol. Microbiol. *21*, 799–809.

Bron, P.A., Tomita, S., Mercenier, A., and Kleerebezem, M. (2013). Cell surface-associated compounds of probiotic lactobacilli sustain the strain-specificity dogma. Curr. Opin. Microbiol. *16*, 262–269. https://doi.org/10.1016/j.mib.2013.06.001

Buck, B.L., Altermann, E., Svingereud, T., Klaenhammer, T.R. (2005). Functional analysis of putative adhesion factors in *Lactobacillus acidophillus* NCFM. Appl. Environ. Mocrobiol. *71*, 8344–8351. https://doi.org/10.1128/AEM.00030-10

Buck, B.L., Azcarate-Peril M.A., and Klaenhammer, T.R. (2009). Role of autoinducer-2 on the adhesion ability of *Lactobacillus acidophilus*. J. Appl. Microbiol.,*107*, 269–279. https://doi.org/10.1111/j.1365-2672.2009.04204.x

Bush, K. (2012). Antimicrobial agents targeting bacterial cell walls and cell membranes. Rev. - Off. Int. Epizoot. *31*, 43–56.

Claes, I.J., Schoofs, G., Regulski, K., Courtin, P., Chapot-Chartier, M.P., Rolain, T., Hols, P., von Ossowski, I., Reunanen, J., de Vos, W.M., *et al.* (2012). Genetic and biochemical characterization of the cell wall hydrolase activity of the major secreted protein of *Lactobacillus rhamnosus* GG. PLOS ONE *7*, e31588. https://doi.org/10.1371/journal.pone.0031588

Corcoran, B.M., Stanton, C., Fitzgerald, G.F., and Ross, R.P. (2005). Survival of probiotic lactobacilli in acidic environments is enhanced in the presence of metabolizable sugars. Appl. Environ. Microbiol. *71*, 3060–3067. https://doi.org/10.1128/AEM.71.6.3060-3067.2005

Corcoran, B.M., Stanton, C., Fitzgerald, G., and Ross, R.P. (2008). Life under stress: the probiotic stress response and how it may be manipulated. Curr. Pharm. Des. *14*, 1382–1399.

Csonka, L.N and Hanson, A.D. (1991). Prokaryotic osmorregulation: genetics and physiology. Ann. Rev. Microbiol 45, 569–606.

Curatolo, W. (1987). The physical properties of glycolipids. Biochim. Biophys. Acta *906*, 111–136.

De Angelis, M., and Gobbetti, M. (2004). Environmental stress responses in *Lactobacillus*: a review. Proteomics *4*, 106–122. https://doi.org/10.1002/pmic.200300497

De Vuyst, L., and Degeest, B. (1999). Heteropolysaccharides from lactic acid bacteria. FEMS Microbiol. Rev. 23, 153–177.

Delcour, J., Ferain, T., Deghorain, M., Palumbo, E., and Hols, P. (1999). The biosynthesis and functionality of the cell wall of lactic acid bacteria. Antonie Van Leeuwenhoek *76*, 159–184.

Derzelle, S., Hallet, B., Francis, K.P., Ferain, T., Delcour, J., and Hols, P. (2000). Changes in *cspL*, *cspP*, and *cspC* mRNA abundance as a function of cold shock and growth phase in *Lactobacillus plantarum*. J. Bacteriol. *182*, 5105–5113.

Dieterle, M.E., Bowman, C., Batthyany, C., Lanzarotti, E., Turjanski, A., Hatfull, G., and Piuri, M. (2014). Exposing the secrets of two well-known *Lactobacillus casei* phages, J-1 and PL-1, by genomic and structural analysis. Appl. Environ. Microbiol. *80*, 7107–7121. https://doi.org/10.1128/AEM.02771-14

Dieterle, M.E., Spinelli, S., Sadovskaya, I., Piuri, M., and Cambillau, C. (2017). Evolved distal tail carbohydrate binding modules of *Lactobacillus* phage J-1: a novel type of anti-receptor widespread among lactic acid bacteria phages. Mol. Microbiol. *104*, 608–620. https://doi.org/10.1111/mmi.13649

Diplock, A.T., Aggett, P.J., Ashwell, M., Bornet, F., Fern, E.B., and Roberfroid, M. (1999). Scientific concepts of functional foods in Europe: consensus document. Br, J. Nutrition *81*, S1- S27.

du Toit E., Vesterlund, S., Gueimonde, M., and Salminen, S. (2013). Assessment of the effect of stress-tolerance acquisition on some basic characteristics of specific probiotics. Int. J. Food Microbiol. *165*, 51–56.

Duary, R.K., Batish, V.K., and Grover, S. (2012). Relative gene expression of bile salt hydrolase and surface proteins in two putative indigenous *Lactobacillus plantarum* strains under in vitro gut conditions. Mol. Biol. Rep. *39*, 2541–2552. https://doi.org/10.1007/s11033-011-1006-9

Fernández Murga, M.L., Bernik, D., Font de Valdez, G., and Disalvo, A.E. (1999). Permeability and stability properties of membranes formed by lipids extracted from *Lactobacillus acidophilus* grown at different temperatures. Arch. Biochem. Biophys. *364*, 115–121. https://doi.org/10.1006/abbi.1998.1093

Fernández Murga M.L., Cabrera, G.M., De Valdez G.F., Disalvo, A., and Seldes, A.M.(2000). Influence of growth temperature on cryotolerance and lipid composition of *Lactobacillus acidophilus*. J Appl. Microbiol. *88*, 342–348.

Ferreira, A.B., De Oliveira, M.N., Freitas, F.S., Alfenas-Zerbini, P., Da Silva, D.F., De Queiroz, M.V., Borges, A.C., and De Moraes, C.A. (2013). Increased expression of *clp* genes in *Lactobacillus delbrueckii* UFV H2b20 exposed to acid stress and bile salts. Benef. Microbes *4*, 367–374. https://doi.org/10.3920/BM2013.0022

Forde, A., and Fitzgerald, G.F. (1999). Bacteriophage defence systems in lactic acid bacteria. Antonie Van Leeuwenhoek *76*, 89–113.

Fozo, E.M., Kajfasz, J.K., and Quivey, R.G. (2004). Low pH-induced membrane fatty acid alterations in oral bacteria. FEMS Microbiol. Lett. *238*, 291–295. https://doi.org/10.1016/j.femsle.2004.07.047

Frece, J., Kos, B., Svetec, I.K., Zgaga, Z., Mrsa, V., and Susković, J. (2005). Importance of S-layer proteins in probiotic activity of *Lactobacillus acidophilus* M92. J. Appl. Microbiol. *98*, 285–292.

Ghuysen, J.M. (1991). Serine beta-lactamases and penicillin-binding proteins. Annu. Rev. Microbiol. *45*, 37–67. https://doi.org/10.1146/annurev.mi.45.100191.000345

Goffin, P., Deghorain, M., Mainardi, J.L., Tytgat, I., Champomier-Vergès, M.C., Kleerebezem, M., and Hols, P. (2005). Lactate racemization as a rescue pathway for supplying D-lactate to the cell wall biosynthesis machinery in *Lactobacillus plantarum*. J. Bacteriol. *187*, 6750–6761. https://doi.org/10.1128/JB.187.19.6750-6761.2005

Grosu-Tudor, S.S., Brown, L., Hebert, E.M., Brezeanu, A., Brinzan, A., Fadda, S., Mozzi, F., and Zamfir, M. (2016). S-layer production by *Lactobacillus acidophilus* IBB 801 under environmental stress conditions. Appl. Microbiol. Biotechnol. *100*, 4573–4583. https://doi.org/10.1007/s00253-016-7355-5

Guerzoni, M.E., Lanciotti, R., and Cocconcelli, P.S. (2001). Alteration in cellular fatty acid composition as a response to salt, acid, oxidative and thermal stresses in *Lactobacillus helveticus*. Microbiology *147*, 2255–2264. https://doi.org/10.1099/00221287-147-8-2255

Henderson, A.J., Kumar, A., Barnett, B., Dow, S.W., and Ryan, E.P. (2012). Consumption of rice bran increases mucosal immunoglobulin a concentrations and numbers of intestinal *Lactobacillus* spp, J. Med. Food, *15*, 469–475. https://doi.org/10.1089/jmf.2011.0213

Hirano, J., Yoshida, T., Sugiyama, T., Koide, N., Mori, I., and Yokochi, T. (2003). The effect of *Lactobacillus rhamnosus* on enterohemorrhagic *Escherichia coli* infection of human intestinal cells in vitro, Microbiol. Immunol. *47*, 405– 409.

Hynönen, U., Kant, R., Lähteinen, T., Pietilä, T.E., Beganović, J., Smidt, H., Uroić, K., Avall-Jääskeläinen, S., and Palva, A. (2014). Functional characterization of probiotic surface layer protein-carrying *Lactobacillus amylovorus* strains. BMC Microbiol. *14*, 199. https://doi.org/10.1186/1471-2180-14-199

Jacobsen, C.N., Rosenfeldt Nielsen, V., Hayford, A.E., Møller, P.L., Michaelsen, K.F., Paerregaard, A., Sandström, B., Tvede, M., and Jakobsen, M. (1999). Screening of probiotic activities of forty-seven strains of *Lactobacillus spp.* by in vitro techniques and evaluation of the colonization ability of five selected strains in humans. Appl. Environ. Microbiol. *65*, 4949–4956.

Johnson, B., Selle, K., O'Flaherty S., Goh, Y.J., and Klaenhammer, T. (2013). Identification of extracellular surface-layer associated proteins in *Lactobacillus acidophilus* NCFM. Microbiology *159*, 2269–2282. https://doi.org/10.1099/mic.0.070755-0

Jones, B.V., Begley, M., Hill, C., Gahan, C.G., and Marchesi, J.R. (2008). Functional and comparative metagenomic analysis of bile salt hydrolase activity in the human gut microbiome. Proc. Natl. Acad. Sci. U.S.A. *105*, 13580–13585. https://doi.org/10.1073/pnas.0804437105

Jordan, S., Hutchings, M.I., and Mascher, T. (2008). Cell envelope stress response in Gram-positive bacteria. FEMS Microbiol. Rev. *32*, 107–146. https://doi.org/10.1111/j.1574-6976.2007.00091.x

Kandler, O. (1983). Carbohydrate metabolism in lactic acid bacteria. Antonie Van Leeuwenhoek *49*, 209–224.

Kankaanpää, P., Yang, B., Kallio, H., Isolauri, E., and Salminen, S. (2004). Effects of polyunsaturated fatty acids in growth medium on lipid composition and on physicochemical surface properties of lactobacilli. Appl. Environ. Microbiol. *70*, 129–136.

Khaleghi, M., Kermanshahi, R.K., Yaghoobi, M.M., Zarkesh-Esfahani, S.H., and Baghizadeh, A. (2010). Assessment of bile salt effects on s-layer production, *slp* gene expression and some physicochemical properties of *Lactobacillus acidophilus* ATCC 4356. J. Microbiol. Biotechnol. *20*, 749–756.

Khaleghi, M., and Kermanshahi, R.K. (2012). Effect of environmental stresses on S-layer production in *Lactobacillus acidophilus* ATCC 4356. In: Advances in Applied Biotechnology, Petre, M., ed. (Intech Open), Chapter 11. https://doi.org/10.5772/28334

Koch, S., Oberson, G., Eugster-Meier, E., Meile, L., and Lacroix, C. (2007). Osmotic stress induced by salt increases cell yield, autolytic activity, and survival of lyophilization of *Lactobacillus delbrueckii* subsp. *lactis*. Int. J. Food Microbiol. *117*, 36–42. https://doi.org/10.1016/j.ijfoodmicro.2007.01.016

Lacroix, C., and Yildirim, S. (2007). Fermentation technologies for the production of probiotics with high viability and functionality. Curr. Opin. Biotechnol. *18*, 176-183. https://doi.org/10.1016/j.copbio.2007.02.002

Lebeer, S., Claes, I.J., and Vanderleyden, J. (2012). Anti-inflammatory potential of probiotics: lipoteichoic acid makes a difference. Trends Microbiol. *20*, 5–10. https://doi.org/10.1016/j.tim.2011.09.004

Linares, D.M., Del Río, B., Ladero, V., Martínez, N., Fernández, M., Martín, M.C., and Alvarez, M.A. (2012). Factors influencing biogenic amines accumulation in dairy products. Front. Microbiol. *3*, 180. https://doi.org/10.3389/fmicb.2012.00180

Machado, M.C., López, C.S., Heras, H., and Rivas, E.A. (2004). Osmotic response in *Lactobacillus casei* ATCC 393: biochemical and biophysical characteristics of membrane. Arch. Biochem. Biophys. *422*, 61–70.

Martínez, M.G., Prado Acosta, M., Candurra, N.A., and Ruzal, S.M. (2012). S-layer proteins of *Lactobacillus acidophilus* inhibits JUNV infection. Biochem. Biophys. Res. Commun. *422*, 590–595. https://doi.org/10.1016/j.bbrc.2012.05.031

Maukonen, J., Matto, J., Suihko, M.L., and Saarela, M. (2008). Intra- individual diversity and similarity of salivary and faecal micro- biota. J. Med. Microbiol. *57*, 1560– 1568. https://doi.org/10.1099/jmm.0.47352-0

Meng, J., Gao, S.M., Zhang, Q.X., and Lu, R.R. (2015). Murein hydrolase activity of surface layer proteins from *Lactobacillus acidophilus* against *Escherichia coli*. Int. J. Biol. Macromol. *79*, 527–532. https://doi.org/10.1016/j.ijbiomac.2015.03.057

Meneghel, J., Passot, S., Dupont, S., and Fonseca, F. (2017). Biophysical characterization of the *Lactobacillus delbrueckii subsp. bulgaricus* membrane during cold and osmotic stress and its relevance for cryopreservation. Appl. Microbiol. Biotechnol. *101*, 1427–1441. https://doi.org/10.1007/s00253-016-7935-4

Mills, S., Stanton, C., Fitzgerald, G.F., and Ross, R.P. (2011). Enhancing the stress responses of probiotics for a lifestyle from gut to product and back again. Microb. Cell Fact. 10 (Suppl. 1), S19. https://doi.org/10.1186/1475-2859-10-S1-S19

Mohamadzadeh, M., Pfeiler, E.A., Brown, J.B., Zadeh, M., Gramarossa, M., Managlia, E., Bere, P., Sarraj, B., Khan, M.W., Pakanati, K.C., *et al.* (2011). Regulation of induced colonic inflammation by *Lactobacillus acidophilus* deficient in lipoteichoic acid. Proc. Natl. Acad. Sci. U.S.A. 108 (Suppl. 1), 4623–4630. https://doi.org/10.1073/pnas.1005066107

Muller, J.A., Ross, R.P., Sybesma, W.F., Fitzgerald, G.F., and Stanton, C. (2011). Modification of the technical properties of Lactobacillus johnsonii NCC 533 by supplementing the growth medium with unsaturated fatty acids. Appl. Environ. Microbiol. 77, 6889–6898. https://doi.org/10.1128/AEM.05213-11

Neuhaus, F.C., and Baddiley, J. (2003). A continuum of anionic charge: structures and functions of D-alanyl-teichoic acids in gram-positive bacteria. Microbiol. Mol. Biol. Rev. *67*, 686–723.

Neville, B.A., Forde, B.M., Claesson, M.J., Darby, T., Coghlan, A., Nally, K., Ross, R.P., and O'Toole, P.W. (2012). Characterization of pro-inflammatory flagellin proteins produced by *Lactobacillus ruminis* and related motile Lactobacilli. PLOS ONE 7, e40592. https://doi.org/10.1371/journal.pone.0040592

Nikolic, M., López, P., Strahinic, I., Suárez, A., Kojic, M., Fernández-García, M., Topisirovic, L., Golic, N., and Ruas-Madiedo, P. (2012). Characterisation of the exopolysaccharide (EPS)-producing *Lactobacillus paraplantarum* BGCG11 and its non-EPS producing derivative strains as potential probiotics. Int. J. Food Microbiol. *158*, 155–162. https://doi.org/10.1016/j.ijfoodmicro.2012.07.015

Palomino, M.M., Sanchez-Rivas, C., and Ruzal, S.M. (2009). High salt stress in *Bacillus subtilis*: involvement of PBP4* as a peptidoglycan hydrolase. Res. Microbiol. *160*, 117–124. https://doi.org/10.1016/j.resmic.2008.10.011

Palomino, M.M., Allievi, M.C., Gründling, A., Sanchez Rivas C., and Ruzal, S.M. (2013). Osmotic stress adaptation in *Lactobacillus casei* BL23 leads to structural changes in the cell wall polymer lipoteichoic acid. Microbiology *159*, 2416–2426. https://doi.org/10.1099/mic.0.070607-0

Palomino, M.M., Waehner, P.M., Fina Martin, J., Ojeda, P., Malone, L., Sánchez Rivas, C., Prado Acosta, M., Allievi, M.C., and Ruzal, S.M. (2016). Influence of osmotic stress on the profile and gene expression of surface layer proteins in *Lactobacillus acidophilus* ATCC 4356. Appl. Microbiol. Biotechnol. *100*, 8475–8484. https://doi.org/10.1007/s00253-016-7698-y

Perly, B., Smith, I.C., and Jarrell, H.C. (1985). Acyl chain dynamics of phosphatidylethanolamines containing oleic acid and dihydrosterculic acid: 2H NMR relaxation studies. Biochemistry *24*, 4659–4665.

Piuri, M., Sanchez-Rivas, C., and Ruzal, S.M. (2003). Adaptation to high salt in *Lactobacillus*: role of peptides and proteolytic enzymes. J. Appl. Microbiol. *95*, 372–379.

Piuri, M., Sanchez-Rivas C., and Ruzal, S.M. (2005). Cell wall modifications during osmotic stress in *Lactobacillus casei*. J. Appl. Microbiol. *98*, 84–95 https://doi.org/10.1111/j.1365-2672.2004.02428.x

Poolman, B., and Glaasker, E. (1998). Regulation of compatible solute accumulation in bacteria. Mol. Microbiol. *29*, 397–407.

Popham, D.L., and Young, K.D. (2003). Role of penicillin-binding proteins in bacterial cell morphogenesis. Curr. Opin. Microbiol. *6*, 594–599.

Pouwels, P., Kolen, C.P.A.M., and Boot, H.J. (1997). S-layer protein genes in *Lactobacillus*. FEMS Microbiol Rev 20,78–82

Prado Acosta, M., Mercedes Palomino, M., Allievi, M.C., Sanchez Rivas, C., and Ruzal, S.M. (2008). Murein hydrolase activity in the surface layer of *Lactobacillus acidophilus* ATCC 4356. Appl. Environ. Microbiol. *74*, 7824–7827. https://doi.org/10.1128/AEM.01712-08

Prasad, J., McJarrow, P., and Gopal, P. (2003). Heat and osmotic stress responses of probiotic *Lactobacillus rhamnosus* HN001 (DR20) in relation to viability after drying. Appl. Environ. Microbiol. *69*, 917–925.

Regulski, K., Courtin, P., Meyrand, M., Claes, I.J., Lebeer, S., Vanderleyden, J., Hols, P., Guillot, A., and Chapot-Chartier, M.P. (2012). Analysis of the peptidoglycan hydrolase complement of *Lactobacillus casei* and characterization of the major γ-D-glutamyl-L-lysyl-endopeptidase. PLOS ONE 7, e32301. https://doi.org/10.1371/journal.pone.0032301

Remus, D.M., Kleerebezem, M., and Bron, P.A. (2011). An intimate tête-à-tête - how probiotic lactobacilli communicate with the host. Eur. J. Pharmacol. 668 (Suppl. 1), S33–42. https://doi.org/10.1016/j.ejphar.2011.07.012

Salazar, N., Gueimonde, M., de los Reyes-Gavilán C.G., Ruas-Madiedo P. (2015), Exopolysaccharides produced by lactic acid bacteria and bifidobacteria as fermentable substrates by the intestinal microbiota, Crit. Rev. Food

Sci. Nutr. *12*, 10. https://doi.org/10.1080/10408398.2013.770728

Sánchez, B., Ruiz, L., Gueimonde, M., Ruas-Madiedo, P., and Margolles, A. (2013). Adaptation of bifidobacteria to the gastrointestinal tract and functional consequences. Pharmacol. Res. *69*, 127–136. https://doi.org/10.1016/j.phrs.2012.11.004

Sanders, T.A. (1999). Food production and food safety. BMJ *318*, 1689–1693.

Sára, M., and Sleytr, U.B. (2000). S-Layer proteins. J. Bacteriol. *182*, 859–868.

Schleifer, K.H., and Kandler, O. (1972). Peptidoglycan types of bacterial cell walls and their taxonomic implications. Bacteriol. Rev. *36*, 407–477.

Silva, P.D.L., Bezerra, M.F., Santos, K.M.O., and Correia, R.T.P. (2015). Potentially probiotic ice cream from goat's milk: characterization and cell viability during processing, storage and simulated gastrointestinal conditions. LWT - Food Science and Technology, *62*, 452–457.

Spano, G., Capozzi, V., Vernile, A., and Massa, S. (2004). Cloning, molecular characterization and expression analysis of two small heat shock genes isolated from wine *Lactobacillus plantarum*. J. Appl. Microbiol. *97*, 774–782. https://doi.org/10.1111/j.1365-2672.2004.02359.x

Stackebrandt, E., and Teuber, M. (1988). Molecular taxonomy and phylogenetic position of lactic acid bacteria. Biochimie *70*, 317–324.

Taranto, M.P., Fernandez Murga, M.L., Lorca, G., and de Valdez, G.F. (2003). Bile salts and cholesterol induce changes in the lipid cell membrane of *Lactobacillus reuteri*. J. Appl. Microbiol. *95*, 86–91.

Trip, H., Mulder, N.L., and Lolkema, J.S. (2012). Improved acid stress survival of *Lactococcus lactis* expressing the histidine decarboxylation pathway of *Streptococcus thermophilus* CHCC1524. J. Biol. Chem. *287*, 11195–11204. https://doi.org/10.1074/jbc.M111.330704

van de Guchte, M., Serror, P., Chervaux, C., Smokvina, T., Ehrlich, S.D., and Maguin, E. (2002). Stress responses in lactic acid bacteria. Antonie Van Leeuwenhoek *82*, 187–216.

Vidgrén, G., Palva, I., Pakkanen, R., Lounatmaa, K., and Palva, A. (1992). S-layer protein gene of *Lactobacillus brevis*: cloning by polymerase chain reaction and determination of the nucleotide sequence. J. Bacteriol. *174*, 7419–7427.

Vinogradov, E., Valence, F., Maes, E., Jebava, I., Chuat, V., Lortal, S., Grard, T., Guerardel, Y., and Sadovskaya, I. (2013). Structural studies of the cell wall polysaccharides from three strains of *Lactobacillus helveticus* with different autolytic properties: DPC4571, BROI, and LH1. Carbohydr. Res. *379*, 7–12. https://doi.org/10.1016/j.carres.2013.05.020

Vinogradov, E., Sadovskaya, I., Grard, T., and Chapot-Chartier, M.P. (2016). Structural studies of the rhamnose-rich cell wall polysaccharide of *Lactobacillus casei* BL23. Carbohydr. Res. *435*, 156–161. https://doi.org/10.1016/j.carres.2016.10.002

Wasko, A., Polak-Berecka M., and Gustaw, W. (2013). Increased viability of probiotic *Lactobacillus rhamnosus* after osmotic stress. Acta Aliment. Hung. *42*, 520–528. https://doi.org/10.1016/j.anaerobe.2013.11.004

Yasuda, E., Serata, M., and Sako, T. (2008). Suppressive effect on activation of macrophages by *Lactobacillus casei* strain Shirota genes determining the synthesis of cell wall-associated polysaccharides. Appl. Environ. Microbiol. *74*, 4746–4755. https://doi.org/10.1128/AEM.00412-08

Zuljan, F.A., Repizo, G.D., Alarcon, S.H., and Magni, C. (2014). α-Acetolactate synthase of *Lactococcus lactis* contributes to pH homeostasis in acid stress conditions. Int. J. Food Microbiol. *188*, 99–107. https://doi.org/10.1016/j.ijfoodmicro.2014.07.017

S-Layer Proteins from Lactobacilli: Biogenesis, Structure, Functionality and Biotechnological Applications

6

Mariano Malamud[1], Patricia A. Bolla[2], Paula Carasi[1], Esteban Gerbino[3], Andrea Gómez-Zavaglia[3], Pablo Mobili[3] and María de los Angeles Serradell[1]*

[1]Chair of Microbiology, Department of Biological Sciences, Faculty of Exact Sciences, National University of La Plata, La Plata, Argentina.

[2]Center for Research and Development in Applied Sciences "Dr. Jorge J. Ronco" (CINDECA), National Council of Scientific and Technological Research (CONICET), La Plata, Argentina.

[3]Center for Research and Development in Food Cryotechnology (CIDCA), National Council of Scientific and Technological Research (CONICET), La Plata, Argentina.

*Correspondence: maserr@biol.unlp.edu.ar

https://doi.org/10.21775/9781910190890.06

Abstract

The S-layer is a (glyco)-proteinaceous envelope made up of subunits that self-assemble to form a two-dimensional lattice that covers the surface of different species of Bacteria and Archaea. The S-layer (glyco)-proteins have been shown to possess exceptional physicochemical properties which make them unique organizational structures with high potential application in different areas of biotechnology. It represents a very interesting model system for studying the processes involved in the synthesis, secretion and assembly of extracellular proteins. Among bacteria, its presence has been demonstrated in many species of lactobacilli, some of which could be considered as probiotic microorganisms. In this chapter, we will discuss the most relevant and updated concepts regarding genetics, structural features, cell wall- and self-assembly, functionality and biotechnological applications of the S-layer proteins from lactobacilli.

Introduction

In the course of evolution, nature has developed simple and fascinating solutions to face different challenges. In this sense, most of the microorganisms have had to adapt in order to survive in highly competitive and complex habitats. Consequently, the diversity observed in the molecular architecture of the bacterial cell surface, particularly the structure of the outermost layer, reflects the evolutionary adaptation of the organism to specific ecological and environmental conditions.

A rigid protein envelope, known as S-layer, is considered the most ancient biological membrane, and has been maintained throughout the evolution both in Bacteria and Archaea (Claus *et al.*, 2005). This macromolecular two-dimensional lattice is a crystal-like proteinaceous structure discovered by Houwink in 1953 in *Spirillum serpens* (Houwink, 1953), but it was not being until the 70´s that began to attract the attention of the scientific community.

These proteins are account for approximately 10–15% of total bacterial cellular proteins thus its synthesis represents a great effort for the microorganism. Regarding that the biomass of prokaryotic microorganisms surpasses the biomass of the eukaryotic organisms, these surface proteins can be considered as one of the most abundant biopolymers on our planet (Sleytr *et al.*, 2014). Considering the ubiquitous presence of S-layer-carrying microorganisms and the abundance of the S-layer proteins (SLP), it is evident that these structures reflect the evolutionary adaptations of the organisms to natural habitats and must have provided them with advantages in specific environmental and ecological conditions (Zhu *et al.*, 2017).

S-layers are composed of numerous identical protein or glycoprotein subunits, 40–200 kDa in molecular weight, which form a two-dimensional, regular and highly porous array (unit cell sizes in the range of 3 to 30 nm, and thicknesses of roughly 10nm) with oblique (p1, p2), square (p4) or hexagonal (p3, p6) symmetry. Electronic microscopy-based techniques are the methods of choice for detecting the presence of an S-layer on the bacterial cell surface (Fig. 6.1) (Pavkov-Keller *et al.*, 2011). To obtain information about the 3D structure and surface topology of S-layer assemblies, methods such as atomic force microscopy (AFM) and small-angle X-ray scattering (SAXS) have been applied among others. However, since conventional methods for high-resolution structural analysis such as X-ray crystallography cannot be easily applied to S-layers, little information is available regarding the atomic structure of SLPs and the spatial arrangement of domains and secondary structure elements within the S-layer lattice (Pavkov-Keller *et al.*, 2011).

The subunits are held together and attached to the underlying cell surface by non-covalent interactions (Hynönen and Palva, 2013; Pum *et al.*, 2013). In Gram-negative and Gram-positive bacteria, they are found to be attached to different supramolecular envelope structures. In the former group of bacteria, the S-layer adheres to the lipopolysaccharide (LPS) of the outer membrane via ionic, carbohydrate–carbohydrate, protein–carbohydrate interactions and/or protein–protein interactions. On the other hand, in Gram-positive bacteria the binding can occur to the peptidoglycan layer or to a secondary cell wall polymer (SCWP) (Sleytr *et al.*, 2014).

The subunit proteins are typically rich in hydrophobic amino acids but low in sulphur-containing amino acids (Sára and Sleytr, 2000). Amino acid analysis of SLPs of organisms from all phylogenetic branches revealed a rather similar overall composition, however, despite all these common features, comparative studies of SLPs on microorganisms from different taxonomic affiliations revealed that homologies between non-related organisms are very low. In this sense, it could be considered that from the evolutionary point of view, S-layers had contributed more to the diversification rather than to conservation, in order to help the microorganisms to gain advantages in selection (Zhu *et al.*, 2017).

SLPs were the first glycoproteins detected in prokaryotes and still are among the best-studied

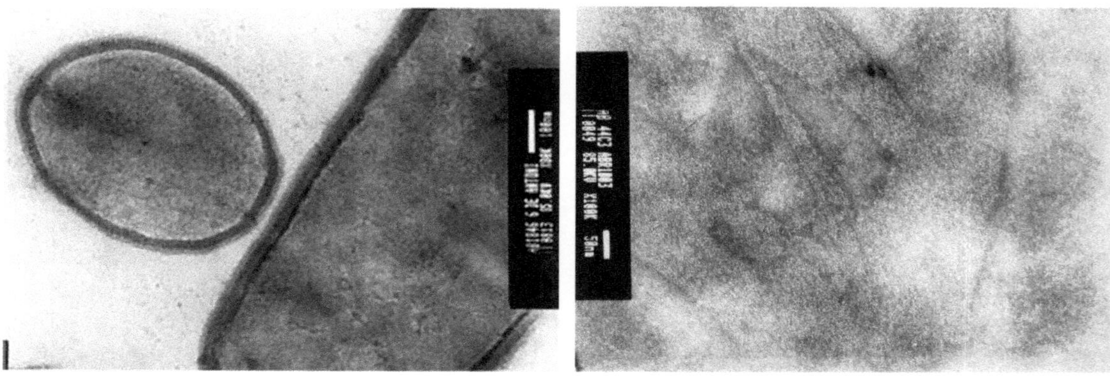

Figure 6.1 Transmission electron micrographs of thin section (A) and (B) negative staining of *Lactobacillus kefiri* CIDCA 8321. Adapted figure from Garrote, G., Delfederico, L., Bibiloni, R., Abraham, A., Pérez, P., Semorile, L., and De Antoni, G. (2004). Lactobacilli isolated from kefir grains: evidence of the presence of S-layer proteins. J. Dairy Res., *71*(2), 222–230 © Proprietors of Journal of Dairy Research, published by Cambridge University Press (with permission).

examples of glycosylated prokaryotic proteins. Indeed, glycosylation is the most frequent post-translational modification of these proteins undergo (Ristl *et al.*, 2011). In bacteria, N-glycosylation is a rare event, so bacterial SLPs were usually characterized as O-glycosylated and the degree of glycosylation varies generally between 2% and 10% (w/w) (Messner *et al.*, 2008).

As it was mentioned above, genes encoding for SLPs are highly expressed. Sleytr and colleagues demonstrated that a bacterial cell needs to synthesize, translocate to the cell surface, and incorporate into the existing S-layer lattice around 500 subunits per second at high growth rates. Moreover, considering the average-size of a lactobacilli cell it has been calculated that at least 5×10^5 S-layer protein subunits are needed during each cell generation (Sleytr and Beveridge, 1999). The mechanisms involved in this highly efficient expression and secretion phenotype are mostly related with the long half-lives of the S-layer gene transcripts (Boot *et al.*, 1997) and the presence of strong and very efficient promoters (Kahala *et al.*, 1997; Narita *et al.*, 2006). The presence of several SLPs genes in the genome of a single strain has been described. However, they are not necessarily expressed at the same time. The existence of silent genes, antigenic variation (Boot and Pouwels, 1996; Sára and Sleytr, 2000; Thompson, 2002), alternative expression *in vivo* or *ex vivo* (Fouet, 2009), sequential expression during growth (Mignot *et al.*, 2004) and, rarely, superimposed S-layers (Cerquetti *et al.*, 2000) or S-layers composed of two different polypeptides (Rothfuss *et al.*, 2006; Fagan *et al.*, 2009; Goh *et al.*, 2009; Sekot *et al.*, 2012; Palomino *et al.*, 2016) have been described.

The information about biological functions of SLPs have been increased in the last decades. No common function for all S-layers has emerged, but instead, many different functional properties have been proposed for this kind of proteins. Thus far, they include, e.g. determination or maintenance of cell shape, functions as a molecular sieve, as a binding site for large molecules, ions or phages, and as a mediator of microbial adhesion (Sára and Sleytr, 2000), protection of microbial cells from adverse environmental conditions, protection for presence of antimicrobial peptides (de la Fuente-Núñez *et al.*, 2012), protection for exposition to radiation (Kotiranta *et al.*, 1999), changes in environmental pH (Gilmour *et al.*, 2000), bacteriophages (Howard and Tipper, 1973), bacterial or eukaryotic microbial predators (Koval and Hynes, 1991; Tarao *et al.*, 2009) or bacteriolytic enzymes (Lortal *et al.*, 1992), and also modulation of immune responses (Konstantinov *et al.*, 2008; Sekot *et al.*, 2011; Taverniti *et al.*, 2013; Collins *et al.*, 2014). Further, in pathogenic bacteria, S-layers could contribute to virulence through several mechanisms including adhesion, coaggregation (Shimotahira *et al.*, 2013), antigenic variation (Spigaglia *et al.*, 2011), protection from complement killing or phagocytosis (Thompson, 2002) or immunomodulation (Ryan *et al.*, 2011; Sekot *et al.*, 2011; Settem *et al.*, 2013). Also, some SLPs have shown the potential to act as enzymes (Calabi *et al.*, 2001; Ahn *et al.*, 2006; Prado Acosta *et al.*, 2008), or even to be involved in motility (McCarren *et al.*, 2005).

The special attention that these proteins have caught from researchers in the last years becomes from their exceptional physicochemical properties which make them a unique organizational structure with high application potential in the field of modern nanobiotechnology. Most techniques for isolation and purification of SLPs involve the employment of detergents (Triton X100 or sodium dodecyl sulphate) or chaotropic agents (such as urea, guanidine hydrochloride, or lithium chloride) (Pum *et al.*, 2013; Schuster and Sleytr, 2013). After extraction from bacterial surface, the SLPs have the intrinsic capability to form free-floating self-assembled products in solution (e.g. flat sheets, tubes, vesicles), to reassemble into mono- and double layers at solid supports, at the air–water interface, at lipid films, and to cover liposomes, nanocapsules, and nanoparticles completely (Fig. 6.2) (Pum *et al.*, 2013). This extraordinary property is one of the key features of SLPs and, in combination with the precise repetitive exposition of functional groups, established the bases for the development of carriers for biomolecules, diagnostic devices, biosensors, biocatalysts, etc (Sleytr *et al.*, 2007; Ilk *et al.*, 2011; Pum *et al.*, 2013; Wang *et al.*, 2015). Thus, this unique capability can be exploited in different areas such as medicine, pharmacology, bioengineering, food industry among others.

The presence of S-layer has been described in many bacterial species, including some of the genus *Lactobacillus* (Fig. 6.1) (Hynönen and Palva, 2013). For these lactic acid bacteria, frequently

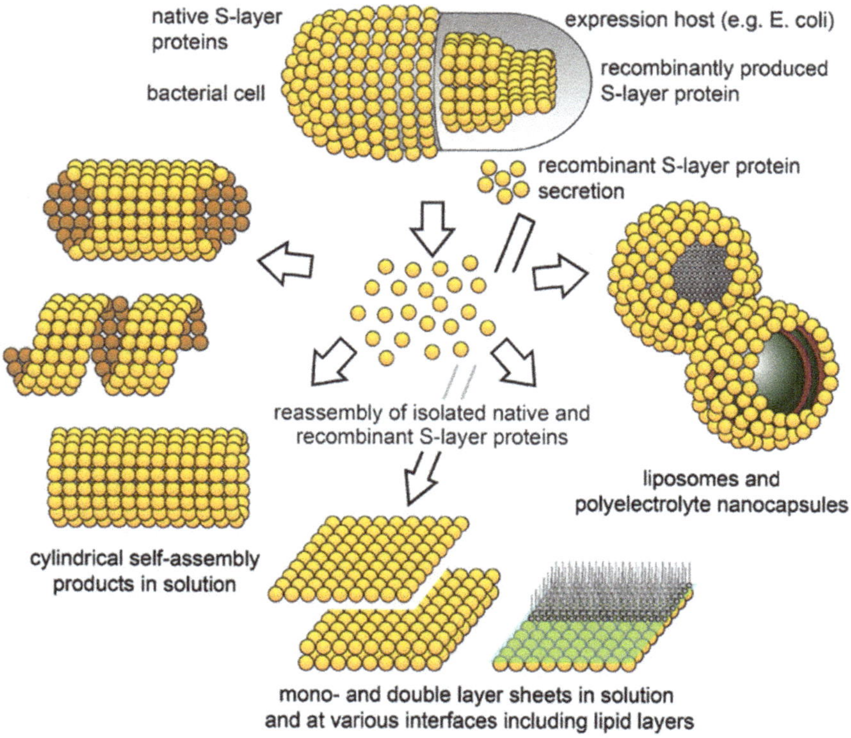

Figure 6.2 Schematic drawing of the reassembly of S-layer proteins. From Pum, D., Toca-Herrera, J., and Sleytr, U. (2013) S-layer protein self-assembly. Int. J. Mol. Sci., *14*, 2484–2501, copyright by MDPI.

found in different natural ecosystems such as plants, foodstuffs, silages, wastewater, and both the gastrointestinal and genital tract of humans and animals, the presence of S-layer becomes a relevant adaptive feature to deal with the adverse conditions of those environments. Moreover, since the S-layer constitutes the outermost surface structure of these bacterial cells, it could mediate the interaction of the microorganisms with different eukaryotic cells (i.e. intestinal epithelial cells, antigen presenting cells) as well as with both extracellular matrix and mucus components, which has been particularly studied for lactobacilli strains characterized as 'probiotics' (Hynönen and Palva, 2013; Fijan, 2014). The reported sizes of SLPs of lactobacilli ranges between 25 and 71 kDa and the predicted pI values are high (9.4–10.4) when compared with SLPs from other bacteria. Only oblique or hexagonal type of lattice symmetry has been described for *Lactobacillus* (Lortal *et al.*, 1992; Smit *et al.*, 2001; Åvall-Jääskeläinen and Palva, 2005; Anzergruber *et al.*, 2014) and glycosylation is a post-transductional modification that has been found and characterized in only a few species (Messner *et al.*, 2008; Malamud *et al.*, 2017; Cavallero *et al.*, 2017).

Recent evidence has also shown that there are additional proteins which non-covalently co-localize to the outermost stratum of the cell surface with the S-layer, called S-layer associated proteins (SLAPs) (Johnson *et al.*, 2013, 2016). These SLAPs were first characterized in *L. acidophilus* NCFM (Johnson *et al.*, 2013), but have since been described in different strains of *L. helveticus*, *L. crispatus*, *L. amylovorus*, and *L. gallinarum* (Johnson *et al.*, 2016) which are S-layer forming members of the *L. acidophilus*–*L. delbrueckii* homology group. Preliminary functional analyses in *L. acidophilus* NCFM have revealed that SLAPs have a broad spectrum of both cellular and probiotic functionality, including cell division and autolysin activity, immunomodulation, and adhesion to extracellular matrices (Johnson *et al.*, 2013; Hymes *et al.*, 2016; Johnson and Klaenhammer, 2016).

In this chapter, we will discuss the most relevant and updated concepts regarding genetics, structural features, cell wall- and self-assembly, functionality and biotechnological applications of the S-layer proteins from different species of lactobacilli.

Occurrence and primary structure of *Lactobacillus* S-layer proteins

In the genus *Lactobacillus*, the presence of an S-layer has been described in several but not all species. Lactobacilli species with an S-layer include *L. buchneri, L. diolivorans, L. hilgardii, L. kisonensis, L. otakiensis, L. parafarraginis, L. acidophilus, L. crispatus, L. plantarum, L. brevis, L. fermentum, L. bulgaricus, L. amylovorus, L. gallinarum, L. helveticus, L. johnsonii, L. gasseri, L. kefiranofaciens, L. parakefiri* and *L. kefiri*, among others (Hynönen and Palva, 2013).

The overall primary structure of lactobacilli S-layers tend to be poorly conserved and can differ between related species and even when more than one SLP gene is present in the same strain the sequences may be different (Jakava-Viljanen *et al.*, 2002; Åvall-Jääskeläinen and Palva, 2005; Hagen *et al.*, 2005; Malamud *et al.*, 2017). However, all the described SLPs have a signal peptide of 25 to 30 amino acids, mostly 19–34, at the N-terminus indicating secretion through the general secretory pathway. It is mostly cleaved off after translocation through the plasma membrane by a signal peptidase (Sará and Sleytr 2000; Antikainen *et al.*, 2002; Zhu *et al.*, 2017).

Generally, the amino acid composition of lactobacilli SLPs is characterized by the absence of cysteine residues, the low presence of sulphur-containing amino acids and a percentage of hydrophobic and hydroxylated amino acids that varies from 31% to 39% and from 23% to 33% respectively. The amount of positively charged amino acid residues ranges from 9.5% to 12% and is always higher than the amount of negatively charged residues (Åvall-Jääskeläinen and Palva, 2005; Dohm *et al.*, 2011; Wásko *et al.*, 2014; Malamud *et al.*, 2017). Messner *et al.* (2008) compared several sequences of lactobacilli S-layer proteins and found that the number of identical amino acids varies from 7.2% to 100%.

Åvall-Jääskeläinen and Palva (2005) evaluated whether the phylogenetic trees built based on SLP sequences and the 16S rRNA gene sequences of the corresponding *Lactobacillus* species revealed a similar overall clustering of strains. They found that *L. brevis* strains clustered clearly separated from strains belonging to the *L. acidophilus* group. However, in this last group different species clustered in different branches even although their 16S RNA genes were almost identical. These results indicate that the rates of evolutionary substitution in the SLPs are higher than in the 16S RNA gene which indicates that there is a strong selective pressure that drives the diversification of SLP genes at least in some group of lactobacilli. Interestingly, the use of LC-MS/MS analysis of SLPs has been proposed for typing strains within the *L. acidophilus* group (Podlesny *et al.*, 2011) and this technique has also been useful recently to distinguish different groups among *L. kefiri* strains (Malamud *et al.*, 2017).

In order to locate the regions with sequence similarities in the *Lactobacillus* SLPs, several authors have aligned amino acid sequences. It has been shown that the SLPs of *L. acidophilus, L. crispatus* and *L. helveticus* are highly similar in the C-terminal region, whereas the N-terminal region of these SLPs is more variable (Åvall-Jääskeläinen and Palva, 2005). However, *L. buchneri, L. parabuchneri, L. hilgardii, L. diolivorans, L. parakefiri* and *L. kefiri* show the highest similarity in the N-terminal region (Fig. 6.3) (Malamud *et al.*, 2017).

Posttranslational modifications in *Lactobacillus* S-layer proteins: glycosylation

Although glycan structures have been identified from the SLPs of several Gram-positive bacteria, most of the SLPs of lactobacilli appear to be non-glycosylated. However, up to now, glycosylation is the only post-translational modification that has been described in *Lactobacillus* SLPs. The glycoprotein nature of lactobacilli SLPs has been reported for *L. buchneri* strains, *L. helveticus* ATCC 12046, *L. acidophilus* NCFM, *L. plantarum* 41021/252 and several *L. kefiri* strains (Möschl *et al.*, 1993; Mozes and Lortal, 1995; Konstantinov *et al.*, 2008; Mobili *et al.*, 2009a; Anzengruber *et al.*, 2014; Malamud *et al.*, 2017). To date, detailed SLP glycan structures have been determined just for three strains: *L. buchneri* 41021/251 and CD034 (Auzengruber *et al.*, 2014) and *L. kefiri* CIDCA 83111 (Cavallero *et al.*, 2017). In both species *O*-glycosidic structures have been found however N-glycosylation structures have only been described for *L. kefiri* CIDCA

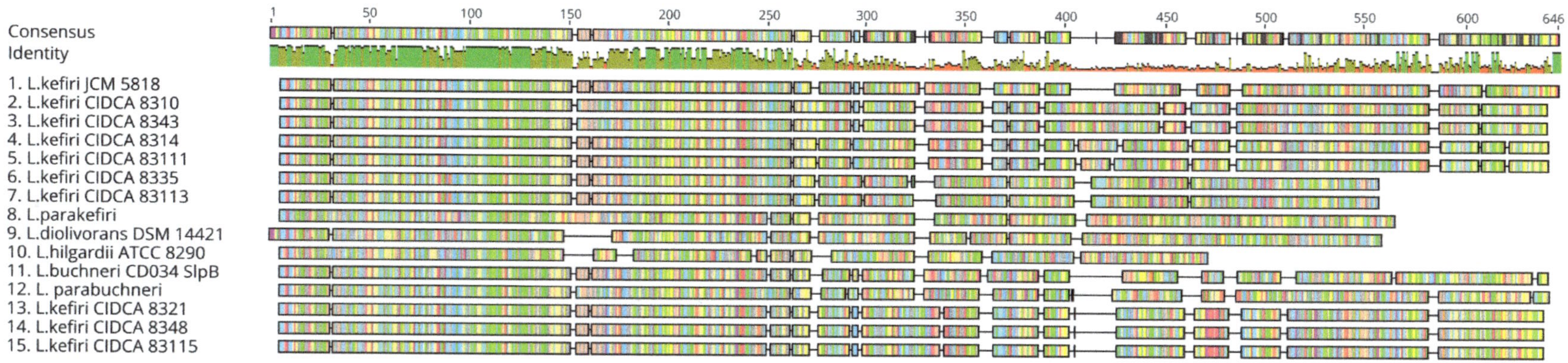

Figure 6.3 Amino acid sequence alignment of S-layer proteins from phylogenetically related *Lactobacillus* species. The highest similarity is observed in the N-terminal region.

83111. Regarding *O*-glycosidic chains, SlpB of *L. buchneri* CD034 and SlpN of *L. buchneri* NRRL B-30929 are both glycosylated at four serine residues within the sequence S_{152}-A-S_{154}-S_{155}-A-S_{157} with, on average, seven Glc(α1-6) residues, each (Anzengruber *et al.*, 2014). More recently, two *O*-glycosylated peptides were detected in *L. kefiri* CIDCA 83111: the peptide sequence SSASSASSA already identified as a signature glycosylation motif in *L. buchneri,* substituted on average with eight glucose residues and decorated with galacturonic acid, and another *O*-glycosylated site on peptide 471–476 with a $Glc_{5-8}GalA_2$ structure (Cavallero *et al.*, 2017). It is interesting to notice that there is a high amino acid sequence homology in the putative *O*-glycosylation sites not only for *L. buchneri* and *L. kefiri* but also for related species such as *L. parafarraginis, L. sunkii, L. senioris, L. kisonensis, L. otakiensis, L. parabuchneri* and *L. parakefiri.* Moreover, these sites are mostly located in the N-terminal region of the proteins (Fig. 6.4).

On the other hand, the presence of ten characteristic sequons (Asn-X-Ser/Thr) SLP amino acid sequence as well as the release of *N*-linked oligosaccharides upon digestion with PNGase F were reported for *L. kefiri* CIDCA 83111 SLP. Analysis of released material by HPAEC-PAD showed a main oligosaccharide migrating near the classical *N*-glycosidic core $GlcNAc_2Man_3$ (Cavallero *et al.*, 2017).

Expression of *Lactobacillus* S-layer protein genes

As mentioned above, the half-lives of the SLP gene transcripts are exceptionally long. It has been described that S-layer gene transcript of *L. brevis* ATCC 8287 is stable for 14 min (Boot *et al.*, 1996). For *L. acidophilus* ATCC 4356 the calculated mRNA half-life is 15 min and it has been demonstrated this is due to the long 5′ untranslated region (UTR) of the transcript which forms a loop that stabilizes the mRNA secondary structure (Kahala *et al.*, 1997; Narita *et al.*, 2006). The same observations were obtained for the strain *L. crispatus* K313 (Sun *et al.*, 2013).

Regarding S-layer proteins promoters, it has been described for *L. acidophilus* ATCC 4356 that its efficiency is twice as that preceding the gene encoding lactate dehydrogenase (Narita *et al.*, 2006). Same high efficiency profile was demonstrated by Sun and colleagues (2013) for the S-layer gene promoter of *L. crispatus* K313. Another important feature of S-layer genes promoters is that two or more of them have also been described and they are likely to be required for the efficient transcription of S-layer genes enhancing and/or regulating gene expression (Boot *et al.*, 1996; Jakava-Viljanen *et al.*, 2002; Hynönen *et al.*, 2010). The SlpA gene of *L. brevis* ATCC 8287 contains two subsequent promoters and both are active during bacterial growth (Kahala *et al.*, 1997; Hynönen *et al.*, 2010). However, in the case of *L. acidophilus* ATCC 4356 only one of the promoters was observed to be functional in the evaluated growth conditions (Boot *et al.*, 1996).

Several *Lactobacillus* strains possess multiple SLP genes in the same strain: *L. brevis* ATCC 14869, *L. acidophilus* NCFM, *L. acidophilus* ATCC 4356, *L. crispatus* ZJ001 and *L. crispatus* K313 and the genomes of *L. gallinarum* strains, *L. brevis* ATCC 367 and *L. buchneri* CD034 and the contigs of *L. kefiri* JCM 5818 (Jakava-Viljanen *et al.*, 2002; Hagen *et al.*, 2005; Konstantinov *et al.*, 2008; Chen *et al.*, 2009; Heinl *et al.*, 2012; Sun *et al.*, 2013; Malamud *et al.*, 2017). However, it has only been proven for some strains the simultaneous expression of more than one S-layer gene under experimental conditions (Jakava-Viljanen *et al.*, 2002; Goh *et al.*, 2009; Sun *et al.*, 2013; Palomino *et al.*, 2016).

Expression of different SLPs, or changed expression levels of SLPs, can occur both at the DNA level and the transcriptional level, or mediated by environmental factors. Chromosomal rearrangements have been studied for *L. acidophilus* ATCC 4356 and *L. acidophilus* NCFM. An active *slpA* gene and a silent *slpB* gene are located in opposite orientations in the bacterial chromosome. The translocation of SlpB behind the active *slpA* promoter via an inversion of a chromosomal segment by site-specific recombination system at the 5′ homologous region has been described (Boot *et al.*, 1996; Buck *et al.*, 2005; Konstantinov *et al.*, 2008).

Stress-mediated and growth stage variations in S-layer expression not involving variation at the DNA level have been also studied in lactobacilli. *L. brevis* ATCC 14869 differentially express *slpB* and *slpD* genes depending on the oxygen content of the growth medium and the growth stage. Under aerated conditions, both transcripts have been

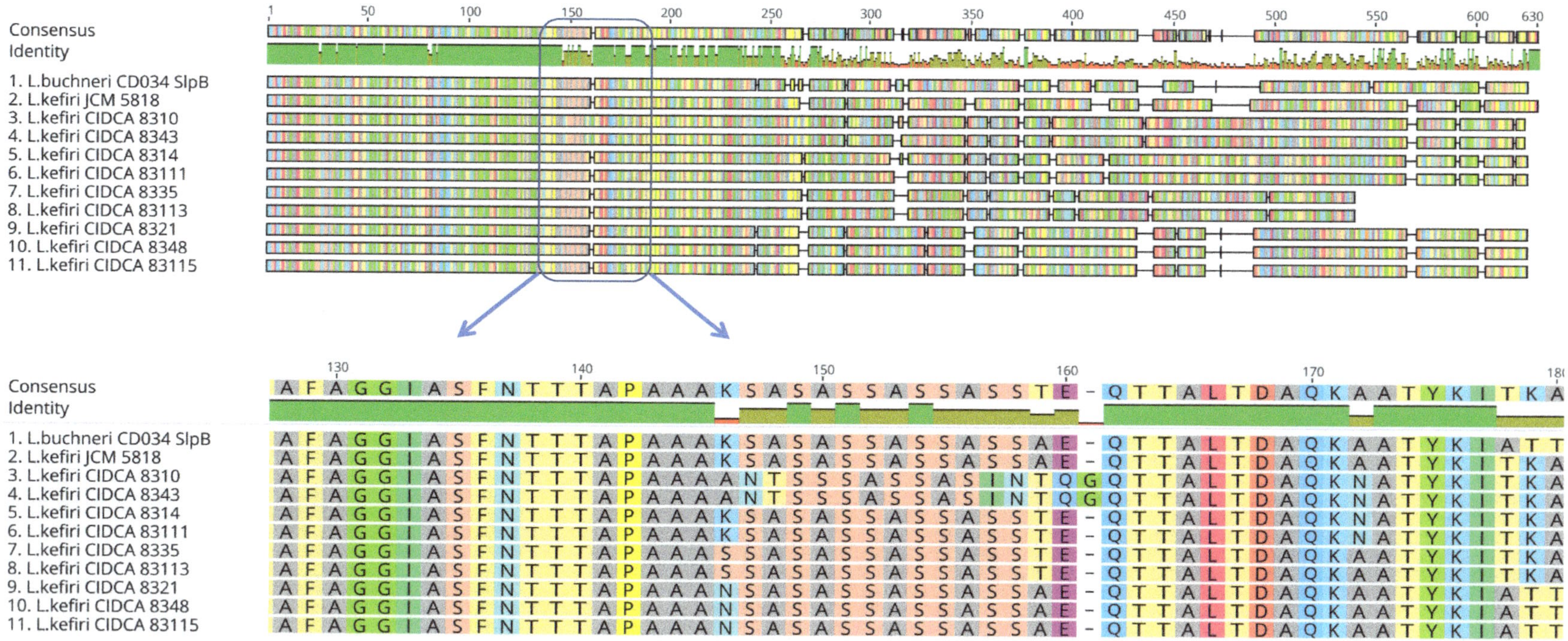

Figure 6.4 Amino acid sequence alignment of S-layer proteins from *L. buchneri* CD034 and ten different strains of *L. kefiri*. Zoom-in on the main *O*-glycosylation site.

detected; whereas under anaerobic conditions only *slpB* mRNA has been observed. Promoter analysis suggests that the variations on expression are mediated by a soluble cytoplasmic factor induced by the environmental changes (Jakava-Viljanen *et al.*, 2002). Another example is the differential SLP expression profiles that have been observed for *L. acidophilus* ATCC 4356 in exponential or stationary phase of growth at two salt concentrations (osmotic stress). At high-salt concentrations the expression of two of the three SLPs present in that strain are increased mostly in the stationary stage (Palomino *et al.*, 2016). The responses for other stress conditions have been evaluated for several *Lactobacillus* strains. In the case of *L. acidophilus* NCC 2628, growth media with limited protein content induced the expression of the SLP gene (Schär-Zammaretti *et al.*, 2005). For *L. acidophilus* ATCC 4356, the expression of *slpA* was not affected by sublethal concentrations of penicillin G even although protein quantities in the bacterial surface were higher (Khaleghi *et al.*, 2011). Additionally, *slpA* expression was increased in the presence of 0.01–0.05% w/v bile salts but decreased at 10-fold higher concentration for the same strain (Khaleghi *et al.*, 2010). In contrast, no changes of S-layer gene expression were observed for *L. acidophilus* NCFM during the passage through an *in vitro* gastrointestinal tract model (Weiss and Jespersen, 2010). In the case of *L. brevis* ATCC 8287, neither bile, pancreatin nor an uncommon carbon source had an impact on the quantities of SlpA expressed on the bacterial cell surface (Hynönen *et al.*, 2010). In the case of *L. acidophilus* IBB 801, an increase of SLP synthesis was detected when the strain was grown at 42°C or in the presence of 0.05% w/v bile salts or 2.0% w/v NaCl (Grosu-Tudor *et al.*, 2016). The described changes in S-layer expression associated with stress factors reinforce the proposed idea that the S-layer plays a role as a protective sheath, functioning as protective coats and helping in the maintenance of cell shape (Avall-Jääskeläinen and Palva, 2005; Hynönen and Palva, 2013; Gerbino *et al.*, 2015). However, this does not seem to be the case for all the studied *Lactobacillus* strains since no changes of S-layer gene expression were observed for *L. acidophilus* NCFM during the passage through an *in vitro* gastrointestinal tract model (Weiss and Jespersen, 2010) or for *L. brevis* ATCC 8287 grown in the presence of bile or pancreatin (Hynönen *et al.*, 2010).

Cell wall binding and self-assembly regions in *Lactobacillus* S-layer proteins

SLPs exhibit mostly two separated morphological regions: one responsible for cell wall binding and the other required for self-assembly. In Gram-positive bacteria, the rigid cell wall is composed of peptidoglycan and accessory (secondary) cell wall polymers (SCWP) such as teichoic acids, lipoteichoic acids, lipoglycans or teichuronic acids embedded in the peptidoglycan matrix. The SLPs of Gram-positive bacteria are anchored to the cell surface via non-covalent interactions with SCWP (Avall-Jääskeläinen and Palva, 2005; Hynönen and Palva, 2013). In general, two conserved Gram-positive S-layer-anchoring modules have been identified, which use either an S-layer homologous (SLH) domain or the cell wall binding 2 (CWB2) domain. Both modules use three domains, which are located either at the N-terminal or C-terminal region of the protein. However, since no SLH domains were found in SLPs from lactobacilli (Sun *et al.*, 2013), different N-terminal or C-terminal regions have been proposed as cell wall binding domains.

It has been demonstrated for the strains *L. acidophilus* ATCC 4356 (Smit *et al.*, 2001) and *L. crispatus* JCM 5810, ZJ001, K313 and K2-4-3 (Antikainen *et al.*, 2002; Chen *et al.*, 2009; Hu *et al.*, 2011; Sun *et al.*, 2013) that the C-terminal region is the most conserved among these strains and that it is responsible for anchoring the SLP to the cell wall. Due to the high homology found between the C-terminal region of the already mentioned species and *L. amylovorus* and *L. helveticus* strains, the same location of the anchoring domain has been proposed for these species (Hynönen *et al.*, 2014; Wasko *et al.*, 2014). Notably, the most conserved region for *L. brevis* ATCC 8287 (Åvall-Jääskeläinen *et al.*, 2008) and *L. hilgardii* B706 (Dohm *et al.*, 2011) is also responsible for the attachment of the SLP to the cell wall however these domains are located in the N-terminal part of the protein. Regarding the mechanism of these interactions, for all the mentioned strains the cell wall binding domains have a similar charge distribution with a high predicted pI thus it has been proposed that electrostatic interactions occur between the cell wall binding regions and the negatively charged SCWP (Smit *et al.*, 2001; Antikainen *et al.*, 2002; Avall-Jääskeläinen *et al.*, 2008; Chen *et al.*, 2009; Dohm *et al.*, 2011; Hu *et al.*, 2011; Sun *et al.*, 2013). The cell wall ligands to which the *Lactobacillus* SLPs attach might be (lipo)-teichoic acids or neutral polysaccharides. SLPs of *L. acidophilus* ATCC 4356, *L. crispatus* JCM 5810 and K313 interact with teichoic acid (Antikainen *et al.*, 2002; Smit and Pouwels, 2002; Sun *et al.*, 2013) while *L. brevis* ATCC 8287 and *L. hilgardii* B706 do not. In fact, Masuda and Kawata (1985) described that the SLPs of *L. brevis* and *L. buchneri* recognize a neutral polysaccharide as a binding site.

On the other hand, the self-assembling region of *Lactobacillus* SLPs, responsible for the self-assembly of the monomers to a periodic S-layer lattice, has been demonstrated to be located in the variable part of the protein. This region is located in the N-terminal part for *L. acidophilus* ATCC 4356 and *L. crispatus* JCM 5810 and K313 and in the C-terminal part for *L. brevis* ATCC 8287 (Antikainen *et al.*, 2002; Smit *et al.*, 2002; Åvall-Jääskeläinen *et al.*, 2008; Chen *et al.*, 2009). It is important to notice that conserved regions and regions predicted to form secondary structures in SLPs are necessary for the formation of a regular lattice (Smit *et al.*, 2002).

Secondary and tertiary structure of *Lactobacillus* S-layer proteins

Early determinations of secondary-structure content in SLPs, based on circular-dichroism measurements, showed about 40% of β sheets and 20% of α helices, with most α-helical segments located in the N-terminal portion of the protein according to secondary-structure predictions based on protein sequence data (Sára and Sleytr, 2000). First estimations of secondary structure content in *Lactobacillus* SLPs were performed by predictions based on known sequences of unprocessed SLPs (Vidgrén *et al.*, 1992; Sillanpää *et al.*, 2000; Ventura *et al.*, 2002). Predictions suggested *L. brevis* SlpA contained α-helix (27%) and β-sheet structures (56%) as the main components (Vidgrén *et al.*, 1992), while a lower but still high content of β-sheet was predicted for CbsA of *L. crispatus* (37%) (Sillanpää *et al.*, 2000) and for Afp1 of *L. gasseri* (29%) and *L. johnsonii* (33%) (Ventura *et al.*, 2002). To avoid comparisons of data generated by secondary structure prediction using different analysis algorithms, Avall-Jääskeläinen and Palva (2005) employed the PHD program (Rost *et al.*, 1994) to predict the secondary structure content of the unprocessed SLPs from different lactobacilli. They found that the SLPs from different *L. brevis* strains consisted of alternating short stretches of α-helices, extended strands and random coils, but located in different positions in each strain, except for the α-helices located at the signal sequences of the proteins. On the other hand, the SLPs from *L. acidophilus*, *L. crispatus* and *L. helveticus* shared a high degree of secondary-structure similarity, with a α-helix at their signal sequences and two other α-helices in the middle of the protein (for a total content of α-helices of about 14%), and similar content of random coils (over 44% in each) and extended strands (over 34% in each) (Avall-Jääskeläinen and Palva, 2005). When the same program was employed to predict the secondary structure of SlpA from *L. brevis* ATCC 8287 in its mature form (without signal peptide), a total content of 55.6% β-sheet and 0% α-helix was found, in agreement with Avall-Jääskeläinen and Palva (2005), which described the α-helices as located at the signal sequences of this protein (Mobili *et al.*, 2009b). A prediction of secondary structure mature SLPs from ten different strains *L. kefiri* was recently performed using the Psipred software (Malamud *et al.*, 2017). No major differences were found among strains, except for SLPs from the strains *L. kefiri* CIDCA 8335 and CIDCA 83113 which showed a higher percentage of α-helix (14.1 and 15.3% respectively) and a lower percentage of random coil (58.9 and 55.8% respectively) than the other proteins. These results are similar to those obtained for the SLP of *L. buchneri* CD034 using the same software.

A few years ago, a Fourier transform infrared spectroscopy (FT-IR) study was performed for the SLPs extracted from *L. brevis* ATCC 8287 and five strains of *L. kefiri*. The SLP from *L. brevis* presented 0% α-helix, 50% β-sheet and 50% other structures, including β-turns and random coils. The SLP from *L. kefiri* strains presented β-sheet contents ranging from 23 to 42% and α-helix contents ranging from

13 to 21%, with a lower content of β-sheet in the aggregating strains than in non-aggregating ones (around 25% and 40% respectively) (Mobili *et al.*, 2009b). These differences were not then observed by software-based prediction analysis; likely due to the glycan residues (that were not included in the predictive analysis) could influence the secondary structure of the whole proteins (Malamud *et al.*, 2017).

In recent years, circular dichroism (CD) measurements in the far-UV region performed in extracted SLPs indicated a content of α-helix, β-sheet and random structures of 26%, 5% and 36% respectively for *L. acidophilus* ATCC 4356 (Eslami *et al.*, 2013), 4%, 41% and 36% respectively for *L. salivarius* 16 (Lighezan *et al.*, 2016), and 6%, 45% and 43% respectively for *L. brevis* KM3 and KM7 (Mobarak Qamsari *et al.*, 2017). In contrast, a lower percentage of β-sheet (at about 12%) as well as a higher percentage of α–helix (at about 34%) were reported by Meng *et al.* (2014) for the SLPs isolated from *L. helveticus* fb213, *L. acidophilus* fb214 and *L. acidophilus* fb116.

It is known that secondary structure of proteins could be influenced by different environmental physicochemical conditions (i.e. pH, temperature, ionic force). In a previous study by Lighezan *et al.* (2012), the far-UV CD spectra taken at different temperatures had been indicated that the thermal denaturation of the secondary structure of the SLP from *L. salivarius* 16 takes place in the temperature range between 40 and 80°C and is partially reversible, suggesting that the high content of β-sheet could take a main role in the thermal stability of this protein. On the other hand, the secondary structure of SLP of *L. acidophilus* ATCC 4356 showed to be stable in simulated gastrointestinal conditions (pH 3.2) whereas incubation at lower pH induced changes that involve a significant increase of β-sheet structure fraction (from 5% to 30%) (Eslami *et al.*, 2013).

The near-UV CD spectrum is sensitive to the tertiary structure of the proteins. For the SLP of *L. salivarius* 16, the tertiary structure is determined by a high content of hydrophobic amino acids, such as Trp, Tyr and Phe, bound into a local chiral environment, which tend to compact the protein's tertiary structure and could be related to the stability to form dimers or oligomers (Lighezan *et al.*, 2012).

Up to now, all the available data provided by predictive or experimental analysis suggests that elucidation of the secondary and tertiary structure of SLPs is a subject of discussion due to the great variability between *Lactobacillus* strains as well as between analytical techniques.

Self-assembly of *Lactobacillus* S-layer proteins in solution

It is known that after extraction from bacterial surface, the SLPs have the intrinsic capability to form free-floating self-assembled products in solution. This reassembly starts with a quick formation of oligomeric precursors, and follows with a slow rearrangement, leading to extended lattices (Jaenicke *et al.*, 1985). Depending on the morphology and bonding properties of the subunits, flat sheets, open-ended cylinders, or closed vesicles can be the final products of the assembly process. To date, the literature about the self-assembly of *Lactobacillus* SLPs is scarce.

Dynamic light scattering (DLS) can be used to evaluate the size and oligomerization of various particles like proteins, polymers, micelles, carbohydrates and nanoparticles. In this sense, this technique was employed to determine the monodispersity of monomeric SLPs from two different strains of *L. brevis*. The results indicated that the main population of soluble SLP molecules in the tested strains were composed of monomers with expected diameter close to 10 nm (Mobarak Qamsari *et al.*, 2017).

On the other hand, small-angle X-ray scattering (SAXS) is a powerful method for determining the low-resolution shape of proteins or to study aggregation of proteins in solution. Up to now, there is only one report about the use of SAXS to study the auto-assembly of a lactobacilli´s SLP in solution. According to SAXS results, two truncated forms (rSlpA$_{167-435}$ and SlpA$_{179-435}$) from *L. brevis* ATCC 8287 form stable globular oligomers of about ten monomers in aqueous solution, likely through hydrophobic interactions (Jääskeläinen *et al.*, 2010).

These results could be useful for further work to identify specific structural features and kinetic of self-assembly process as well as for the application of SLPs in the development of new biotechnological tools.

Functionality of S-layer proteins on *Lactobacillus* cells

As it was previously mentioned, no common function for all S-layers has been found hereto, but instead, different functional properties have been proposed for several SLPs. As the outermost layer in different species of lactobacilli, the S-layer is in direct contact with bacterial environment and thus may be involved in many of their surface properties, including the interaction with eukaryotic cells as well as other kind of microorganisms.

The investigation of surface properties of lactobacilli bearing S-layers and their correlation with specific probiotic properties represent a way to understand both the mechanisms involved in beneficial properties and the functions of the S-layers in these bacteria.

Lactobacillus SLPs on bacterial adherence to different substrates

Several *Lactobacillus* species have a status generally regarded as safe (GRAS), thus they can be used as carriers of live oral vaccines or important effector molecules in the intestine (Hynönen *et al.*, 2002; Mercenier *et al.*, 2003). The ability of *Lactobacillus* to adhere to epithelial and sub-epithelial tissues is a key to its successful persistent colonization of mammalian intestines and other tissue sites. *Lactobacillus* species exhibiting strong adherence to intestinal epithelium cells, could produce an inhibition of the binding of pathogens to host tissues (Chen *et al.*, 2007; Khang *et al.*, 2009). Indeed, different studies revealed that lactobacilli´s SLPs mediate bacterial aggregation as well as adhesion to epithelial cells and to intestinal components such as mucus or extracellular matrix (ECM) proteins (Hynönen and Palva, 2013).

Auto-aggregation and co-aggregation with other microorganisms

Bacterial adhesion is initially based on non-specific physical interactions between two surfaces, which then enable specific interactions between adhesins (usually proteins) and complementary receptors (Kos *et al.*, 2003). Auto-aggregation of probiotic bacterial strains appeared to be necessary for adhesion to intestinal epithelial cells; meanwhile co-aggregation abilities may contribute to form a barrier that prevents colonization by pathogenic microorganisms.

Physicochemical characteristics of the cell surface such as hydrophobicity may affect auto-aggregation as well as adhesion of bacteria to different surfaces (Kos *et al.*, 2003). Determination of these properties on bacterial strains before and after the removal of SLPs may confirm the role of such proteins in those interactions. In fact, the removal of SLP with 5 M LiCl reduces the auto-aggregation ability of different lactobacilli strains such as *L. acidophilus* ATCC 4356 (Kos *et al.*, 2003), *L. kefiri* CIDCA 8321 (Garrote *et al.*, 2004), *L. helveticus* M92 (Beganović *et al.*, 2011) and *L. helveticus* T159 (Waśko *et al.*, 2014). However, the strong auto-aggregating phenotype of *L. crispatus* ZJ001 could be related to other cell surface components than SLP, since it is not markedly affected by treatment with LiCl (Chen *et al.*, 2007).

The ability to aggregate with other microorganisms (coaggregation) could be seen as a part of competitive exclusion mechanism which contributes to the reduction of the pathogenic load during infections. The role of SLP in the interaction with *Salmonella typhimurium* FP1 has been demonstrated for *L. helveticus* M92 (Beganović *et al.*, 2011) and *L. brevis* D6 (Uroić *et al.*, 2016) since its removal produces a decrease significantly the coaggregation with this pathogen. On the other hand, SLPs from auto-aggregative *L. kefiri* strains mediates co-aggregation with *Saccharomyces lypolitica*, since the effect was abrogated by bacterial treatment with 5 M LiCl. This interaction may contribute to the maintenance of structure and composition of a complex microbial ecosystem as the kefir grain (Golowczyc *et al.*, 2009).

Adhesion to intestinal epithelial cells, gastrointestinal mucus and extracellular matrix components

Numerous targeting strategies to identify the mechanisms of bacterial colonization of their host have been employed because of the multiple cell surface-associated factors expressed by lactobacilli. Adhesion to intestinal epithelial cells is considered the first step in the persistent colonization of the host by *Lactobacillus* strains. High-affinity adhesion promotes the residence of lactobacilli in the gut, excludes pathogens and protects epithelial cells. Several difficulties involved in the study of bacterial adhesion *in vivo*, especially in humans, have led to the development of different *in vitro* models such as colonocyte-like epithelial cell lines (Caco-2, HT-29, HT-29 MTX) or mucin/mucus-binding systems (Deepika and Charalampopoulos, 2010).

The S-layer is one of the first bacterial components that could interact with the gastrointestinal surface of the mammalian host. Moreover, diverse bacterial cell surface-associated factors such as carbohydrates, lipoteichoic acids as well as proteinaceous factors LPXTG-like protein could mediate specific adhesion and may act as adhesins (Ventura *et al.*, 2002; Lebeer *et al.*, 2008). Therefore, adhesion likely does not require a unique and ubiquitous mechanism.

Regarding the interaction with the host cells, several studies have shown that the loss of the SLPs from the bacterial surface caused by chemical treatment decreases adhesion to different target cells, indicating that these proteins may be one of the most important factors that mediates bacterial adherence to eukaryotic cells. Depletion of the SlpA from the bacterial surface greatly reduced the adhesion of *L. brevis* ATCC 8287 to the human intestinal cell lines Caco-2 and Intestine 407, the endothelial cell line EA-hy926, and the urinary bladder cell line T24 (Hynönen *et al.*, 2002). The SLP of *L. acidophilus* M92 was proposed to mediate binding to porcine ileal epithelial cells, since removal of the SLP by LiCl decreased significantly the adhesion of this strain to intestinal epithelium (Frece *et al.*, 2005). In a similar way, adhesion ability to the human colorectal adenocarcinoma cell line HT-29 decreased by 61%, 65% and 92% for *L. helveticus* fb213, *L. acidophilus* fb116 and *L. acidophilus* fb214, respectively, after removal of SLP from these strains (Meng *et al.*, 2014). On the contrary, removal of SLP from *L. kefiri* and *L. parakefiri* strains did not produce changes in bacterial adhesion to the intestinal epithelial cell line Caco-2 (Garrote *et al.*, 2004), suggesting the involvement of other bacterial surface components in the interaction with those cells. Recently, Wang *et al.* (2017) showed that the SLP choline-binding protein A (CbpA) of *L. salivarius* REN is involved in the high affinity binding to HT-29 cells, since adhesion of a CbpA deletion mutant was significantly reduced compared with that of the wild-type.

Since mucus is the first barrier that covers the epithelial cells of the gastrointestinal tract, a possible mechanism for bacterial adherence and colonization involves the binding of microbial cell-surface molecules to the protective mucus layer. Despite mucus adhesion ability has been reported by several lactobacilli strains (Tuomola *et al.*, 2000; MacKenzie *et al.*, 2010), the involvement of SLP in that process was only studied in a few cases. Some years ago, Rojas *et al.* (2002) reported the purification and characterization of a LiCl-extractable surface protein from *L. fermentum* 104R with the ability to bind to porcine small intestinal mucus and gastric mucin. More recently, Carasi *et al.* (2014) reported that *L. kefiri* strains treated with NaOH and LiCl have a lower capacity to adhere to porcine gastric mucin and mucus components extracted from piglet small intestine and colon. Moreover, the addition of soluble SLPs increased *L. kefiri* adhesion to these substrates, suggesting that the presence of these proteins in the gastrointestinal tract could enhance the bacterial adhesion to mucus layer, improving the interaction with the epithelium (Carasi *et al.*, 2014).

Adhesion to the subepithelial constituents of the extracellular matrix (ECM), such as glycoproteins like fibronectin and laminin as well as to different types of collagen, could be an important event to the successful colonization of mammalian intestine and other tissue sites. In this sense, some years ago, Sillanpää and colleagues (2000) described that adhesiveness of strain *L. crispatus* JCM 5810 to human sub-intestinal extracellular matrix, to mouse basement membrane and to human collagens was mediated by an adhesive surface structure preliminary characterized as a SLP, which was named CbsA (for 'collagen-binding S-layer protein A'). Binding to immobilized types I and IV collagen was observed with a truncated form of SlpB (1–379) from *L. crispatus* K313, suggesting that location of the ECM domain was located in the N-terminal region of the protein (Sun *et al.*, 2013). On the other hand, the involvement of SLPs in the strong adhesion to fibronectin and laminin was reported for different strains of *L. brevis, L. crispatus* and *L. amylovorus* species (Jakava-Vijanen and Palva, 2007), as well as for *L. brevis* D6 and *L. helveticus* M92 (Uroić *et al.*, 2016), since removal of SLPs with guanidine hydrochloride abolished the adhesiveness of those bacteria. Indeed, in 2006, de Leeuw and colleagues were able to demonstrate the direct interaction of the SlpA of *L. brevis* ATCC 8287 with fibronectin, laminin, fibrinogen and collagen using surface plasmon resonance technology. SlpA was found to interact with high affinity to fibronectin and laminin, whereas the interaction with collagen and fibrinogen was found to be much lower affinity.

Considering all the scientific evidence, it is quite clear that adhesion of lactobacilli to the intestinal surface is a result of multifactorial specific and non-specific interactions. Among the factors responsible for that interaction, SLPs seem to play a critical role in several strains, although other surface molecules such as moonlighting proteins, adhesins, lipoteichoic acids, and exopolysaccharides could be involved.

Lactobacillus SLPs as a mechanical barrier to protect bacteria in harmful environments

To date, some other functions than adhesion have been shown or proposed for *Lactobacillus* SLPs. The gastrointestinal tract (GIT) is an environment full of stressful conditions, including low pH, proteolytic enzymes, digestive juices, etc. Therefore, different authors investigated the role of SLPs in protecting microorganisms in these and other harmful conditions.

The survival of *L. brevis* D6 in different stress conditions, particularly in simulated GIT conditions and during freeze-drying were better before removal of SLP from the cell surface (Uroić *et al.*, 2016). In this sense, the viability of *L. acidophilus* M92 without SLP decreases in simulated gastric juice at low pH and simulated pancreatic juice with bile salts (Frece *et al.*, 2005). SLP of *L. acidophilus* M92 showed to be resistant to pepsin and pancreatin. However, the treatment with proteinase K leads to a significant proteolysis of SLP.

The expression of SLPs in different conditions of fermentation medium was studied for *L. acidophilus* NCC2628 by Schär-Zammaretti and colleagues (2005). They observed that in the absence of peptones, the overall protein content of cell wall is significantly lower than in complete medium, whereas the expression of SLP is strongly enhanced. This suggests that SLP could be preferentially expressed in conditions that are not optimal for bacterial growth. These results agree with those reported by Khaleghi and coworkers who demonstrated that SLP gene expression were altered by different stress conditions in the probiotic strain *L. acidophilus* ATCC 4356. First, the amount of SLP extracted from bacterial cells increased in presence of 0.01–0.1% bile, and the SlpA encoding gene expression showed a similar behaviour (Khaleghi *et al.*, 2010). This is in concurrence with the recent observation made by Palomino and colleagues (2016) that showed that SLPs are involved in the adaptation of *L. acidophilus* to osmotic stress. They observed an increased expression of the SlpX protein due to high-salt conditions, accompanying the predominant SlpA. Moreover, it has been shown that other certain stress culture conditions like high incubation temperature, presence of bile salts or NaCl, and acidic pH induce SLP production by *L. acidophilus* IBB 801, which most probably helped the strain to maintain cell viability under detrimental culture conditions (Grosu-Tudor *et al.*, 2016).

A potential protective role of the S-layer of *L. helveticus* ATCC 12046 against muramidases has been studied by Lortal and colleagues (1992). Although LiCl treatment of bacterial cells did not increase sensitivity to lysozyme, mutanolysin had a more rapid lytic effect on cells devoid of S-layer. Since the presence of the S-layer could not protect completely the peptidoglycan layer from enzymatic activities, it can be concluded that the pores in the oblique S-layer lattice of *L. helveticus* ATCC 12046 allow the passage of these enzymes through it.

On the other hand, Prado Acosta and colleagues (2008) observed that the C-terminal part of the SLP of *L. acidophilus* ATCC 4356 have murein hydrolase (endopeptidase) activity against the cell wall of, for example, *Salmonella enterica*. This proteolytic activity could provide S-layer-bearing lactobacilli strains with an additional means to succeed and survive in a competitive habitat such as GIT.

Finally, although the capacity of lactic acid bacteria to interact with different metals has been addressed, the role of SLP in metal biosorption has only been reported for *L. kefiri*. Gerbino and colleagues (2015b) addressed the capacity of *L. kefiri* strains to sequester Pb^{2+} before and after removing SLPs with proteinase K. They found that bacteria without SLP increased their sequestrant capacity. The removal of SLP had no significant effect on bacterial viability in control conditions, but microorganisms without SLP were more prone to the detrimental effect of Pb^{2+}, thus suggesting that the S-layer acts as a protective rather than as a sequestrant layer in this strain. In this sense, *L. hilgardii* wine isolate B706 without S-layer was more susceptible to wine-related conditions like the presence of copper sulphate or tannic acid (Dohm *et al.*, 2011).

Immunomodulatory properties of *Lactobacillus* SLPs

Considering that certain lactobacilli are autochthonous microorganisms of the gastrointestinal tract of animals and humans, and some of them are considered as probiotics, the study of the interaction of *Lactobacillus* cells with the mucosal immune system has increased in recent years. The cross talk between the host and intestinal bacteria is due to the ability of host cell to recognize specific bacterial components. Recognition of microbial components by host immune cells is mediated by different pattern recognition receptors (PRRs), including Toll-like receptors (TLRs) (Wan *et al.*, 2016) and C-type lectins receptors (CLRs) (Mayer *et al.*, 2017). After recognition of microbial components by PRRs, antigen-presenting cells like macrophages and dendritic cells (DC), internalize microorganisms and present microbe-derived antigens on major histocompatibility complex (MHC) molecules to T cells, thereby initiating adaptive immune responses (Geijtenbeek and Gringhuis, 2016). To identify the microbial components that act as effectors of the immune system is a key point in the understanding of the complex interaction between bacteria and host. In this sense, the S-layer emerges as a potential target for bacterial recognition since it is in direct contact with the bacterial environment.

One of the cellular models most used for the study of the interaction of lactobacilli and its components with the host are the intestinal epithelial cells lines. Ashida and colleagues (2011), showed that the adhesive abilities of *L. acidophilus* strains to the human Caco-2 cells correlated closely to the amount of SlpA in bacteria and the productivity of IL-12, an inflammatory cytokine produced by DC. On the contrary, using the same cell line, Taverniti and colleagues (2013) showed that the SlpA from *L. helveticus* MIMLh5 exerts an anti-inflammatory effect by reducing the activation of NF-kB (a transcriptional factor involved in the expression of several proinflammatory cytokines). In line with Taverniti and colleagues, other researchers demonstrated that SLPs from *L. amylovorus* DSM 16698^{T} could activate protective immune regulation on Caco-2 cells in response to F4+ enterotoxigenic *Escherichia coli* through regulation of NF-kB, since the co-treatment of the pathogen with cell wall fractions coated with SLPs (S-CWF) inhibited NF-kB activation (Roselli *et al.*, 2016). However, Hynönen and colleagues (2014), using S-CWF from the same *L. amylovorus* strain have found a low level of adhesion to porcine intestinal epithelial cells (IPEC-1), showing that the same stimulus could have different modes of action depending on the cell type, origin and polarization. (Habil *et al.*, 2011). Consistent with results showing an anti-inflammatory profile of SLPs on intestinal cells, Gao and colleagues (2017), showed that the SLP from *L. rhamnosus* GG, inhibited p38MAPK signalling in porcine IPEC-J2 cells after LPS-stimulation resulting in decreased transcription of inflammatory cytokines. Also, it has shown that treatment with SLPs from *L. acidophilus* reduced *Salmonella typhimurium*-induced IL-8 secretion on Caco-2 cells (Li *et al.*, 2011).

Since glycosylation is the post-translational modification most frequently found in SLPs, several studies have been carried out to evaluate the role of carbohydrates present in the SLPs of different *Lactobacillus* species in their immunomodulatory properties. Konstantinov and colleagues (2008) reported by first time the critical role of SlpA from *L. acidophilus* NFCM in the interaction of bacterial cells with human monocyte-derived DCs by specific binding to the DC-specific intercellular adhesion molecule 3 (ICAM-3)-grabbing nonintegrin (DC-SIGN), since a SlpA-knockout mutant showed a significantly reduced binding to this receptor. The binding of SlpA from *L. acidophilus* NCFM on DCs was not sufficient to induce a strong maturation, but the combination with LPS could induce a higher secretion of IL-10 compared with LPS alone without affecting the secretion of IL-6 or TNF-α. This indicate that SlpA was responsible for the anti-inflammatory cytokine profile observed for *L. acidophilus* NCFM (Konstantinov *et al.*, 2008). The SLPs-DC-SIGN engagement was also demonstrated by Liu and colleagues (2011), who have shown that S-layer like proteins from *L. plantarum* bind to DC-SIGN and induce the maturation of DCs, and also has been suggested for the SLP from *L. kefiri* JCM 5818 (Prado Acosta *et al.*, 2016). Recently, it has been demonstrated that SLP from *L. kefiri* CIDCA 8348 (SLP-8348) is internalized by murine macrophages RAW 264.7 cells in a process mediated by carbohydrate–receptor interactions since it was inhibited by glucose, mannose or EGTA, a Ca^{+2} chelating agent (Malamud *et al.*, 2018). Furthermore, while SLP-8348 was not able to induce macrophage´s activation by itself, it enhanced the LPS-induced response, increasing the expression of surface cell markers MHC-II, CD86 and CD40, as well as in IL-6 and IL-10 expression at both transcript and protein levels, in comparison with LPS-stimulated cells. The presence of EGTA completely abrogated this synergistic effect, suggesting the involvement of both glycosidic residues and Ca^{+2} ions in the recognition of SLP-8348 by cellular receptors. In line with these results, purified SlpA of *L. helveticus* MIMLh5, a strain closely related to *L. acidophilus*, exerted proinflammatory effects by inducing high levels of TNF-α and COX-2 and lower levels of IL-10 in the human macrophage cell line U937 via recognition through Toll-like receptor 2 (TLR2) (Taverniti *et al.*, 2013). A similar response was observed following stimulation of murine macrophages isolated from bone marrow or peritoneal cavity. Although the simultaneous stimulation with *L. helveticus* MIMLh5 SlpA and LPS did not result in an additive or synergistic effect on the proinflammatory response.

The involvement of the SLPs in the interaction between *Lactobacillus* and host, led to several *in vivo* experiments focusing on the immunomodulatory effects of SLPs. In this regard, it was reported that oral administration of *L. helveticus* M92 to mice induced higher levels of total serum IgA, IgG and IgM than those evoked by *L. helveticus* M92 without SlpA, whereas purified SlpA did not induced a specific humoral immune response after oral application (Beganović *et al.*, 2011). Interestingly, the immunogenicity of SlpA from *L. acidophilus* was demonstrated by Kajikawa and colleagues (2015), who showed that repeated, high dose immunization with *L. acidophilus* evoked SLP-specific antibodies and cytokines responses. Splenocytes isolated from mice immunized with *L. acidophilus* were re-stimulated with purified SLPs and the production of IFN-γ and IL-17 were markedly up-regulated. Recently, it was demonstrated that *L. acidophilus* NCK2187, a strain which solely expresses SlpA, and its purified SlpA binds to the C-type lectin SIGNR3 to trigger regulatory signals that result in mitigation of colitis, maintenance of healthy gastrointestinal microbiota, and protection of gut mucosal barrier function on mice, since this protective effect was not observed in Signr3$^{-/-}$ mice (Lightfoot *et al.*, 2015). These authors performed an experimentally infectious model using *Citrobacter rodentium*, and both the *L. acidophilus* NCK2187 and its SlpA accelerated pathogen clearance, resulting in a decreased colonic IL-1β expression. They also performed an irritant-induced colitis model with dextran sulfate sodium, and neither the *L. acidophilus* NCK2187 nor its SlpA prevent colitis in the absence of SIGNR3-signalling. Considering that of the eight murine homologues of DC-SIGN, SIGNR3 exhibits the most biochemical similarity to human DC-SIGN, these results agree with those previously obtained by Konstantinov *et al.* (2008). A step forward in the use of SlpA from *L. acidophilus* NCK2187 for the treatment of intestinal disorders in humans was presented by Sahay and colleagues (2015) who used NaCl 5 M to improve the process of SlpA isolation and purification, and demonstrated that using this protocol, the protein does not elicit potential toxicity when administered orally to C57BL/6 mice.

Considering all the evidence, the SLPs could be essential to determine the interaction of lactobacilli with the host immune system, as well as to contribute to the shaping of the immunological response to these lactic acid bacteria. Moreover, the immunomodulatory properties of isolated *Lactobacillus* SLPs encourage further studies to evaluate the ability of these proteins to act as new adjuvant or carrier for vaccine antigens.

Antimicrobial activity of *Lactobacillus* SLPs

It is known that different strains of *Lactobacillus* have an antagonistic effect on various intestinal pathogenic virus and bacteria. In several cases, the involvement of SLPs in the antimicrobial activity has been studied.

As was mentioned above, DC-SIGN is a transmembrane protein C-type calcium-dependent lectin that can capture antigens for processing and presentation. It is known that several viruses including HIV, Ebola, Junín (JUNV), hepatitis C, and Dengue, and important bacteria such as *Mycobacterium tuberculosis*, use DC-SIGN receptor to enhance their ability to infect host cells. In agreement with the results obtained for *L. acidophilus* NCFM (Konstantinov *et al.*, 2008), *L. plantarum* (Liu *et al.*,

2011) and *L. kefiri* JCM 5818 (Prado Acosta *et al.*, 2016), Martínez and colleagues (2012) described the interaction involving the DC-SIGN and the SLP presented on the bacterial cell-envelope of *L. acidophilus* ATCC 4356, an inhibitor of both viral and bacterial infection. The SLP of this strain inhibited JUNV invasion into 3T3 cells over-expressing DC-SIGN, in the early stages of viral infection. However, this inhibition does not involve the classic recognition of mannose by this C-type lectin as the SLP of this strain showed no evidence to be glycosylated. More recently, Gao and colleagues (2017) demonstrated that mouse DCs treated with SLP from *L. acidophilus* ATCC 4356 were less susceptible to H9N2 virus infection than untreated cells. Indeed, SLP-treated DCs were more activated and secreted a significantly higher amount of IL-10 than DCs infected with only H9N2 virus, which could aid in the control of the exacerbated inflammation caused by H9N2 infection.

The involvement of surface associated components of *L. brevis* CD2 in the antiviral activity against Herpes virus type-2 (HSV-2) was demonstrated (Mastromarino *et al.*, 2011). However, in this case, the resistance of the inhibitory activity to proteases and heating suggested that SLPs are not responsible for this effect.

Some lactobacilli's SLPs have shown activity against pathogenic bacteria. Pre-treatment of pathogen bacteria with SLPs from different *Lactobacillus* species reduced bacteria viability but also prevent infection by three enterobacteria, *Escherichia coli*, *Salmonella enterica* serovar *Typhi* and *Klebsiella pneumoniae*, and *Mycobacterium smegmatis*, a non-pathogenic model for *Mycobacterium* infection (Prado Acosta *et al.*, 2016). In other cases, such as *L. paracasei* subsp. *paracasei* M5-L, *L. rhamnosus* J10-L and *L. casei* Q8-L the involvement of SLPs in the inhibition of *Shigella sonnei* adhesion to HT-29 cells was suggested since the inhibitory activity of these lactobacilli decrease after removal of their SLP (Zhang *et al.*, 2010).

The SLP from *L. plantarum* and *L. acidophilus* CICC 6074 has been reported to protect intestinal epithelial cells injuries induced by enteropathogenic *Escherichia coli* (Liu *et al.*, 2011; Zhang *et al.*, 2017). The SLPs from different strains of kefir-isolated *L. kefiri* showed to inhibit the invasion of *Salmonella enterica* serovar *Enteritidis* to Caco-2 cells (Golowczyc *et al.*, 2007) as well as to antagonize the cytopathic effects of toxins from *Clostridium difficile in vitro* likely through a direct interaction SLP-toxins (Carasi *et al.*, 2012). On the other hand, the murein hydrolase activity shown by the SLP of *L. acidophilus* ATCC 4356 against the cell wall of, for example, *Salmonella enterica* serovar *Newport* (Prado Acosta *et al.*, 2008) and *E. coli* (Meng *et al.*, 2015), provides this strain with an additional mechanism against Gram-positive bacterial pathogens, such as *Staphylococcus aureus* and *Bacillus cereus* (Prado-Acosta *et al.*, 2010).

Biotechnological applications of *Lactobacillus* S-layer proteins

The fundamental knowledge explained in the previous sections about structure, genetics, chemistry, morphogenesis, and functions of SLPs represents a strong support for both biological and non-biological applications. The capacity of SLPs to reassembly into lattices, either in solution or at interfaces after being removed from bacteria opens up several biophysical applications not directly related with the biological functions of SLPs on bacteria (Pum *et al.*, 2013). The reassembly of SLPs occurs in solution, on mica and silicon substrates, on self-assembled monolayers, on polyelectrolyte layers, on lipid interfaces, and on liposomes and nanocapsules. As SLPs have certain affinity to biopolymers (in particular to secondary cell wall polymers), which are controlled although carbohydrate–protein interactions, reassembly experiments involving SLPs aim at engineering biomimetic surfaces. This way, cationic and anionic polyelectrolyte layers demonstrated to be suitable substrates for S-layer reassembly (Toca-Herrera *et al.*, 2005).

Although there are numerous reports about biotechnological applications of bacterial SLPs, the studies using *Lactobacillus* SLPs are limited.

Reassembly of *Lactobacillus* SLPs on liposomes

The reassembly of SLPs at the air/water interface of lipid surfaces, and on lipid films, opened a broad spectrum of applications in basic and applied membrane research (Schuster *et al.*, 2009). In this regard, SLPs attached to lipid bilayers provide them a much higher mechanical strength and life-time and a decreased tendency to be disrupted (Schuster *et* al., 2008). Since pI values of *Lactobacillus* SLPs are generally high (9.4–10.4), they can reassemble on positively charged particles surfaces, however, the reassembly on negatively charged lipid monolayer was reported for SLP from *L. acidophilus* ATCC 4356 (Smit *et al.*, 2001).

Artificial lipid vesicles such as liposomes are widely used for enhancing the delivery of different biologically active molecules into cells and tissues. SLPs from *L. brevis* ATCC 14869 and *L. kefiri* JCM 5818 adsorbed on the surface of positively charged to liposomes composed by dipalmitoylphosphatidylcholine (DPPC) or soybean lecithin, with the addition of similar ratios of cholesterol and stearylamine. Such strategy protected liposomes from adverse environmental conditions such as low pH, high temperatures or simulated gastro-intestinal conditions (Hollmann *et al.*, 2007, 2010). Similar results were recently reported by Wang and colleagues (2017b) who showed that novel positively charged liposomes composed by soybean lecithin, Eudragit1RL100 and cholesterol, and then coated by the SLPs from *L. helveticus* CGMCC1.1877 were more stable than naked liposomes. Moreover, *in vivo* experiments performed in mice indicated that these SLP-coated liposomes could adhere to the gastrointestinal tract, suggesting their perspective as oral vaccine vectors.

Besides, SAXS measurements revealed a marked change in the lattice constants for SlpA of *L. brevis* ATCC 8287 reassembled on cationic liposomes compared to those determined for SlpA on native CWF (Kontro *et al.*, 2014). These findings indicate that despite its stability, SlpA reassembles into a different structure on liposomes.

Lactobacillus SLPs expression and/or secretion signals in heterologous gene expression

Taking advantage of the high efficiency of the SLP promoters, the expression and/or secretion signals of *Lactobacillus* SLP genes have been used for some biotechnological applications such as the expression of intra or extracellular proteins in *Lactobacillus* and *Lactococcus* (Hynönen and Palva, 2013). Using the expression and secretion signals of SlpA from *L. brevis* ATCC 8287, high secretion levels of β-lactamase have been achieved even although differences between the recognition efficiency in distinct hosts were reported (Savijoki *et al.*, 1997). Similar performances were observed for the promoter region of the SLP gene of *L. acidophilus* ATCC 4356 (Hynönen and Palva, 2013). Furthermore, the addition of the signal peptide encoding sequence of *slpA* from *L. brevis* ATCC 8287 upstream of the 5′end of the human interferon alpha gene increased the secretion efficiency of the cytokine in *Lactococcus lactis* (Zhang *et al.*, 2010).

Heterologous expression of *Lactobacillus* SLPs and application to vaccine development

The display of heterologous proteins on the cell surface of lactic acid bacteria is a very interesting and promising research area due to its high applicability to the development of a variety of biotechnological tools, including live vaccine delivery systems, diagnostic devices, peptide library screening and whole-cell biocatalysts (Hu *et al.*, 2011). Regarding this, and considering the structural features of SLPs, the construction of fusion proteins between the desired functional molecules and SLPs has been reported. Some years ago, the green fluorescent protein (GFP) or β-galactosidase (Gal) were fused to the C-terminal region of SlpB (LcsB) from *L. crispatus* K2-4-3, and the fused proteins were successfully produced in *E. coli* (Hu *et al.*, 2011). After mixing them with the non-genetically modified lactic acid bacteria cells, the fused GFP–LcsB and Gal–LcsB were functionally associated with the cell surface of various lactic acid bacteria tested. Furthermore, when the fused DNA fragment gfp:lcsB was inserted into *Lactococcus lactis* expression vector pSec:Leiss:Nuc, the GFP was successfully expressed onto the bacterial surface with the aid of the LcsB anchor (Hu *et al.*, 2011).

Delivery of antigens to mucosal surfaces by lactic acid bacteria is considered as a safe alternative to live attenuated pathogens because of their food grade status. As surface components that frequently mediate specific interactions with host cells, several experiments focusing on the use of SLP-bearing microorganisms or isolated SLPs as antigen/hapten carriers, as adjuvants, or as part of vaccination vesicles have been conducted by different research groups (Sleytr *et al.*, 2014). This interest mainly relies on the fact that the antigen display using a SLP as a carrier results in the simultaneous expression of foreign peptides as multiple copies regularly arranged, and also that both the *Lactobacillus* cells as well as SLPs have intrinsic adjuvant properties (Seegers, 2002; Beganović *et al.*, 2011; Hynonen *et al.*, 2013). In 2002, Åvall-Jääskeläinen and colleagues reported the development of an inducible expression system for production, secretion and surface display of the 11-amino acid-long epitope c-myc from the human c-myc proto-oncogene inserted at different sites of the SlpA from *L. brevis* ATCC 8287 (Åvall-Jääskeläinen *et al.*, 2002). More recently, Kajikawa *et al.* (2015) constructed genetically modified *L. acidophilus* strains expressing the membrane proximal external region (MPER) from human immunodeficiency virus type 1 (HIV-1) within the context of the SlpA. Intragastric immunization of mice with the recombinants, induced a specific antibody response against both the MPER and SlpA in serum and mucosal secretions.

In a different approach, a fusion protein of SlpB from *L. buchneri* CD034 and the most potent peanut allergen Ara h-2 derived peptide, resulted to be highly immunogenic (Anzengruber *et al.*, 2017), however more than one allergen-peptide will be needed to induce a broader protection of patients allergic to Ara h2.

Bioremediation of heavy metals

SLPs play a key role in the interactions between the bacteria and metal substrates, although most of the reported examples have been developed in non-lactic acid bacteria (Pollmann *et al.*, 2005; Patel *et al.*, 2010). The special capabilities of S-layers offer an interesting alternative in the bioremediation processes of heavy metals, which can be applied to decontaminate waste waters, to recover precious metals from wastes of the electronic industry, and even to produce metal nanoclusters (Sleytr *et al.*, 2014). SLPs may also be used for the formation of inorganic nanocrystal superlattices (e.g. CdS, Au, Ni, Pt, or Pd) as required for molecular electronics and non-linear optics (Sleytr *et al.*, 1997).

Among lactic acid bacteria, different species of lactobacilli have been successfully used to remove Cu, Pb, Cd, Zn and Ni (*L. buchneri*, *L. brevis*, *L. kefiri*, *L. rhamnosus* GG) (Halttunen *et al.*, 2007; Gerbino *et al.*, 2011; Schut *et al.*, 2011). In most cases, metal/bacteria interaction was shown to be a fast metabolism-independent surface process, suggesting that binding occurred passively to the surface of bacteria (biosorption) rather than by accumulation inside the cell (bioaccumulation) (Halttunen *et al.*, 2007). *L. kefiri* strains bearing SLPs can successfully remove Pb^{2+}, Cd^{2+}, Zn^{2+} and Ni^{2+} from the growth medium either in growing and in non-growing conditions (Gerbino *et al.*, 2011, 2015b).

The interaction of metal ions with SLPs involves certain functional groups from naturally occurring amino acid residues, posttranslational modifications of the proteins or modifications introduced through chemical or genetic engineering of the proteins (Fahmy *et al.*, 2006) and these interactions may be used as an initial step in cluster formation. In this regard, the SLPs from *L. kefiri* interact with metals mainly through coordination with the side chain carboxyl groups of Asp and Glu residues and additional coordination involving NH groups from the peptide backbone (Gerbino *et al.*, 2011).

In spite of the scarce information about the capacity of SLPs from *Lactobacillus* to bind metals, the existing background supports novel applications involving the development of sensors or the removing of trace amounts of heavy metals from food related products. Considering the huge existing background on SLPs, although not necessarily on those of lactic acid bacteria, such information could be applied for highly innovative applications.

Conclusions and future trends

The S-layer proteins are one of the most abundant biopolymers in our planet, as well as the simplest biological membranes developed during evolution. Years and years of basic and applied research have resulted in a deep understanding of the structure, genetics, and functions of these special prokaryotic proteins. Considering that SLPs are the outermost surface components of S-layer-bearing bacteria, a particular interest in studying different structural and functional characteristics of *Lactobacillus* SLPs has been emerged, relying mostly on the GRAS status of these lactic acid bacteria as well as the probiotic properties of some lactobacilli strains. In this context, the S-layer-carrying lactobacilli will gain relevance for health-related applications such as live oral vaccines.

Moreover, the special ability of SLPs to reassemble, either in solution or at different surfaces, have led to an astonishing spectrum of applications in nano(bio)technology, synthetic biology and biomimetics. In this sense, *Lactobacillus* SLPs could be excellent candidates as carriers of antigens or other medically important molecules.

To resume, the structural and functional characterization of *Lactobacillus* SLPs is not only essential to get deep insight into the beneficial and technological properties of these microorganisms, but also encourage us to evaluate their potential in the development of new nano(bio)technological tools in the life and non-life sciences.

References

Ahn, J.S., Chandramohan, L., Liou, L.E., and Bayles, K.W. (2006). Characterization of CidR-mediated regulation in *Bacillus anthracis* reveals a previously undetected role of S-layer proteins as murein hydrolases. Mol. Microbiol. *62*, 1158–1169.

Antikainen, J., Anton, L., Sillanpää, J., and Korhonen, T.K. (2002). Domains in the S-layer protein CbsA of *Lactobacillus crispatus* involved in adherence to collagens, laminin and lipoteichoic acids and in self-assembly. Mol. Microbiol. *46*, 381–394.

Anzengruber, J., Pabst, M., Neumann, L., Sekot, G., Heinl, S., Grabherr, R., Altmann, F., Messner, P., and Schäffer, C. (2014). Protein O-glucosylation in *Lactobacillus buchneri*. Glycoconj. J. *31*, 117–131. https://doi.org/10.1007/s10719-013-9505-7

Ashida, N., Yanagihara, S., Shinoda, T., and Yamamoto, N. (2011). Characterization of adhesive molecule with affinity to Caco-2 cells in *Lactobacillus acidophilus* by proteome analysis. J. Biosci. Bioeng. *112*, 333–337. https://doi.org/10.1016/j.jbiosc.2011.06.001

Åvall-Jääskeläinen, S., Kylä-Nikkilä, K., Kahala, M., Miikkulainen-Lahti, T., and Palva, A. (2002). Surface display of foreign epitopes on the *Lactobacillus brevis* S-layer. Appl. Environ. Microbiol. *68*, 5943–5951.

Åvall-Jääskeläinen, S., and Palva, A. (2005). *Lactobacillus* surface layers and their applications. FEMS Microbiol. Rev. *29*, 511–529.

Åvall-Jääskeläinen, S., Hynönen, U., Ilk, N., Pum, D., Sleytr, U.B., and Palva, A. (2008). Identification and characterization of domains responsible for self-assembly and cell wall binding of the surface layer protein of *Lactobacillus brevis* ATCC 8287. BMC Microbiol. *8*, 165. https://doi.org/10.1186/1471-2180-8-165

Beganović, J., Frece, J., Kos, B., Leboš Pavunc, A., Habjanič, K., and Sušković, J. (2011). Functionality of the S-layer protein from the probiotic strain *Lactobacillus helveticus* M92. Antonie Van Leeuwenhoek *100*, 43–53. https://doi.org/10.1007/s10482-011-9563-4

Boot, H.J., and Pouwels, P.H. (1996). Expression, secretion and antigenic variation of bacterial S-layer proteins. Mol. Microbiol. *21*, 1117–1123.

Boot, H.J., Kolen, C.P., and Pouwels, P.H. (1996). Interchange of the active and silent S-layer protein genes of *Lactobacillus acidophilus* by inversion of the chromosomal slp segment. Mol. Microbiol. *21*, 799–809.

Boot, H.J., Kolen, C.P., Andreadaki, F.J., Leer, R.J., and Pouwels, P.H. (1996). The *Lactobacillus acidophilus* S-layer protein gene expression site comprises two consensus promoter sequences, one of which directs transcription of stable mRNA. J. Bacteriol. *178*, 5388–5394.

Buck, B.L., Altermann, E., Svingerud, T., and Klaenhammer, T.R. (2005). Functional analysis of putative adhesion factors in *Lactobacillus acidophilus* NCFM. Appl. Environ. Microbiol. *71*, 8344–8351.

Calabi, E., Ward, S., Wren, B., Paxton, T., Panico, M., Morris, H., Dell, A., Dougan, G., and Fairweather, N. (2001). Molecular characterization of the surface layer proteins from *Clostridium difficile*. Mol. Microbiol. *40*, 1187–1199.

Carasi, P., Ambrosis, N.M., De Antoni, G.L., Bressollier, P., Urdaci, M.C., and Serradell, M.d.e.L. (2014). Adhesion properties of potentially probiotic *Lactobacillus kefiri* to gastrointestinal mucus. J. Dairy Res. *81*, 16–23. https://doi.org/10.1017/S0022029913000526

Cavallero, G.J., Malamud, M., Casabuono, A.C., Serradell, M.L.Á., and Couto, A.S. (2017). A glycoproteomic approach reveals that the S-layer glycoprotein of *Lactobacillus kefiri* CIDCA 83111 is O- and N-glycosylated. J. Proteomics *162*, 20–29.

Cerquetti, M., Molinari, A., Sebastianelli, A., Diociaiuti, M., Petruzzelli, R., Capo, C., and Mastrantonio, P. (2000). Characterization of surface layer proteins from different *Clostridium difficile* clinical isolates. Microb. Pathog. *28*, 363–372. https://doi.org/10.1006/mpat.2000.0356

Chen, X., Chen, Y., Li, X., Chen, N., and Fang, W. (2009). Characterization of surface layer proteins in *Lactobacillus crispatus* isolate ZJ001. J. Microbiol. Biotechnol. *19*, 1176–1183.

Chen, X., Xu, J., Shuai, J., Chen, J., Zhang, Z., and Fang, W. (2007). The S-layer proteins of *Lactobacillus crispatus* strain ZJ001 is responsible for competitive exclusion against *Escherichia coli* O157:H7 and *Salmonella* typhimurium. Int. J. Food Microbiol. *115*, 307–312.

Claus, H., Akça, E., Debaerdemaeker, T., Evrard, C., Declercq, J.P., Harris, J.R., Schlott, B., and König, H. (2005). Molecular organization of selected prokaryotic S-layer proteins. Can. J. Microbiol. *51*, 731–743.

Collins, L.E., Lynch, M., Marszalowska, I., Kristek, M., Rochfort, K., O'Connell, M., Windle, H., Kelleher, D., and Loscher, C.E. (2014). Surface layer proteins isolated from *Clostridium difficile* induce clearance responses in macrophages. Microbes Infect. *16*, 391–400. https://doi.org/10.1016/j.micinf.2014.02.001

de la Fuente-Núñez, C., Mertens, J., Smit, J., and Hancock, R.E. (2012). The bacterial surface layer provides protection against antimicrobial peptides. Appl. Environ. Microbiol. *78*, 5452–5456. https://doi.org/10.1128/AEM.01493-12

Deepika, G., and Charalampopoulos, D. (2010). Surface and adhesion properties of lactobacilli. Adv. Appl. Microbiol. *70*, 127–152. https://doi.org/10.1016/S0065-2164(10)70004-6

Dohm, N., Petri, A., Schlander, M., Schlott, B., König, H., and Claus, H. (2011). Molecular and biochemical properties of the S-layer protein from the wine bacterium *Lactobacillus hilgardii* B706. Arch. Microbiol. *193*, 251–261. https://doi.org/10.1007/s00203-010-0670-9

Eslami, N., Kermanshahi, R.K., and Erfan, M. (2013). Studying the Stability of S-Layer Protein of *Lactobacillus acidophilus* ATCC 4356 in simulated gastrointestinal fluids using SDS-PAGE and circular dichroism. Iran. J. Pharm. Res. *12*, 47–56.

Fagan, R.P., Albesa-Jové, D., Qazi, O., Svergun, D.I., Brown, K.A., and Fairweather, N.F. (2009). Structural insights into the molecular organization of the S-layer from *Clostridium difficile*. Mol. Microbiol. *71*, 1308–1322. https://doi.org/10.1111/j.1365-2958.2009.06603.x

Fahmy, K., Merroun, M., Pollmann, K., Raff, J., Savchuk, O., Hennig, C., and Selenska-Pobell, S. (2006). Secondary structure and Pd(II) coordination in S-layer proteins from *Bacillus sphaericus* studied by infrared and X-ray absorption spectroscopy. Biophys. J. *91*, 996–1007.

Fijan, S. (2014). Microorganisms with claimed probiotic properties: an overview of recent literature. Int. J. Environ. Res. Public Health *11*, 4745–4767. https://doi.org/10.3390/ijerph110504745

Fouet, A. (2009). The surface *of Bacillus anthracis*. Mol. Aspects Med. *30*, 374–385. https://doi.org/10.1016/j.mam.2009.07.001

Frece, J., Kos, B., Svetec, I.K., Zgaga, Z., Mrsa, V., and Susković, J. (2005). Importance of S-layer proteins in probiotic activity of *Lactobacillus acidophilus* M92. J. Appl. Microbiol. *98*, 285–292.

Gao, K., Wang, C., Liu, L., Dou, X., Liu, J., Yuan, L., Zhang, W., and Wang, H. (2017). Immunomodulation and signalling mechanism of *Lactobacillus rhamnosus* GG and its components on porcine intestinal epithelial cells stimulated by lipopolysaccharide. J. Microbiol. Immunol. Infect. *50*, 700–713. https://doi.org/10.1016/j.jmii.2015.05.002

Garrote, G.L., Delfederico, L., Bibiloni, R., Abraham, A.G., Pérez, P.F., Semorile, L., and De Antoni, G.L. (2004). Lactobacilli isolated from kefir grains: evidence of the presence of S-layer proteins. J. Dairy Res. *71*, 222–230.

Geijtenbeek, T.B., and Gringhuis, S.I. (2016). C-type lectin receptors in the control of T helper cell differentiation. Nat. Rev. Immunol. *16*, 433–448. https://doi.org/10.1038/nri.2016.55

Gerbino, E., Carasi, P., Araujo-Andrade, C., Tymczyszyn, E.E., and Gómez-Zavaglia, A. (2015). Role of S-layer proteins in the biosorption capacity of lead by *Lactobacillus kefir*. World J. Microbiol. Biotechnol. *31*, 583–592. https://doi.org/10.1007/s11274-015-1812-7

Gerbino, E., Carasi, P., Mobili, P., Serradell, M.A., and Gómez-Zavaglia, A. (2015). Role of S-layer proteins in bacteria. World J. Microbiol. Biotechnol. *31*, 1877–1887. https://doi.org/10.1007/s11274-015-1952-9

Gerbino, E., Mobili, P., Tymczyszyn, E., Fausto, R., and Gómez-Zavaglia, A. (2011). FTIR spectroscopy structural analysis of the interaction between *Lactobacillus kefir* S-layers and metal ions. J. Mol. Struct. *987*, 186–192. https://doi.org/10.1016/j.molstruc.2010.12.012

Gilmour, R., Messner, P., Guffanti, A.A., Kent, R., Scheberl, A., Kendrick, N., and Krulwich, T.A. (2000). Two-dimensional gel electrophoresis analyses of pH-dependent protein expression in facultatively alkaliphilic *Bacillus pseudofirmus* OF4 lead to characterization of an S-layer protein with a role in alkaliphily. J. Bacteriol. *182*, 5969–5981.

Goh, Y.J., Azcárate-Peril, M.A., O'Flaherty, S., Durmaz, E., Valence, F., Jardin, J., Lortal, S., and Klaenhammer, T.R. (2009). Development and application of a upp-based counterselective gene replacement system for the study of the S-layer protein SlpX of *Lactobacillus acidophilus* NCFM. Appl. Environ. Microbiol. *75*, 3093–3105. https://doi.org/10.1128/AEM.02502-08

Golowczyc, M.A., Mobili, P., Garrote, G.L., Abraham, A.G., and De Antoni, G.L. (2007). Protective action of *Lactobacillus kefir* carrying S-layer protein against *Salmonella enterica* serovar *Enteritidis*. Int. J. Food Microbiol. *118*, 264–273. https://doi.org/10.1016/j.ijfoodmicro.2007.07.042

Golowczyc, M.A., Mobili, P., Garrote, G.L., de Los Angeles Serradell, M., Abraham, A.G., and De Antoni, G.L. (2009). Interaction between *Lactobacillus kefir* and *Saccharomyces lipolytica* isolated from kefir grains: evidence for lectin-like activity of bacterial surface proteins. J. Dairy Res. *76*, 111–116. https://doi.org/10.1017/S0022029908003749

Grosu-Tudor, S.S., Brown, L., Hebert, E.M., Brezeanu, A., Brinzan, A., Fadda, S., Mozzi, F., and Zamfir, M. (2016). S-layer production by *Lactobacillus acidophilus* IBB 801 under environmental stress conditions. Appl. Microbiol. Biotechnol. *100*, 4573–4583. https://doi.org/10.1007/s00253-016-7355-5

Habil, N., Al-Murrani, W., Beal, J., and Foey, A.D. (2011). Probiotic bacterial strains differentially modulate macrophage cytokine production in a strain-dependent and cell subset-specific manner. Benef. Microbes *2*, 283–293. https://doi.org/10.3920/BM2011.0027

Hagen, K.E., Guan, L.L., Tannock, G.W., Korver, D.R., and Allison, G.E. (2005). Detection, characterization, and in vitro and in vivo expression of genes encoding S-proteins in *Lactobacillus gallinarum* strains isolated from chicken crops. Appl. Environ. Microbiol. *71*, 6633–6643.

Halttunen, T., Salminen, S., and Tahvonen, R. (2007). Rapid removal of lead and cadmium from water by specific lactic acid bacteria. Int. J. Food Microbiol. *114*, 30–35.

Heinl, S., Wibberg, D., Eikmeyer, F., Szczepanowski, R., Blom, J., Linke, B., Goesmann, A., Grabherr, R., Schwab, H., Pühler, A., *et al.* (2012). Insights into the completely annotated genome of *Lactobacillus buchneri* CD034, a strain isolated from stable grass silage. J. Biotechnol. *161*, 153–166. https://doi.org/10.1016/j.jbiotec.2012.03.007

Hollmann, A., Delfederico, L., De Antoni, G., Semorile, L., and Disalvo, E.A. (2010). Interaction of bacterial surface layer proteins with lipid membranes: synergysm between surface charge density and chain packing. Colloids Surf. B Biointerfaces 79, 191–197. https://doi.org/10.1016/j.colsurfb.2010.03.046

Hollmann, A., Delfederico, L., Glikmann, G., De Antoni, G., Semorile, L., and Disalvo, E.A. (2007). Characterization of liposomes coated with S-layer proteins from lactobacilli. Biochim. Biophys. Acta *1768*, 393–400.

Houwink, A.L. (1953). A macromolecular mono-layer in the cell wall of *Spirillum* spec. Biochim. Biophys. Acta *10*, 360–366.

Howard, L., and Tipper, D.J. (1973). A polypeptide bacteriophage receptor: modified cell wall protein subunits in bacteriophage-resistant mutants of *Bacillus sphaericus* strain P-1. J. Bacteriol. *113*, 1491–1504.

Hu, S., Kong, J., Sun, Z., Han, L., Kong, W., and Yang, P. (2011). Heterologous protein display on the cell surface of lactic acid bacteria mediated by the s-layer protein. Microb. Cell Fact. *10*, 86. https://doi.org/10.1186/1475-2859-10-86

Hymes, J.P., Johnson, B.R., Barrangou, R., and Klaenhammer, T.R. (2016). Functional analysis of an S-layer-associated fibronectin-binding protein in *Lactobacillus acidophilus* NCFM. Appl. Environ. Microbiol. *82*, 2676–2685. https://doi.org/10.1128/AEM.00024-16

Hynönen, U., and Palva, A. (2013). *Lactobacillus* surface layer proteins: structure, function and applications. Appl. Microbiol. Biotechnol. *97*, 5225–5243. https://doi.org/10.1007/s00253-013-4962-2

Hynönen, U., Avall-Jääskeläinen, S., and Palva, A. (2010). Characterization and separate activities of the two promoters of the *Lactobacillus brevis* S-layer protein gene. Appl. Microbiol. Biotechnol. *87*, 657–668. https://doi.org/10.1007/s00253-010-2500-z

Hynönen, U., Kant, R., Lähteinen, T., Pietilä, T.E., Beganović, J., Smidt, H., Uroić, K., Avall-Jääskeläinen, S., and Palva, A. (2014). Functional characterization of probiotic surface layer protein-carrying *Lactobacillus amylovorus* strains. BMC Microbiol. *14*, 199. https://doi.org/10.1186/1471-2180-14-199

Hynönen, U., Westerlund-Wikström, B., Palva, A., and Korhonen, T.K. (2002). Identification by flagellum display of an epithelial cell- and fibronectin-binding function in the SlpA surface protein of *Lactobacillus brevis*. J. Bacteriol. *184*, 3360–3367.

Ilk, N., Schumi, C.T., Bohle, B., Egelseer, E.M., and Sleytr, U.B. (2011). Expression of an endotoxin-free S-layer/allergen fusion protein in gram-positive *Bacillus subtilis* 1012 for the potential application as vaccines for immunotherapy of atopic allergy. Microb. Cell Fact. *10*, 6. https://doi.org/10.1186/1475-2859-10-6

Jääskeläinen, P., Engelhardt, P., Hynönen, U., Torkkeli, M., Palva, A., and Serimaa, R. (2010). Small angle X-ray scattering and transmission electron microscopy study of the *Lactobacillus brevis* S-layer protein. J. Phys.: Conf. Ser. *247*, 012017. http://doi.org/10.1088/1742-6596/247/1/012017

Jaenicke, R., Welsch, R., Sára, M., and Sleytr, U.B. (1985). Stability and self-assembly of the S-layer protein of the cell wall of *Bacillus stearothermophilus*. Biol. Chem. Hoppe-Seyler *366*, 663–670.

Jakava-Viljanen, M., Avall-Jääskeläinen, S., Messner, P., Sleytr, U.B., and Palva, A. (2002). Isolation of three new surface layer protein genes (slp) from *Lactobacillus brevis* ATCC 14869 and characterization of the change in their expression under aerated and anaerobic conditions. J. Bacteriol. *184*, 6786–6795.

Johnson, B.R., and Klaenhammer, T.R. (2016). AcmB Is an S-layer-associated β-N-acetylglucosaminidase and functional autolysin in *Lactobacillus acidophilus* NCFM. Appl. Environ. Microbiol. *82*, 5687–5697. https://doi.org/10.1128/AEM.02025-16

Johnson, B.R., Hymes, J., Sanozky-Dawes, R., Henriksen, E.D., Barrangou, R., and Klaenhammer, T.R. (2016). Conserved S-layer-associated proteins revealed by exoproteomic survey of S-layer-forming Lactobacilli. Appl. Environ. Microbiol. *82*, 134–145. https://doi.org/10.1128/AEM.01968-15

Johnson, B., Selle, K., O'Flaherty, S., Goh, Y.J., and Klaenhammer, T. (2013). Identification of extracellular surface-layer associated proteins in *Lactobacillus acidophilus* NCFM. Microbiology *159*, 2269–2282. https://doi.org/10.1099/mic.0.070755-0

Kahala, M., Savijoki, K., and Palva, A. (1997). In vivo expression of the *Lactobacillus brevis* S-layer gene. J. Bacteriol. *179*, 284–286.

Kajikawa, A., Zhang, L., LaVoy, A., Bumgardner, S., Klaenhammer, T.R., and Dean, G.A. (2015). Mucosal immunogenicity of genetically modified *Lactobacillus acidophilus* expressing an HIV-1 epitope within the surface layer protein. PLOS ONE *10*, e0141713. https://doi.org/10.1371/journal.pone.0141713

Khaleghi, M., Kermanshahi, R.K., Yaghoobi, M.M., Zarkesh-Esfahani, S.H., and Baghizadeh, A. (2010). Assessment of bile salt effects on s-layer production, slp gene expression and some physicochemical properties of *Lactobacillus acidophilus* ATCC 4356. J. Microbiol. Biotechnol. *20*, 749–756.

Khaleghi, M., Kasra Kermanshahi, R., and Zarkesh-Esfahani, S.H. (2011). Effects of penicillin G on morphology and certain physiological parameters of *Lactobacillus acidophilus* ATCC 4356. J. Microbiol. Biotechnol. *21*, 822–829.

Khang, Y.H., Park, H.Y., Jeong, Y.S., Kim, J.A., and Kim, Y.H. (2009). Recombinant S-layer proteins of *Lactobacillus brevis* mediating antibody adhesion to calf intestine alleviated neonatal diarrhea syndrome. J. Microbiol. Biotechnol. *19*, 511–519.

Konstantinov, S.R., Smidt, H., de Vos, W.M., Bruijns, S.C., Singh, S.K., Valence, F., Molle, D., Lortal, S., Altermann, E., Klaenhammer, T.R., *et al.* (2008). S layer protein A of *Lactobacillus acidophilus* NCFM regulates immature dendritic cell and T cell functions. Proc. Natl. Acad. Sci. U.S.A. *105*, 19474–19479. https://doi.org/10.1073/pnas.0810305105

Kontro, I., Wiedmer, S.K., Hynönen, U., Penttilä, P.A., Palva, A., and Serimaa, R. (2014). The structure of *Lactobacillus brevis* surface layer reassembled on liposomes differs from native structure as revealed by SAXS. Biochim. Biophys. Acta *1838*, 2099–2104. https://doi.org/10.1016/j.bbamem.2014.04.022

Kos, B., Susković, J., Vuković, S., Simpraga, M., Frece, J., and Matosić, S. (2003). Adhesion and aggregation ability of probiotic strain *Lactobacillus acidophilus* M92. J. Appl. Microbiol. *94*, 981–987.

Kotiranta, A.K., Ito, H., Haapasalo, M.P., and Lounatmaa, K. (1999). Radiation sensitivity of *Bacillus cereus* with and without a crystalline surface protein layer. FEMS Microbiol. Lett. *179*, 275–280.

Lebeer, S., Vanderleyden, J., and De Keersmaecker, S.C. (2008). Genes and molecules of lactobacilli supporting probiotic action. Microbiol. Mol. Biol. Rev. *72*, 728–764, Table of Contents. https://doi.org/10.1128/MMBR.00017-08

de Leeuw, E., Li, X., and Lu, W. (2006). Binding characteristics of the *Lactobacillus brevis* ATCC 8287 surface layer to extracellular matrix proteins. FEMS Microbiol. Lett. *260*, 210–215.

Li, P., Yu, Q., Ye, X., Wang, Z., and Yang, Q. (2011). *Lactobacillus* S-layer protein inhibition of *Salmonella*-induced reorganization of the cytoskeleton and activation of MAPK signalling pathways in Caco-2 cells. Microbiology *157*, 2639–2646. https://doi.org/10.1099/mic.0.049148-0

Lighezan, L., Georgieva, R.,and Neagu, A. (2012). A study of the thermal denaturation of the S-layer protein from *Lactobacillus salivarius*. Physica Scripta, *86*, 035801.

Lighezan, L., Georgieva, R., and Neagu, A. (2016). The secondary structure and the thermal unfolding parameters of the S-layer protein from *Lactobacillus salivarius*. Eur. Biophys. J. *45*, 491–509. https://doi.org/10.1007/s00249-016-1117-2

Lightfoot, Y.L., Selle, K., Yang, T., Goh, Y.J., Sahay, B., Zadeh, M., Owen, J.L., Colliou, N., Li, E., Johannssen, T., *et al.* (2015). SIGNR3-dependent immune regulation by *Lactobacillus acidophilus* surface layer protein A in colitis. EMBO J. *34*, 881–895. https://doi.org/10.15252/embj.201490296

Liu, Z., Ma, Y., Shen, T., Chen, H., Zhou, Y., Zhang, P., Zhang, M., Chu, Z., and Qin, H. (2011). Identification of DC-SIGN as the receptor during the interaction of *Lactobacillus plantarum* CGMCC 1258 and dendritic cells. World J. Microbiol. Biotechnol. *27*, 603–611. https://doi.org/10.1007/s11274-010-0495-3

Lortal, S., Van Heijenoort, J., Gruber, K., and Sleytr, U.B. (1992). S-layer of *Lactobacillus helveticus* ATCC 12046: isolation, chemical characterization and re-formation after extraction with lithium chloride. Microbiology *138*, 611–618.

MacKenzie, D.A., Jeffers, F., Parker, M.L., Vibert-Vallet, A., Bongaerts, R.J., Roos, S., Walter, J., and Juge, N. (2010). Strain-specific diversity of mucus-binding proteins in the adhesion and aggregation properties of *Lactobacillus reuteri*. Microbiology *156*, 3368–3378. https://doi.org/10.1099/mic.0.043265-0

Malamud, M., Carasi, P., Bronsoms, S., Trejo, S.A., and Serradell, M.L. (2017). *Lactobacillus kefiri* shows inter-strain variations in the amino acid sequence of the S-layer proteins. Antonie Van Leeuwenhoek *110*, 515–530. https://doi.org/10.1007/s10482-016-0820-4

Malamud, M., Carasi, P., Freire, T., and Serradell, M.L.A. (2018). S-layer glycoprotein from *Lactobacillus kefiri* CIDCA 8348 enhances macrophages response to LPS in a Ca+2-dependent manner. Biochem. Biophys. Res. Commun. *495*, 1227–1232.

Martínez, M.G., Prado Acosta, M., Candurra, N.A., and Ruzal, S.M. (2012). S-layer proteins of *Lactobacillus acidophilus* inhibits JUNV infection. Biochem. Biophys. Res. Commun. *422*, 590–595. https://doi.org/10.1016/j.bbrc.2012.05.031

Mastromarino, P., Cacciotti, F., Masci, A., and Mosca, L. (2011). Antiviral activity of *Lactobacillus brevis* towards herpes simplex virus type 2: role of cell wall associated components. Anaerobe *17*, 334–336. https://doi.org/10.1016/j.anaerobe.2011.04.022.

Masuda, K., and Kawata, T. (1985). Reassembly of a regularly arranged protein in the cell wall of *Lactobacillus buchneri* and its reattachment to cell walls: chemical modification studies. Microbiol. Immunol. *29*, 927–938.

Mayer, S., Raulf, M.K., and Lepenies, B. (2017). C-type lectins: their network and roles in pathogen recognition and immunity. Histochem. Cell Biol. *147*, 223–237. https://doi.org/10.1007/s00418-016-1523-7

McCarren, J., Heuser, J., Roth, R., Yamada, N., Martone, M., and Brahamsha, B. (2005). Inactivation of swmA results in the loss of an outer cell layer in a swimming *Synechococcus* strain. J. Bacteriol. *187*, 224–230.

Meng, J., Zhu, X., Gao, S.M., Zhang, Q.X., Sun, Z., and Lu, R.R. (2014). Characterization of surface layer proteins and its role in probiotic properties of three *Lactobacillus* strains. Int. J. Biol. Macromol. 65, 110–114. https://doi.org/10.1016/j.ijbiomac.2014.01.024

Mercenier, A., Pavan, S., and Pot, B. (2003). Probiotics as biotherapeutic agents: present knowledge and future prospects. Curr. Pharm. Des. 9, 175–191.

Messner, P., Steiner, K., Zarschler, K., and Schäffer, C. (2008). S-layer nanoglycobiology of bacteria. Carbohydr. Res. *343*, 1934–1951. https://doi.org/10.1016/j.carres.2007.12.025

Mignot, T., Couture-Tosi, E., Mesnage, S., Mock, M., and Fouet, A. (2004). In vivo *Bacillus anthracis* gene expression requires PagR as an intermediate effector of the AtxA signalling cascade. Int. J. Med. Microbiol. *293*, 619–624.

Mobarak Qamsari, E., Kasra Kermanshahi, R., Erfan, M., Ghadam, P., Sardari, S., and Eslami, N. (2017). Characteristics of surface layer proteins from two new and native strains of *Lactobacillus brevis*. Int. J. Biol. Macromol. *95*, 1004–1010.

Mobili, P., Serradell, M.A., Trejo, S.A., Avilés Puigvert, F.X., Abraham, A.G., and De Antoni, G.L. (2009a). Heterogeneity of S-layer proteins from aggregating and non-aggregaing *Lactobacillus kefir* strains. Antonie Van Leeuwenhoek 95, 363–372. doi: 10.1007/s10482-009-9322-y

Mobili, P., Londero, A., Maria, T.M.R., Eusébio, M.E.S., De Antoni, G.L., Fausto, R., and Gómez-Zavaglia, A. (2009b). Characterization of S-layer proteins of *Lactobacillus* by FTIR spectroscopy and differential scanning calorimetry. Vibr. Spectrosc. *50*, 68–77. https://doi.org/10.1016/j.vibspec.2008.07.016

Möschl, A., Schäffer, C., Sleytr, U.B., Messner, P., Christian, R., and Schulz, G. (1993). Characterization of the S-layer glycoproteins of two lactobacilli. In Advances in Bacterial Paracrystalline Surface Layers. NATO ASI Series (Series A: Life Sciences), Beveridge, T.J., and Koval, S.F. eds. (Springer, Boston), pp. 281–284. https://doi.org/10.1007/978-1-4757-9032-0_27

Mozes, N., and Lortal, S. (1995). X-ray photoelectron spectroscopy and biochemical analysis of the surface of Lactobacillus helveticus ATCC 12046. Microbiology *141*, 11–19.

Narita, J., Ishida, S., Okano, K., Kimura, S., Fukuda, H., and Kondo, A. (2006). Improvement of protein production in lactic acid bacteria using 5'-untranslated leader sequence of slpA from *Lactobacillus acidophilus*. Improvement in protein production using UTLS. Appl. Microbiol. Biotechnol. 73, 366–373. https://doi.org/10.1007/s00253-006-0477-4

Palomino, M.M., Waehner, P.M., Fina Martin, J., Ojeda, P., Malone, L., Sánchez Rivas, C., Prado Acosta, M., Allievi, M.C., and Ruzal, S.M. (2016). Influence of osmotic stress on the profile and gene expression of surface layer proteins in *Lactobacillus acidophilus* ATCC 4356. Appl. Microbiol. Biotechnol. *100*, 8475–8484. https://doi.org/10.1007/s00253-016-7698-y

Patel, J., Zhang, Q., McKay, R.M., Vincent, R., and Xu, Z. (2010). Genetic engineering of *Caulobacter crescentus* for removal of cadmium from water. Appl. Biochem.

Biotechnol. *160*, 232–243. https://doi.org/10.1007/s12010-009-8540-0

Pavkov-Keller, T., Howorka, S., and Keller, W. (2011). The structure of bacterial S-layer proteins. Prog. Mol. Biol. Transl. Sci. *103*, 73–130. https://doi.org/10.1016/B978-0-12-415906-8.00004-2

Podlesny, M., Jarocki, P., Komon, E., Glibowska, A., and Targonski, Z. (2011). LC-MS/MS analysis of surface layer proteins as a useful method for the identification of lactobacilli from the *Lactobacillus acidophilus* group. J. Microbiol. Biotechnol. *21*, 421–429.

Pollmann, K., Raff, J., Schnorpfeil, M., Radeva, G., and Selenska-Pobell, S. (2005). Novel surface layer protein genes in *Bacillus sphaericus* associated with unusual insertion elements. Microbiology *151*, 2961–2973. https://doi.org/10.1099/mic.0.28201-0

Prado Acosta, M., Palomino, M.M., Allievi, M.C., Sánchez Rivas, C., and Ruzal, S.M. (2008). Murein hydrolase activity in the surface layer of *Lactobacillus acidophilus* ATCC 4356. Appl. Environ. Microbiol. *74*, 7824–7827. https://doi.org/10.1128/AEM.01712-08

Prado Acosta, M., Ruzal, S.M., and Cordo, S.M. (2016). S-layer proteins from *Lactobacillus* sp. inhibit bacterial infection by blockage of DC-SIGN cell receptor. Int. J. Biol. Macromol. *92*, 998–1005.

Pum, D., Toca-Herrera, J.L., and Sleytr, U.B. (2013). S-layer protein self-assembly. Int. J. Mol. Sci. *14*, 2484–2501. https://doi.org/10.3390/ijms14022484

Ristl, R., Steiner, K., Zarschler, K., Zayni, S., Messner, P., and Schäffer, C. (2011). The s-layer glycome-adding to the sugar coat of bacteria. Int. J. Microbiol. *2011*, 127870. https://doi.org/10.1155/2011/127870

Rojas, M., Ascencio, F., and Conway, P.L. (2002). Purification and characterization of a surface protein from *Lactobacillus fermentum* 104R that binds to porcine small intestinal mucus and gastric mucin. Appl. Environ. Microbiol. *68*, 2330–2336.

Roselli, M., Finamore, A., Hynönen, U., Palva, A., and Mengheri, E. (2016). Differential protection by cell wall components of *Lactobacillus amylovorus* DSM 16698T against alterations of membrane barrier and NF-kB activation induced by enterotoxigenic F4+ *Escherichia coli* on intestinal cells. BMC Microbiol, *16*, 226. https://doi.org/10.1186/s12866-016-0847-8

Rost, B., and Sander, C. (1994). Combining evolutionary information *and* neural networks to predict protein secondary structure. Proteins *19*, 55–72. https://doi.org/10.1002/prot.340190108

Rothfuss, H., Lara, J.C., Schmid, A.K., and Lidstrom, M.E. (2006). Involvement of the S-layer proteins Hpi and SlpA in the maintenance of cell envelope integrity in *Deinococcus radiodurans* R1. Microbiology *152*, 2779–2787. https://doi.org/10.1099/mic.0.28971-0

Ryan, A., Lynch, M., Smith, S.M., Amu, S., Nel, H.J., McCoy, C.E., Dowling, J.K., Draper, E., O'Reilly, V., McCarthy, C., *et al.* (2011). A role for TLR4 in *Clostridium difficile* infection and the recognition of surface layer proteins. PLOS Pathog. *7*, e1002076. https://doi.org/10.1371/journal.ppat.1002076

Sahay, B., Ge, Y., Colliou, N., Zadeh, M., Weiner, C., Mila, A., Owen, J.L., and Mohamadzadeh, M. (2015). Advancing the use of *Lactobacillus acidophilus* surface layer protein A for the treatment of intestinal disorders in humans. Gut Microbes *6*, 392–397. https://doi.org/10.1080/19490976.2015.1107697

Sára, M., and Sleytr, U.B. (2000). S-Layer proteins. J. Bacteriol. *182*, 859–868.

Savijoki, K., Kahala, M., and Palva, A. (1997). High level heterologous protein production in *Lactococcus* and *Lactobacillus* using a new secretion system based on the *Lactobacillus brevis* S-layer signals. Gene *186*, 255–262.

Schär-Zammaretti, P., Dillmann, M.L., D'Amico, N., Affolter, M., and Ubbink, J. (2005). Influence of fermentation medium composition on physicochemical surface properties of *Lactobacillus acidophilus*. Appl. Environ. Microbiol. *71*, 8165–8173.

Schuster, B., and Sleytr, U.B. (2009). Composite S-layer lipid structures. J. Struct. Biol. *168*, 207–216. https://doi.org/10.1016/j.jsb.2009.03.004

Schuster, B., and Sleytr, U.B. (2013). Nanotechnology with S-layer proteins. In: Protein Nanotechnology. Methods in Molecular Biology (Methods and Protocols), J. Gerrard, ed. (Humana Press, Totowa, NJ), pp. 153175. https://doi.org/10.1007/978-1-62703-354-1_9

Schuster, B., Pum, D., and Sleytr, U.B. (2008). S-layer stabilized lipid membranes (Review). Biointerphases *3*, FA3. https://doi.org/10.1116/1.2889067

Schut, S., Zauner, S., Hampel, G., König, H., and Claus, H. (2011). Biosorption of copper by wine-relevant lactobacilli. Int. J. Food Microbiol. *145*, 126–131. https://doi.org/10.1016/j.ijfoodmicro.2010.11.039

Seegers, J.F. (2002). Lactobacilli as live vaccine delivery vectors: progress and prospects. Trends Biotechnol. *20*, 508–515.

Sekot, G., Posch, G., Messner, P., Matejka, M., Rausch-Fan, X., Andrukhov, O., and Schäffer, C. (2011). Potential of the *Tannerella forsythia* S-layer to delay the immune response. J. Dent. Res. *90*, 109–114. https://doi.org/10.1177/0022034510384622

Sekot, G., Posch, G., Oh, Y.J., Zayni, S., Mayer, H.F., Pum, D., Messner, P., Hinterdorfer, P., and Schäffer, C. (2012). Analysis of the cell surface layer ultrastructure of the oral pathogen *Tannerella forsythia*. Arch. Microbiol. *194*, 525–539. https://doi.org/10.1007/s00203-012-0792-3

Settem, R.P., Honma, K., Nakajima, T., Phansopa, C., Roy, S., Stafford, G.P., and Sharma, A. (2013). A bacterial glycan core linked to surface (S)-layer proteins modulates host immunity through Th17 suppression. Mucosal Immunol. *6*, 415–426. https://doi.org/10.1038/mi.2012.85

Shimotahira, N., Oogai, Y., Kawada-Matsuo, M., Yamada, S., Fukutsuji, K., Nagano, K., Yoshimura, F., Noguchi, K., and Komatsuzawa, H. (2013). The surface layer of *Tannerella forsythia* contributes to serum resistance and oral bacterial coaggregation. Infect. Immun. *81*, 1198–1206. https://doi.org/10.1128/IAI.00983-12

Sillanpää, J., Martínez, B., Antikainen, J., Toba, T., Kalkkinen, N., Tankka, S., Lounatmaa, K., Keränen, J., Höök, M., Westerlund-Wikström, B., *et al.* (2000). Characterization of the collagen-binding S-layer protein CbsA of *Lactobacillus crispatus*. J. Bacteriol. *182*, 6440–6450.

Sleytr, U.B., and Beveridge, T.J. (1999). Bacterial S-layers. Trends Microbiol. *7*, 253–260.

Sleytr, U.B., Bayley, H., Sára, M., Breitwieser, A., Küpcü, S., Mader, C., Weigert, S., Unger, F.M., Messner, P.,

Jahn-Schmid, B., *et al.* (1997). Applications of S-layers. FEMS Microbiol. Rev. *20*, 151–175.

Sleytr, U.B., Egelseer, E.M., Ilk, N., Pum, D., and Schuster, B. (2007). S-Layers as a basic building block in a molecular construction kit. FEBS J. *274*, 323–334.

Sleytr, U.B., Schuster, B., Egelseer, E.M., and Pum, D. (2014). S-layers: principles and applications. FEMS Microbiol. Rev. *38*, 823–864. https://doi.org/10.1111/1574-6976.12063

Smit, E., and Pouwels, P.H. (2002). One repeat of the cell wall binding domain is sufficient for anchoring the *Lactobacillus acidophilus* surface layer protein. J. Bacteriol. *184*, 4617–4619.

Smit, E., Jager, D., Martinez, B., Tielen, F.J., and Pouwels, P.H. (2002). Structural and functional analysis of the S-layer protein crystallisation domain of *Lactobacillus acidophilus* ATCC 4356: evidence for protein-protein interaction of two subdomains. J. Mol. Biol. *324*, 953–964.

Smit, E., Oling, F., Demel, R., Martinez, B., and Pouwels, P.H. (2001). The S-layer protein of *Lactobacillus acidophilus* ATCC 4356: identification and characterisation of domains responsible for S-protein assembly and cell wall binding. J. Mol. Biol. *305*, 245–257. https://doi.org/10.1006/jmbi.2000.4258

Spigaglia, P., Barbanti, F., and Mastrantonio, P. (2011). Multidrug resistance in European *Clostridium difficile* clinical isolates. J. Antimicrob. Chemother. *66*, 2227–2234. https://doi.org/10.1093/jac/dkr292

Sun, Z., Kong, J., Hu, S., Kong, W., Lu, W., and Liu, W. (2013). Characterization of a S-layer protein from *Lactobacillus crispatus* K313 and the domains responsible for binding to cell wall and adherence to collagen. Appl. Microbiol. Biotechnol. *97*, 1941–1952. https://doi.org/10.1007/s00253-012-4044-x

Taverniti, V., Stuknyte, M., Minuzzo, M., Arioli, S., De Noni, I., Scabiosi, C., Cordova, Z.M., Junttila, I., Hämäläinen, S., Turpeinen, H., *et al.* (2013). S-layer protein mediates the stimulatory effect of *Lactobacillus helveticus* MIMLh5 on innate immunity. Appl. Environ. Microbiol. *79*, 1221–1231. https://doi.org/10.1128/AEM.03056-12

Thompson, S.A. (2002). *Campylobacter* surface-layers (S-layers) and immune evasion. Ann. Periodontol. *7*, 43–53. https://doi.org/10.1902/annals.2002.7.1.43

Toca-Herrera, J.L., Krastev, R., Bosio, V., Küpcü, S., Pum, D., Fery, A., Sára, M., and Sleytr, U.B. (2005). Recrystallization of bacterial S-layers on flat polyelectrolyte surfaces and hollow polyelectrolyte capsules. Small *1*, 339–348. https://doi.org/10.1002/smll.200400035

Tuomola, E.M., Ouwehand, A.C., and Salminen, S.J. (2000). Chemical, physical and enzymatic pre-treatments of probiotic lactobacilli alter their adhesion to human intestinal mucus glycoproteins. Int. J. Food Microbiol. *60*, 75–81.

Uroić, K., Novak, J., Hynönen, U., Pietilä, T.E., Pavunc, A.L., Kant, R., Kos, B., Palva, A., and Šušković, J. (2016). The role of S-layer in adhesive and immunomodulating properties of probiotic starter culture *Lactobacillus brevis* D6 isolated from artisanal smoked fresh cheese. LWTFood Sci. Technol. *69*, 623–632. http://doi.org/10.1016/j.lwt.2016.02.013

Ventura, M., Jankovic, I., Walker, D.C., Pridmore, R.D., and Zink, R. (2002). Identification and characterization of novel surface proteins *in Lactobacillus johnsonii* and *Lactobacillus gasseri*. Appl. Environ. Microbiol. *68*, 6172–6181.

Vidgrén, G., Palva, I., Pakkanen, R., Lounatmaa, K., and Palva, A. (1992). S-layer protein gene of *Lactobacillus brevis*: cloning by polymerase chain reaction and determination of the nucleotide sequence. J. Bacteriol. *174*, 7419–7427.

Wan, L.Y., Chen, Z.J., Shah, N.P., and El-Nezami, H. (2016). Modulation of intestinal epithelial defense responses by Probiotic Bacteria. Crit. Rev. Food Sci. Nutr. *56*, 2628–2641. https://doi.org/10.1080/10408398.2014.905450

Wang, R., Jiang, L., Zhang, M., Zhao, L., Hao, Y., Guo, H., Sang, Y., Zhang, H., and Ren, F. (2017). The Adhesion of *Lactobacillus salivarius* REN to a human intestinal epithelial cell line requires S-layer proteins. Sci. Rep. *7*, 44029. https://doi.org/10.1038/srep44029

Wang, W., Shao, A., Feng, S., Ding, M., and Luo, G. (2017b). Physicochemical characterization and gastrointestinal adhesion of S-layer proteins-coating liposomes. Int. J. Pharm. *529*, 227–237.

Wang, X.Y., Wang, D.B., Zhang, Z.P., Bi, L.J., Zhang, J.B., Ding, W., and Zhang, X.E. (2015). A S-Layer protein of *Bacillus anthracis* as a building block for functional protein arrays by in vitro self-assembly. Small *11*, 5826–5832. https://doi.org/10.1002/smll.201501413

Waśko, A., Polak-Berecka, M., Kuzdraliński, A., and Skrzypek, T. (2014). Variability of S-layer proteins in *Lactobacillus helveticus* strains. Anaerobe *25*, 53–60. https://doi.org/10.1016/j.anaerobe.2013.11.004

Weiss, G., and Jespersen, L. (2010). Transcriptional analysis of genes associated with stress and adhesion in *Lactobacillus acidophilus* NCFM during the passage through an in vitro gastrointestinal tract model. J. Mol. Microbiol. Biotechnol. *18*, 206–214. https://doi.org/10.1159/000316421

Zhang, Q., Zhong, J., Liang, X., Liu, W., and Huan, L. (2010). Improvement of human interferon alpha secretion by *Lactococcus lactis*. Biotechnol. Lett. *32*, 1271–1277. https://doi.org/10.1007/s10529-010-0285-x

Zhu, C., Guo, G., Ma, Q., Zhang, F., Ma, F., Liu, J., Xiao, D., Yang, X., and Sun, M. (2017). Diversity in S-layers. Prog. Biophys. Mol. Biol. *123*, 1–15.

Bacteriophages of *Lactobacillus* Species

7

María Eugenia Dieterle[1,2] and Mariana Piuri[1,2*]

[1]Universidad de Buenos Aires, Facultad de Ciencias Exactas y Naturales, Departamento de Química Biológica, Buenos Aires, Argentina.
[2]CONICET – Universidad de Buenos Aires. Instituto de Química Biológica de la Facultad de Ciencias Exactas y Naturales (IQUIBICEN), Buenos Aires, Argentina.

*Correspondence: mpiuri@qb.fcen.uba.ar

https://doi.org/10.21775/9781910190890.07

Abstract
Bacteriophages infecting lactic acid bacteria (LAB) are the main cause of fermentation failures leading to economic losses. Phages that infect *Lactococcus* and *Streptococcus* strains have been widely characterized in the literature, mainly because of their relevance in industry. Despite the above, strains of *Lactobacillus* are particularly of interest because besides their contribution to the organoleptic properties of fermented products, several strains have purported probiotic properties and are part of commercial formulations. The use of these strains is the result of years of research that validated the claimed benefits in food products. Phage attack on these specifically chosen strains is particularly deleterious, as they cannot be easily replaced. Advances in phage genomics, virus–bacteria interactions, and new tools designed to avoid infection are described in this chapter.

Introduction
Bacteriophages, or simply phages, are unable to reproduce independently and use the bacterial machinery for their own replication. They are considered the most abundant biological entities on earth overtaking bacteria by a factor of 10 (Breitbart and Rohwer, 2005). At the end of each infection, hundreds of new phages are released to lyse neighbouring bacteria (Kutter *et al.*, 2005). Consequently, every biotechnology process that relies on the use of bacteria could be disrupted by phages causing great economic losses. Extensive research has been focused in controlling phage population, mainly studying bacteria currently used in dairy industry.

Everyday, immense volumes of milk are inoculated with lactic acid bacteria (LAB), used as starters, to produce fermented products (around 10^{14} LAB are needed to produce one ton of cheese). The starter culture is usually a combination of *Streptococcus thermophilus*, *Lactococcus lactis*, *Leuconostoc* spp. and/or *Lactobacillus* spp. The last two are frequently used as adjunct cultures adding flavour and ripening during cheese production. Contamination in the dairy industry is common, as phages can be present in the substrates for fermentation, on surfaces of vessels and other equipment or recycled ingredients (Garneau and Moineau, 2011). Cultures of starter strains themselves can be phage contaminated, likely due to induction of resident prophages and concomitant cell lysis (Desiere *et al.*, 2002). It is highly recommended to avoid lysogenic LAB, but the nature of starter strains is not always well defined, and in some strains multiple prophages can be present. If a sensitive cell is present in the starter culture, a phage population can increase rapidly (virulent phages disperse quickly with a latent period of 30 minutes and a burst size of 100). A batch is considered compromise when the phage titre is higher than 10^4 UPF/ml.

Phage attack may produce a delay in acidification rates, alter the quality of the product or, in the worst case, the complete loss of the ferment (Brüssow and Desiere, 2001). Early indicators of phage contamination include high pH values, lactose remnant and low lactic acid values. These changes can lead to the alteration of the organoleptic properties of the commercial products making them unreliable. Even worse, this can be an ideal environment for spoilage bacteria development risking human consumption. Nowadays, phages are considered one of the most important factors of economic losses causing between 0.1% and 10% of the problems in fermentation vats (Moineau and Lévesque, 2005). Different strategies to minimize the negative impact of phages on fermentation include milk pasteurization, sanitation of equipment, factory design and rotation of starter strains (Briggiler Marcó *et al.*, 2012).

The most intensely researched LAB-infecting phages are those infecting *L. lactis* and *S. thermophilus*. These LAB are widely used as starter bacteria in fermentation processes, thus big efforts have been made to avoid phage contamination. In the last decade, phages that specifically infect *Lactobacillus* spp. started to have a more relevant role (Capra *et al.*, 2009). *Lactobacillus* spp. are widely employed in fermentation of vegetables (sauerkraut, pickles), meat and dairy products, including cheese and fermented milk. Not only are they used as starter/adjunct cultures, but many strains are part of NSLAB (non-starter lactic acid bacteria). NSLAB survives pasteurization processes and form a significant portion of the microbial flora of most cheese varieties during ripening (Beresford *et al.*, 2001). These pool of unknown bacteria (possibly carrying prophages in their genomes) increases the spectrum of potentially unknown phages.

Most important, probiotic properties of commercial formulations relies on the use of specifically selected *Lactobacillus* strains (FAO/WHO, 2006). The use of these strains is the result of years of research that validated the claimed benefits and safety in food products and led to their approval for human consumption (Douillard *et al.*, 2013) For these reasons, phage attack on these specifically chosen and studied strains, which cannot be easily replaced, is particularly deleterious.

A multidisciplinary research approach to study phage- bacteria interactions in *Lactobacillus* spp. is needed to design new strategies in order to minimize or control phage population in fermentation processes. This chapter aims to highlight scientific advances concerning phages and its interaction with susceptible *Lactobacillus* strains and future innovative tools to overcome fermentation failures.

Genomics

More than 100 species are actually classified in the *Lactobacillus* genus (Claesson *et al.*, 2007). Construction of pangenomes reveals genetic and phenotypic diversity and explains adaptability of lactobacilli to various habitats (Stefanovic *et al.*, 2017). The great plasticity of *Lactobacillus* genomes increased phage heterogeneity and makes hard to find common features to improve taxonomic data (Mahony *et al.*, 2012). The last detailed review concerning *Lactobacillus* phages came from Villion and Moineau (2009). In this review, 231 phages were documented from peer-reviewed articles since 1960: 109 belonging to *Siphoviridae* (phages with long non-contractile tails), 76 to *Myoviridae* (phages with contractile tails) and only one to *Podoviridae* (characterized by short non contractile tails) (Villion and Moineau, 2009). It is noteworthy that only nine complete *Lactobacillus* phage genomes were available for comparisons and evolution studies. The mosaic nature seemed to be the common feature found in comparative studies that reflect the horizontal transfer between different phages. Because of information scarcity and great diversity, actual phage classification considers host first and then morphology (*Siphoviridae, Podoviridae* and *Myoviridae*).

Current phage taxonomic efforts significantly depends on comparative genomic analysis and derived information and improvement of phage resistance of *Lactobacillus* starter cultures (Mahony and van Sinderen, 2014). Low cost and improved technologies, sequencing and annotation have dramatically increased the number of genomes available. In particular, the function associated with each ORF (open reading frame) is not always precise. Phage diversity also shows a high number of genes that do not show similarity to other genes present in databases (Hatfull, 2015). *Siphoviridae* phages show a conserved synteny that helps on prediction when identity is low. In other cases, identity is high with proteins in databases with no apparent or clear function associated with phages but that might be interesting on evolutionary terms. Powerful *in silico* tools, like Pfam or HHpred, helps in annotation when sequence similarity is low but a same folding appear. According to the NCBI database, 200 phages infecting LAB have been sequenced. In March 2017, 39 genome sequences of phages that infect *Lactobacillus* spp. were described (three prophage sequences were not included in the analysis since no induction information was reported). Four sequences corresponded to *Myoviridae* phages and 35 to *Siphoviridae*. *Lactobacillus* phage genomes belonging to the *Siphoviridae* family are between 29 and 49 kbp with exception of Ldl1, infecting *L. delbruekii*, which is extremely large (71 kbp). Most of the phages with available sequences infect *L. casei* (10), followed by *L. delbrueckii* (9) and then between one to six sequences reported from *L. plantarum* (6), *L. fermentum*, *L. gasseri*, *L. helveticus*, *L. brevis* and *L. johnsonii*.

In order to contribute with classification a dot plot analysis of *Lactobacillus* phages belonging to the *Siphoviridae* family is shown. Individual genomes were concatenated in a unique sequence. The most similar were found closer to each other. Then, we compared the nucleotide sequence of the concatemer to each other resulting in a dot plot graphic. This graph represents the simplest interpretation, where genomes with higher identities are together in the same group and those distant from each other are in different groups. Fig. 7.1 shows that genomes of phages infecting *Lactobacillus* spp. are not uniformly diverse. At the top left, a cluster of *L. delbrueckii* phages is observed, showing that, with exception of phiJB, LL-H and Ldl1, are genetically related. Another cluster is observed including those phages that infect bacteria belonging to the *L. casei*

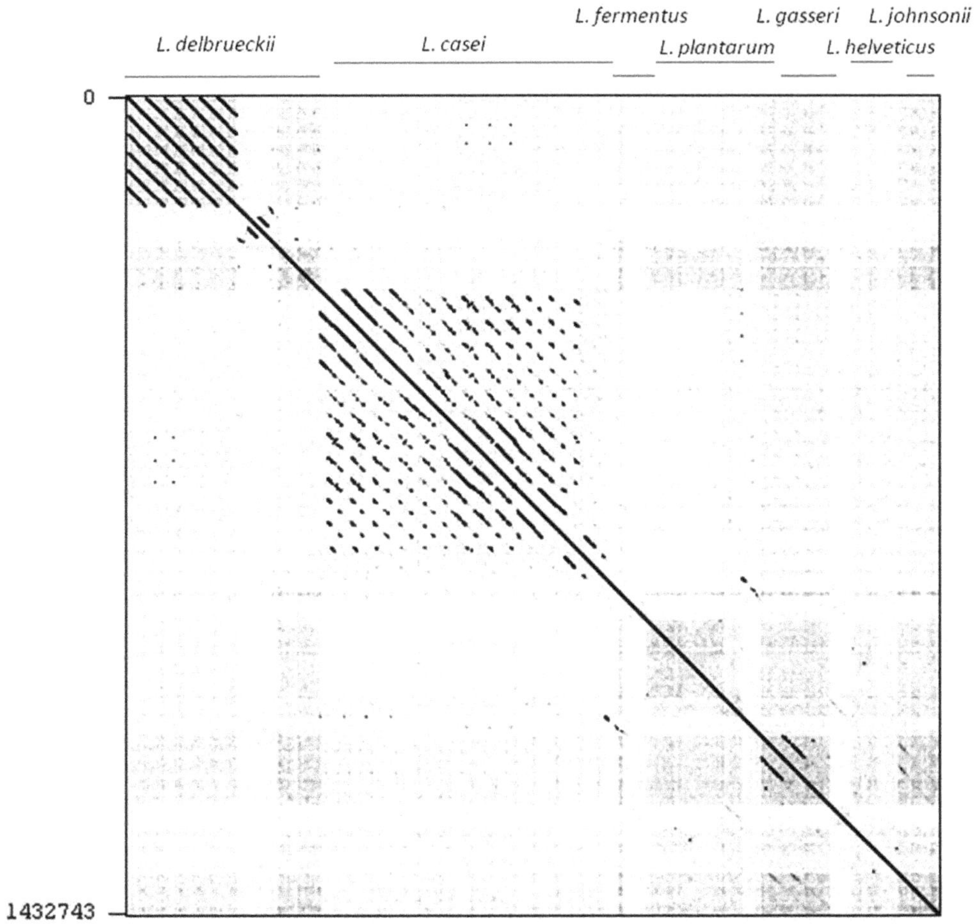

Figure 7.1 Dot plot of 36 genomes of *Lactobacillus* spp phages. Genomes of phages infecting *Lactobacillus* spp. were concatenated and compared to each other using Gepard (Krumsiek *et al.*, 2007). The bacterial host is indicated in the upper part of the plot.

group (*L. casei*, *L. paracasei* and *L. rhamnosus*) but with a higher variability. However, the rest of the analysed genomes are difficult to group. Although some efforts have been done towards sequencing, the high *Lactobacillus* diversity along with a high heterogeneity of phages infecting these species make hard to find a suitable classification as it was previously described (Claesson *et al.*, 2007; Villion and Moineau, 2009). Despite this, *L. delbrueckii* and *L. casei* phages might be possibly be grouped once the number of genomes sequences increase. Using DNA–DNA hybridization and comparative genome analysis, *L. delbrueckii* phages were organized into five distinct groups (a, b, c, d, e) (Casey *et al.*, 2015). In group a (LL-H and phiJB), LL-H is the most studied phage. LL-H was originally isolated in 1972 and its full genome sequence and transcriptional map were determined (Mikkonen *et al.*, 1996). Research on phage–host interactions has contributed to its understanding and is discussed in the next section. Ld17 is the most characterized phage from group b (c5, LL-Ku; Ld3, Ld17, Ld25A and phiLdb), with a defined transcriptome consisting of two transcripts and several structural proteins identified. Little is known concerning group c with only one representative, phage JCL1032. Phages belonging to Groups d and e remain significantly understudied, each represented by a single isolate. Prolate headed phage 0252 infects *L. delbrueckii* subsp. *lactis* but its sequence is not yet available. Phage Ldl1, more related to *Lactobacillus plantarum* phage ATCC 8014-B2, is the singleton of group **e** (Casey *et al.*, 2015).

Phages belonging to the *L. casei* group are shown in detail in Fig. 7.2. Even although all genomes share homology regions, we observed two differential subgroups. Subgroup I (top left) is formed by J-1, PL-1, A2, Lrm1. These phages show a high sequence similarity on terminases and morphological regions but after that module identity drops.

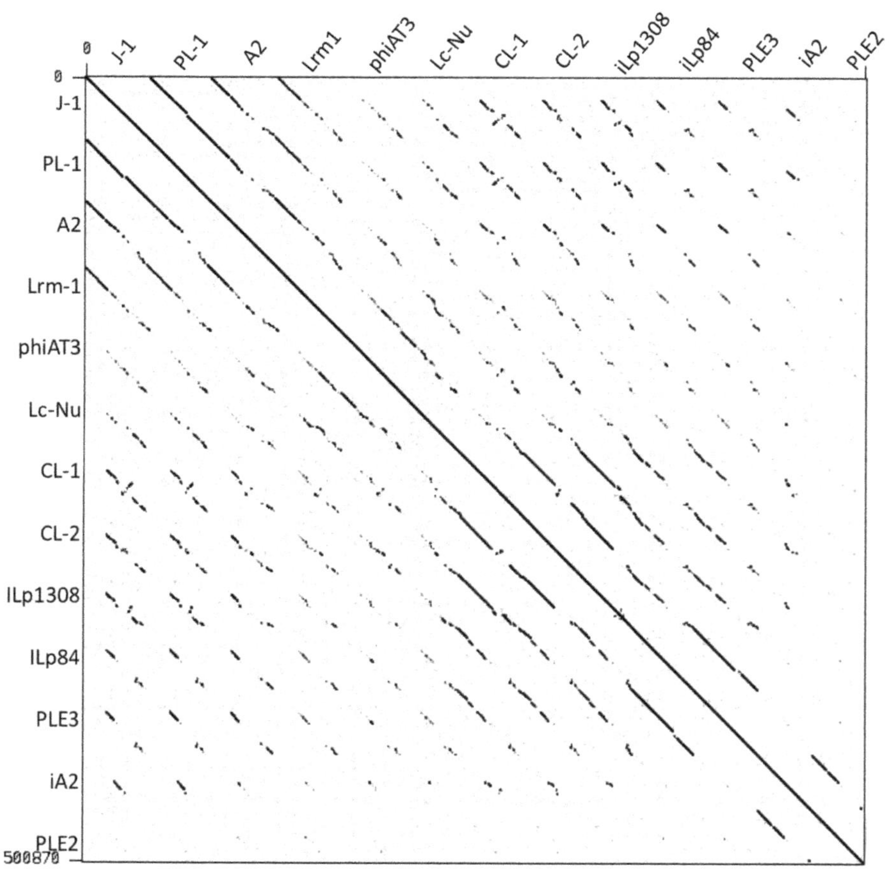

Figure 7.2 Dot plot of phages infecting Lactobacilli of the *L. casei* group. The indicated phage genomes were compared to each other using Gepard (Krumsiek *et al.*, 2007).

Lactobacillus phage J-1 was isolated in 1965 from a failed fermentation of the Japanese beverage Yakult (Yakult®, Minato-ku, Japan). It was the first described phage infecting a *L. casei* (or *L. paracasei*) strain. At the time, an *L. casei* strain resistant to J-1 infection was used instead, but after two years of use a second phage, designated PL-1, was isolated. J-1 and PL-1 are serologically related (28–30) and extensively characterized (reviewed by Sechaud *et al.*, 1988). Phage A2 is another well characterized *L. casei* phage isolated in Spain from whey of a failed home-made blue cheese production using *L. casei* ATCC 393. The second subgroup (subgroup II), contains phages CL1, CL2, Ilp84, Ilp1308 and PLE3. Lc-Nu and phiAT3 are not clustered and shown between these groups. Phage phiAT3 was recovered from *L. casei* ATCC 393 following induction with mitomycin C (Lo *et al.*, 2005). Finally, phages iA2 and PLE2 are part of another group. This analysis is in agreement with the maximum-likelihood tree done by Mercanti *et al.* (2016) using terminase sequences (terminases are considered the most conserved gene among phages and it has been proposed as a phylogenetic classifier according to the DNA encapsidation system) (Mercanti *et al.*, 2016). Phages J-1, PL-1, A2 and Lrm1 (subgroup I), PhiAT3 and Lc-Nu were grouped according to a *cos* packaging system with cohesive ends with 3′ extensions. Phage iA2 was also included in this group. CL1, CL2, iLP84, iLp1308 were grouped based on a full head encapsidation system as P22.

Prophages and the risk of lysogenic strains

Bacteriophages can be found on substrates used for fermentation, surfaces of equipment or in generated aerosol drops, but also can be found as prophages of the bacterial strains used as starters (Durmaz *et al.*, 2008; Garneau and Moineau, 2011; Ventura *et al.*, 2006; Verreault *et al.*, 2011, 2008). The increasing number of sequenced bacterial genomes has revealed the presence of multiple temperate phages and phage remnants in the genomes of *Lactobacillus* spp. and other LAB (Canchaya *et al.*, 2003; Desiere *et al.*, 2002; Douillard *et al.*, 2013; Sechaud *et al.*, 1988; Ventura *et al.*, 2006). Additionally, the presence of phage gene homologues spread in different bacterial strains likely suggests horizontal gene transfer between these related species (Baugher *et al.*, 2014). Because of the risk of induction, starter cultures with specific organoleptic properties that cannot be replaced using strain rotation strategies or probiotic strains, seem to be the most vulnerable targets. Prophage stability should be considered when strains are selected for specific fermentative process (Emond and Moineau, 2007). The presence of prophages or its excision can influence the bacterial phenotype (Feiner *et al.*, 2015; Rabinovich *et al.*, 2012).

It is known that different environmental stress conditions such as heat, osmotic stress, antimicrobials or starvation might activate prophage induction followed by replication and phage release. Under laboratory conditions, prophage induction is usually carried out by treatment of cultures with either mitomycin C (MC), UV light or hydrogen peroxide activating the SOS response. Research experience has shown that at least for *Lactobacillus* strains, MC seems to be the most effective (Capra *et al.*, 2010; Dieterle *et al.*, 2016; Durmaz *et al.*, 2008; Mercanti *et al.*, 2016).

An early study showed that from 148 strains of *Lactobacillus* (15 different species), 27% released phage to the supernatant when exposed to MC (Yokokura *et al.*, 1974). In the genome of the sequenced *L. johnsonii* NCC 533 (Ventura *et al.*, 2004), *Lactobacillus plantarum* commensal WCFS1 (Ventura *et al.*, 2003), *Lactobacillus gasseri* ATCC 33323, *Lactobacillus salivarius* subsp. *salivarius* UCC 118, and *Lactobacillus casei* ATCC 334 (Ventura *et al.*, 2006) many prophages were reported but attempts to induce their excision failed or were not described. In all cases, transcription analysis showed that most of the prophage genome is silent but short regions near the attachment sites were highly transcribed.

Mercanti *et al.* (2011) described phage induction on 30 strains of the *L. casei* group exposed to MC and 25 of them allowed direct recovering of phage DNA. From this study, 10 out of 11 commercial strains tested, contained prophages (Mercanti *et al.*, 2011). Phages CL1 and CL2 were isolated spontaneously from *L. paracasei* A, a strain used commercially. By adding MC, phage iA2 was also induced from the same strain (Capra *et al.*, 2010). Also, using MC as inducer, the temperate bacteriophage phiAT3 was obtained from *L. casei* ATCC 393 (Lo *et al.*, 2005). Phage iLp84 was isolated from *L. paracasei* 84 and Lrm1 induced from *L. rhamnosus* (Durmaz *et al.*, 2008). Moreover, prophage remnants, found in the genome of several *Lactobacillus* strains, are considered as well a phage risk because of recombination with virulent phages (Moineau *et al.*, 1994; Shimizu Kadota *et al.*, 1983).

L. casei BL23 is a widely used laboratory strain for physiological, biochemical and genetic studies (Bourand *et al.*, 2013; Muñoz-Provencio *et al.*, 2012; Piuri *et al.*, 2003; Revilla-Guarinos *et al.*, 2013) and it has been shown to exhibit probiotic properties (Rochat *et al.*, 2007). Despite of its extensive use, the presence of a mobile element in *L. casei* BL23 has been recently described (Mercanti *et al.*, 2016; Dieterle *et al.*, 2016). Three complete prophages (PLE1, PLE2, and PLE3) in the genome of *L. casei* BL23 were identified. All of them are also present in commercial probiotic strains BDII and W56. PLE1 and PLE2 are also present in the patented probiotic strain LCW2. Using quantitative real-time PCR, we have shown that PLE2 and PLE3 can be induced but with different kinetics. Additionally, PLE2 genome can replicate inside the host cell suggesting that this phage could undergo a complete phage lytic cycle. Sequencing, electron microscopy and mass spectrometry analysis showed that at least one of them forms a potential infectious virion (Dieterle *et al.*, 2016). Interestingly, one of these prophages, PLE1, is 99% similar to the sequenced iA2 phage but unlike iA2 was not induced after MC exposure. PLE1 has a small deletion of 5 bp in the integration/immunity region that could account for this different behaviour. Even although these data showed a low level of spontaneous induction in the conditions tested, the multiple prophages found in *L. casei* BL23, their sequence similarities to other phages or prophages present in *Lactobacillus* strains contributes to the idea of high rates of recombination among them.

All these results evidence the widely occurrence of lysogeny in *Lactobacillus* spp. genomes and the potential risk of their use in fermentative processes. However, in most cases complete lysis was not observed and the growth of the culture was not significantly affected showing only a partial arrest after induction. In summary, these prophages appear to be very stable in the tested conditions. Different factors present during fermentation processes such as osmolarity, pH and temperature fluctuations could act as prophage inducers as well and need to be tested in the future to avoid host lysis and consequent fermentation failures.

Prophages are not passive genetic elements and may also play a role in cellular physiology or niche adaptation (Canchaya *et al.*, 2003). As mobile elements, genetic exchange has been reported between modules in prophages and phages (Baugher *et al.*, 2014; Bouchard and Moineau, 2000; Labrie and Moineau, 2007; Lima-Mendez *et al.*, 2011). In *L. gasseri* genomes, many prophages (even present in tandem) show spontaneous induction and their high similarity with prophages and genes present in multiple *Lactobacillus* strains suggest that contribute to horizontal gene transfer in these genera (Baugher *et al.*, 2014; Raya *et al.*, 1989). It has also been suggested that prophages could be involved in cheese ripening. During cheese manufacturing, *L. lactis* lysis increase free amino acids and reduction of bitterness by hydrolysis of hydrophobic peptides. Autolysis of *L. lactis* subsp. *cremoris* strain is correlated with activation of a thermoinducible prophage. A cured *L. lactis* subsp. *cremoris* AM2 strain showed slower lytic events compared with the wild type strain (Lepeuplel *et al.*, 1998; Lortal and Chapot-Chartier, 2005). No similar events were described for *Lactobacillus* spp.

Prophages have also provided a wide range of tools for phage/bacteria genetic manipulation. Although these tools have been mainly used in *Lactococcus* and *Streptococcus*, some advances in *Lactobacillus* have been done recently. Phage integration is catalysed by a phage-encoded recombinase (integrase) and occurs between the *attB* and *attP* attachment sites located on the bacterial and phage genomes, respectively (Groth and Calos, 2004). For example, *L. delbrueckii* prophage mv4 has been extensively studied for site specific recombination (Auvray *et al.*, 1999; Coddeville and Ritzenthaler, 2010; Coddeville *et al.*, 2014; Dupont *et al.*, 2004). Mv4 integrase does not need any accessory host factors for recombination. A non replicative vector (pMC1) carrying an *attP* site and the integrase-coding (*int*) gene of mv4 is able to integrate into the chromosome of *Lactobacillus plantarum*, *Lactobacillus casei*, *Lactococcus lactis* subsp. *cremoris*, *Enterococcus faecalis* and *Streptococcus pneumoniae* (Dupont *et al.*, 2004). In *Lactobacillus rhamnosus* HN001 an integrative vector constructed using phiAT3 *attP* site and its integrase was also successfully tested (Lo *et al.*, 2005). A new optimized integration vector was constructed to stably integrate heterologous reporter genes into the genomes of *L. casei* ATCC 334, *L. casei* ATCC 27139, and *L. fermentum* ATCC 14931 (Lin *et al.*, 2013). A recent work using PLE3 (LCABL_13040-50-60) as a novel λ Red-like recombinase operon allowed recombineering in *L. casei* BL23. Combining cre-lox technology, the authors were able to catalyse markerless deletion, insertion and precision point mutations. Moreover, with the assistance of Redγ, the LCABL_13040-50-60 proteins also showed recombination activity in six other *L. casei* strains, *L. paracasei* and *L. plantarum* WCSF1 (Xin *et al.*, 2003).

Phage–host interaction

Phages belonging to the Siphoviridae family posses a dedicated proteinaceous machinery for bacterial recognition, the first step required for a successful infection.

The high divergence of phage genomes is usually an obstacle to detect homology between proteins and determine phylogenetic relationships among them. However, the high degree of conservation in the order of genes in phages that belong to the *Siphoviridae*, *Podoviridae* and *Myoviridae* families (Duplessis and Moineau, 2001; Vegge *et al.*, 2005), mainly in the morphogenesis module, is striking. Particularly, the canonical organization of tail genes in *Siphoviridae* is as follows: tail terminator protein, major tail protein (MTP), tail chaperones with the classical translational frameshift (Xu *et al.*, 2004), tape measure protein (TMP), the distal tail protein (Dit), tail associated lysins (Tal) and the receptor-binding proteins (RBP) (Veesler and Cambillau, 2011).

The Dit protein is a key piece for baseplate morphogenesis and initiates the assembly of the adsorption complex. The structure of Dit proteins in phages that infect Gram-positive bacteria like SPP1, p2 and TP901–1 is highly conserved and postulates a conserved structural motif in *Siphoviridae* phages. The Dit protein has two domains corresponding to the C- and N- terminal parts of the polypeptide chain. The Dit protein forms an hexamer, with a central channel that allows DNA passage during phage infection (Veesler *et al.*, 2010). One monomer connects to the N- terminal of the next polypeptide through a belt region ensuring the strength of the hexamer. The C- terminal protrudes out of the cylinder and displays a galectin fold. Aminoacidic sequence analysis of Dit, from phages that infect Gram-positive bacteria, shows a higher conservation of the N-terminus of the protein, that interacts with the MTP, and more differences in the C- terminus according to the context of Dit.

The Tal protein is present in phages that are far phylogenetically like T4 (gp27), Mu (gp44), p2 (ORF16) and TP901-1 (ORF47) with a conserved structure and a trimeric assembly (Veesler and Cambillau, 2011). In the Tal protein usually resides a peptidoglycan hydrolytic activity that facilitates DNA passage through the cell wall during infection (Kenny *et al.*, 2004; Vegge *et al.*, 2006).

The receptor-binding proteins are variable and depending on the macromolecular nature of the host receptor, either a protein or carbohydrate, the phage tail tip (that contains the RBPs) appears to display a particular morphology. Phages that interact with a proteinaceous receptor usually have a pointed end while those that interact with saccharidic receptors usually display a more obvious structure (Mahony and van Sinderen, 2012). A good example of these differences is seen in SPP1, where only the tip of the Tal protein (gp21) interacts with the protein YueB of *B. subtilis* (Goulet *et al.*, 2011), in comparison with p2, in which 18 trimeric RBPs recognize the cell wall polysaccharide of *L. lactis* (Sciara *et al.*, 2010).

In bacteriophages J-1 and PL-1 (that infect different strains of *Lactobacillus casei/paracasei*) the morphogenic module that codes the tail genes could be identified. Genes coding for the tail terminator, MTP, chaperones, TMP and the baseplate proteins Dit and Tal could be assigned but surprisingly we were not able to identify the RBP genes (Dieterle *et al.*, 2014a). The lack of a dedicated RBP prompted us to look for other proteins that could fulfil this role in these phages.

Analysis of Dit proteins present in other phages infecting *Lactobacillus* spp. showed that these proteins had a higher molecular weight than classic Dits. Particularly in J-1 and PL-1 the protein presented two insertions, correspondent to two extra domains and we so called them 'evolved' Dits.

The presence of Dit as structural protein of the virion was confirmed by MS analysis and using an electron microscopy assay we could localize it at the tip of the phage tail. Dit could specifically bind to *L.*

casei/paracasei cells and Inhibition of J-1 infection in the presence of anti-Dit antibodies strongly suggested its involvement in host recognition (Dieterle *et al.*, 2014b, 2017).

Using a bioinformatic approach we found homology between the first extra domain of Dit with carbohydrate-binding modules or CBMs but we did not find homology with known proteins for the second extra domain. However, this second domain (further called CBM2) but not CBM1 could specifically bind *L. casei/paracasei* cells. A combination of assays, including phage adsorption inhibition assays, binding of fusion fluorescent proteins visualized by microscopy, binding of fusion proteins in the presence of nanobodies followed by flow cytometry (Dieterle *et al.*, data unpublished) analysis and immunofluorescence assays validated this result concluding that CBM2 was most probably involved in bacterial recognition (Dieterle *et al.*, 2017).

The nature of this domain was confirmed after determination of the structure by X-ray crystallography with a resolution of 1.28 Å. The crystal structure corroborated that this domain was indeed a CBM with homology to lectins. Moreover, after comparison with similar structures we were able to assigned a putative carbohydrate-binding site.

The lack of activity of CBM1 *in vitro* could lead to a wrong assignment of the function of this CBM in nature where infectivity conditions differ from laboratory conditions. Hatfull *et al.* (2015) proposed that in the constant battle between phage and bacteria, a phage could have acquired a particular function in the past that does not provide an advantage in the actual host (Hatfull, 2015). So we cannot discard that CBM1 could be functional in other bacteria with a different cell envelope or in a different environmental context.

Even although the bacterial receptor for these phages is not strictly identified, in a recent study the presence and structure of a cell wall polysaccharide (CWP) fraction in the model *L. casei* BL23 strain, that is susceptible to J-1 infection, was described (Vinogradov *et al.*, 2016). All constituent polysaccharides were rich in rhamnose. We and others have previously reported that this sugar inhibits J-1 adsorption and Dit binding to its host (Dieterle *et al.*, 2014b; Yokokura, 1971). In addition, we have demonstrated that the CWP from *L. casei* BL23 can inhibit Dit and CBM2 binding to this strain and to *L. casei* subsp. *casei* ATCC 27139, strongly suggesting that at least some components of the CWPS could be the receptor for phage J-1 and most probably for PL-1 (Dieterle *et al.*, 2017).

Recently, using electron microscopy, we determined the structure of the baseplate of phage J-1 with a resolution of 20 Å. The baseplate of phage J-1 is ~125–210 Å wide and the length is ~110 Å, much smaller than baseplates of *Lactococcus* phages (Fig. 7.3) and we determined that CBM2 is pointing out of the J-1 baseplate, fully available for efficient interaction with the host CWPS (Fig. 7.4).

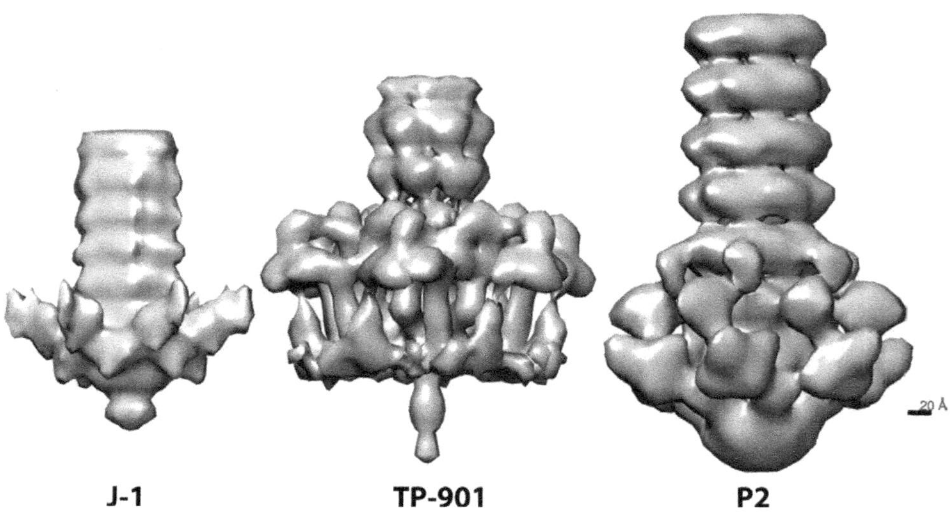

Figure 7.3 3D reconstructions of the baseplate of phages J-1, TP-901 and P2.

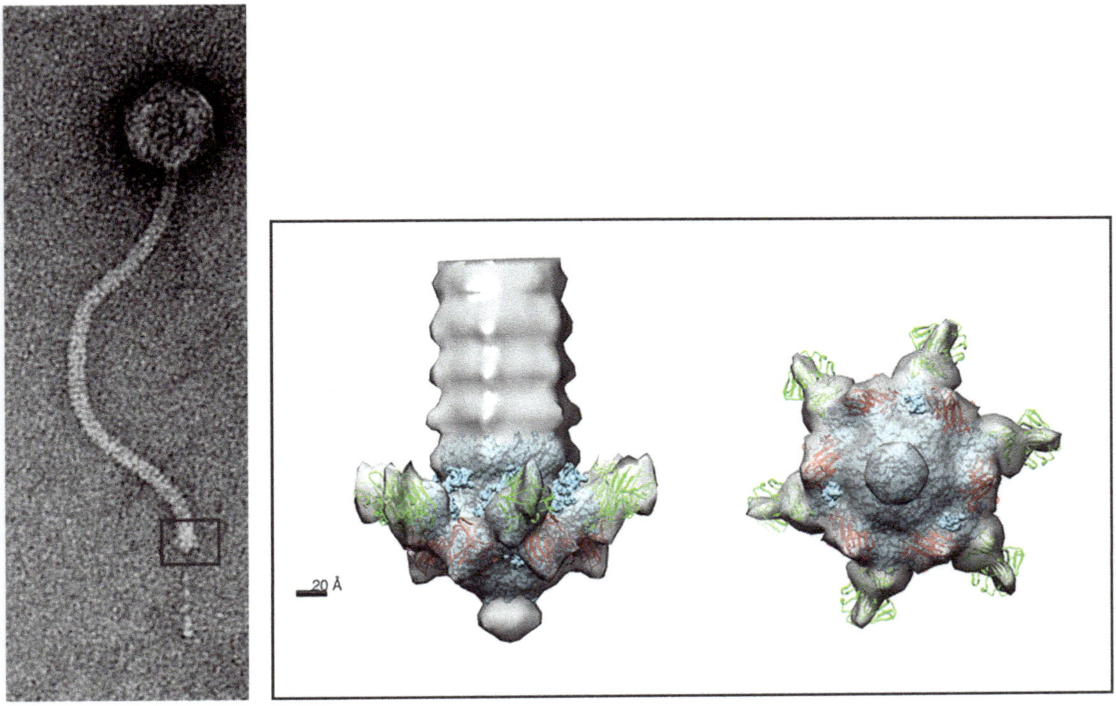

Figure 7.4 Negative staining electron microscopy single particle reconstruction of J-1 virion baseplate. Left. View of phage J-1 virion and its tail tip with the baseplate boxed. Right. Lateral and bottom view of the baseplate with dimensions. Dit core and Tal are blue; CBM2 is green; and 2xom, a topological model of CBM1, is red.

All Together our results suggest that Dit CBM2 is a bona fide receptor-binding protein of phage J-1. We suggest that such Dit inserted CBM domains – putative receptor-binding protein – exist in other species of the siphophage world. This hypothesis is also supported by the presence of putative CBMs in *Lactobacillus* phage PLE3 (Dieterle *et al.*, 2016) and other phages infecting lactobacilli of the *L. casei* group (Table 7.1). Also, in a recent study involving 38 lactococcal phage genomes from the 936 group, 16 Dit proteins showed a length compatible with an evolved Dit type (~450 residues or more) (Murphy *et al.*, 2016). These results introduced a drastic paradigm shift concerning the Dit proteins: evolved Dits should be considered as functional proteins towards the host, and not only passive hubs during baseplate formation. They may serve as unique RBPs or as associated recognition modules involved in the host adhesion process.

In other *Lactobacillus* phages the host recognition apparatus has not been studied at the structural level but the anti-receptor proteins and the corresponding bacterial receptor were identified in a few systems.

Phage LL-H has been used as a model for the infection of thermophilic *Lactobacillus delbrueckii* subsp. *lactis* host strains (Räisänen *et al.*, 2007; Riipinen and Alatossava, 2011). This phage has a small baseplate and a flexible tail fibre at the end of the tail. The tail fibre is a flexible hexapolymer of the anti receptor protein SP58 (coded by gene 71) (Ravin *et al.*, 2002). LL-H recognizes LTA (lipoteichoic acids) in strain ATCC 15808. In the current model for phage–host interaction, the C-terminal end of each anti receptor protein subunit contains one domain responsible for the reversible, specificity-determining binding to the surface end of the LTA, primarily mediated through hydrogen bonds to a glucose moiety. The second domain of the anti receptor protein SP58 ensures the irreversible binding to the negatively charged glycerol phosphate groups close to the surface end of the LTA, possibly mediated through ionic bonds (Munsch-Alatossava and Alatossava, 2013).

L. plantarum phage B2 uses teichoic acids as receptor. A spontaneous B2 resistant strain was found to have a lower glucose content in its teichoic acids, implicating that similar to phage LL-H a

Table 7.1 Putative CBMs present in the baseplate of phages of the *L. casei* group. The list of phages belonging to the *L. casei* group where putative CBMs were found are shown and compared with those identified in *Lactococcus lactis* phages. Presence or absence of CBM is shown in pink or grey, respectively. White indicates absence of the indicated protein as part of the baseplate

Phage	Accession nunber	Dit	Tal	BppU	BppA	RBP	Group	Reference
A2	AJ251789.2						*L. casei/ L. paracasei*	García *et al.* (2003)
CL1	KR905066						*L. casei/ L. paracasei*	Mercanti *et al.* (2016)
CL2	KR905067						*L. casei/ L. paracasei*	Mercanti *et al.* (2016)
IA2	KR905068						*L. casei/ L. paracasei*	Mercanti *et al.* (2016)
ilp1308	KR905070						*L. casei/ L. paracasei*	Mercanti *et al.* (2016)
ILP84	KR905069						*L. casei/ L. paracasei*	Mercanti *et al.* (2016)
J-1	KC171646						*L. casei/ L. paracasei*	Dieterle *et al.* (2014)
Lc-Nu	NC_007501.1						*L. rhamnosus*	Lo *et al.* (2005)
Lrm1	EU246945.1						*L. rhamnosus*	Durmaz *et al.* (2008)
PHIAt3	NC_005893.1						*L. casei/ L. paracasei*	Tuohimaa *et al.* (2005)
PL-1	KC171647						*L. casei/ L. paracasei*	Dieterle *et al.* (2014)
PLE1*	–						*L. casei/ L. paracasei*	Maze *et al.* (2010)
PLE2	KU848187						*L. casei/ L. paracasei*	Dieterle *et al.* (2016)
PLE3	KU848186						*L. casei/ L. paracasei*	Dieterle *et al.* (2016)
p2	GQ979703						*936 L. lactis*	Sciara *et al.* (2010)
TP901-1	NC_002747						P335 *L. lactis*	Veesler *et al.* (2012)
Tuc2009	NC_002703						P335 *L. lactis*	Collins *et al.* (2013)
PhiM1127	KP793132						*936 L. lactis*	Murphy *et al.* (2016)

glucose moiety is involved in host recognition (Nes *et al.*, 1988).

Interestingly, using a similar approach for *L. plantarum* phage B1, phage resistant derivatives of strain ATCC 8014 with a different CWPS composition (lower levels of galactose and rhamnose) compared with the sensitive strain were isolated, suggesting that these sugars could be used as receptors (Douglas and Wolin, 1971).

Finally, for phage CNRZ 832-B1, that infects *L. helveticus*, it was shown that the S-layer proteins were necessary for phage adsorption and several resistant derivatives have point mutations in the S- layer protein encoding gene (Zago *et al.*, 2017).

Future trends

Rational tools to avoid phage infection

As we previously mentioned, bacteriophages can become a huge problem in any fermentative process that employs bacteria. Rather than eradicated, phages can be controlled. Control strategies have been developed in order to fight phages. Reinforcing hygiene and safety procedures used along with appropriate technology in the dairy industry have allowed more accurate environmental conditions to avoid phage contamination. In a review from Brigiler Marcó *et al*, (2012), a list of strategies used nowadays in industrial plants are listed. These actions include physical separation of plant areas, air filters, control of bioaerosols, use of effective sanitizers or physical treatments (UV light irradiation, thermal treatments, high pressure technologies) (Marcó *et al.*, 2012). Other strategies include direct vat inoculation of starters, culture rotation programs or the use of starter bacteria resistant to bacteriophage infection. These actions have improved the everyday production but phages are still a problem. Consequently, great efforts concerning phage bacteria interaction research have been made in order to develop new tools to avoid phage replication. The knowledge of every step in the phage life cycle (adsorption, DNA injection, replication, assembling and host lysis) has been used to design new strategies to block phage propagation Bacteria owning anti phage mechanisms have been strategically adopted to avoid phage contamination (Labrie *et al.*, 2010). Research on *Lactobacillus* phages still remains behind in comparison with all the accumulated data about *Streptococcus* and *Lactococcus* phages. Here again, the increased number of bacteria genomes available along with new molecular technologies helped in the development of new tools on behalf of natural strategies.

Viral infection starts with the recognition of a bacterial receptor. Any modification on its receptor, has the potential to protect bacteria against phages. When bacteria are challenged with phages at high multiplicity of infection, BIMs (Bacteriophage insensitive mutants) can be isolated. This methodology, when is possible, does not involve genetic manipulation and can render strains with no restrictions in food industry application. Mutants with a confirmed highly stable phage-resistant phenotype and maintaining the desired technological properties – are suitable for industrial utilization. The inactivation of PIP protein of *L. lactis*, a known protein receptor of phage c2, inhibits phage c2 adsorption (Valyasevi *et al.*, 1991). This mechanism has been so effective that today type c2 phages are rare in Canada (Samson and Moineau, 2013). Mutants resistant to phage J-1, PL-1 and MLC-A have been isolated (Capra *et al.*, 2011). BIMs are sometimes related to a reduction of adsorption rates (Bouchard and Moineau, 2000, 2001). However, the cell surface properties did not seem to be affected in these cases suggesting that another mechanism is involved. Guglielmotti *et al.* (2007) also obtained phage-resistant derivatives from *L. delbrueckii* strains (Guglielmotti *et al.*, 2007). In *L. lactis* MG1363, It was described that phage sk1 receptors localize in the polysaccharide pellicle that covers the cell surface. Mutations on genes involved in pellicle biosynthesis inhibited infection. A rational design of insensitive strains is desired in food industry. A good example was described by Mahony *et al.* (2013), who conducted an exhaustive study of phages and *Lactococcus* host bacteria that could help to solve problems in dairy industry. In this work, a detailed analysis of the conserved and variable regions from the cluster related to cell wall polysaccharide synthesis was done allowing strain classification in subgroups (CWPS A, B and C). In parallel, a phage panel of *Lactococcus* phages (family 936) was classified according to host range and RBPs. A correlation was observed between the type of pellicule of a strain and the host range (Mahony *et al.*, 2013).

Noteworthy, monitoring of phage genome changes, specifically of those associated with host range could help to determine hosts. Nevertheless, adsorption mechanisms are unknown in most of LAB phages described until now and ways to avoid infection could become complex (Labrie *et al.*, 2010).

Two of the most studied mechanisms that involve bacterial DNA edition are restriction modification systems (R/M) and more recently CRISPR–Cas systems. The first one is based on phage genome degradation once injected inside bacteria. Two complementary enzymatic systems contributes to the mechanism: an endonuclease responsible of hydrolysing DNA after sequence recognition (restriction) and a methylase which modifies bacterial DNA in specific sequences (modification) turning bacteria immune to its own restriction enzyme (Hyman and Abedon, 2010). Depending on the structure and which mechanism is involved, four groups have been described (type I to IV) (Vasu and Nagaraja, 2013). Only I, II and III were found in LAB and are coded either in plasmids or in the chromosome. In *Lactobacillus*, R/M systems were reported in *L. helveticus, L. delbrueckii and L. plantarum* and generally appeared associated with another mechanism of phage resistance (Deng *et al.*, 2018; Guinane *et al.*, 2011; Zago *et al.*, 2017). Phages could counterattack expressing a methyltransferase that blocks putative recognition sites before restriction enzymes cut the DNA. Also phages carrying molecules that inhibit restriction enzymes or have less recognition sites in their genomes have been described (Szczepankowska *et al.*, 2013). BIMs isolation is probably one of the most efficient way to avoid phage propagation. Phage bacteria interaction contributes with BIMs. Probably, CRISPR–Cas system is the most interesting natural tool that has succeeded in dairy industry. DUPONT company started developing BIMs using CRISPR–Cas carrying bacteria since 2007 and CHOOZITSWIFT was the first product derived from this technology available in 2012. CRISPR (clustered regularly interspersed palindromic repeats) loci and their associated Cas proteins provides adaptive and inherited immunity in bacteria against phages and plasmids (Barrangou *et al.*, 2007; Garneau *et al.*, 2010). During the CRISPR–Cas immune response, small fragments from foreign genetic element (protospacers) are incorporated and arranged between direct repeats to originate the CRISPR arrays (acquisition or adaptation). These loci are subsequently transcribed and processed into small interfering RNAs (biogenesis) that guide nucleases for specific cleavage of complementary sequences in plasmid/phage DNA (interference) (Barrangou and Marraffini, 2014) Maturation and interference have been extensively studied during the past years (Deltcheva *et al.*, 2011; Hale *et al.*, 2009; Díez-Villaseñor *et al.*, 2013). CRISPR systems have been identified in several *Lactobacillus* genomes (Horvath *et al.*, 2009) but how each system work remains poorly understood. A type II-A CRISPR–Cas system in the commensal species *L. gasseri* has been incipiently described. Several of the spacers showed similarity to phage and plasmid sequences and in *L. gasseri* JV-V03 and NCK 1342 the system interfered with transforming plasmids containing sequences matching the last acquired CRISPR spacers in each strain (Sanozky-Dawes *et al.*, 2015). On the other hand, in *Lactobacillus helveticus* ATCC 10386 a BIM to phage Lh56 was isolated and the mechanism of resistance was interrogated. Even this strain has a CRISPR–Cas system, two other resistance mechanisms took place. The inhibition of phage adsorption and the up-regulation of Type I R/M genes, explained *L. helveticus* resistance to Lh56 (Zago *et al.*, 2017).

Other efforts that do not imply genome manipulation but require a deeper knowledge of phage molecular mechanisms of infection have been made in order to avoid phage contamination. Mahony and van Sinderen (2015) discussed the possibility of using baseplate macromolecular complex as phage competitors. They claimed that inhibition of adsorption is not complete and the expression of those high molecular weight proteins seem to be struggling and not useful for industrial purposes. From our own experience, this approach does not seem to be appropriate for *Lactobacillus* strains infected by J-1 and PL-1 since in competition assays Dit (one of the baseplate proteins) did not show a complete inhibition of infection (Dieterle *et al.*, 2014b).

VHHs have been also been suggested as an antiphage strategy. VHH or nanobodies against critical phage components have been used successfully and have many advantages for application in the dairy industry: VHHs are easily produced, low concentrations are needed and bacterial growth or acidification is not altered (Ledeboer *et al.*, 2002). As an example, VHHs anti RBP of TP901 were successfully employed to completely neutralize infection (Desmyter *et al.*, 2013). Interestingly, also for phage p2, an anti-RBP VHH was successfully expressed in a GRAS *L. paracasei* strain inhibiting *Lactococcus* cell infection (Hultberg *et al.*, 2007). The nanobody was also effective when it was scaled up in a cheese manufacturing environment (Ledeboer *et al.*, 2002). On the contrary, It needs to be addressed the cost associated with its use in the dairy plant pipeline and the possibility of mutant phages overtaking VHH neutralizing effect. DARPins (Designed Ankyrin Repeat Proteins) that specifically bind to RPBs have been selected to inhibit phage adsorption (Veesler *et al.*, 2009). DARPins binds to proteins with high affinity, are stable and easily produced in *E. coli*. Union sites differs from VHH and could be used in combination to cover many epitopes. This technology, even promising, needs to be tested along with operating and development cost.

Nowadays, there is not an only way to control phage infections. A constant monitoring to evaluate the imminent presence of bacteriophages in combination with the increasing knowledge on the steps of infection could provide a rational anti-phage solution for a particular phage population (Mahony and van Sinderen, 2015).

Even different tools prove to be successful, the biggest problem seems to be the specificity of the target which means that only works with one phage or a very related group of phages. Again, this issue is related with phage plasticity to overcome natural or artificial barriers (Koskella and Brockhurst, 2014).

Phages to avoid undesirable bacteria

Throughout this chapter phages were described as undesirable entities with a tremendous need to be controlled. Even not explored deeply in *Lactobacillus*, phages or its gene products, could be used to avoid bacteria propagation.

S. aureus is known as a spoilage bacteria in dairy products. Phage endolysins have been proposed as a potential antibacterial agent against *S. aureus* as an alternative to antibiotic usage. *Lactobacillus delbrueckii* phage phiLdb encodes endolysin Lysdb, one of the two-components of the cell lysis cassette, and it was shown to be effective against *S. aureus*. Moreover, Guo *et al.* (2016) showed that engineered *L. casei* BL23 was able to constitutively deliver Lysdb during a lab-scale cheese making from raw milk and that *S. aureus* levels were reduced by 10(5)-fold. Additionally, this pathogenic agent counts remain low (10^4 CFU/g) after 6 weeks of ripening (Guo *et al.*, 2016).

Another example of bacteria spoilage occurs during biofuel ethanol fermentation. It is described that around 10^5–10^8 colony forming units (cfu) per millilitre of bacteria may be present in the system (Skinner and Leathers., 2004). LAB are considered a problem in this type of fermentations because they generate by-products that inhibit yeast growth. Consequently, LAB reduces the efficiency of the ethanol fermentation process. Phage cocktails are seen as a promissory alternative to antibiotics to control undesired bacteria population. Two phages (EcoSau and EcoInf) were isolated and added to a *L. fermentum*-contaminated maize mash fermentation model. Levels of ethanol were restored suggesting phages could act as an antibiotic-free strategy to control contaminants in the industry (Liu *et al.*, 2015).

Concluding remarks

Lactobacillus phage research has been relegated in comparison to phages infecting *Lactococcus* and *Streptococcus* strains. But in the past years, mainly because of the increasing number of *Lactobacillus* probiotic strains incorporated in dairy products, a larger knowledge about phages infecting these bacteria has been accumulated.

Understanding of the different steps of infection in combination with structural biology studies would bring a comprehensive picture of these entities in order to design strategies to avoid phage infection but also to design phage-derived tools for biotechnological applications and bacterial manipulation.

Web resources

https://pfam.xfam.org/
https://toolkit.tuebingen.mpg.de/
www.rcsb.org/
www.emdatabank.org/
https://www.ncbi.nlm.nih.gov/genomes/GenomesGroup.cgi taxid=10239&opt=Virus

References

Auvray, F., Coddeville, M., Ordonez, R.C., and Ritzenthaler, P. (1999). Unusual structure of the attB site of the site-specific recombination system of *Lactobacillus delbrueckii* bacteriophage mv4. J. Bacteriol. *181*, 7385–7389.

Barrangou, R., and Marraffini, L.A. (2014). CRISPR-Cas systems: Prokaryotes upgrade to adaptive immunity. Mol. Cell *54*, 234–244. https://doi.org/10.1016/j.molcel.2014.03.011

Barrangou, R., Fremaux, C., Deveau, H., Richards, M., Boyaval, P., Moineau, S., Romero, D.A., and Horvath, P. (2007). CRISPR provides acquired resistance against viruses in prokaryotes. Science *315*, 1709–1712.

Baugher, J.L., Durmaz, E., and Klaenhammer, T.R. (2014). Spontaneously induced prophages in *Lactobacillus gasseri* contribute to horizontal gene transfer. Appl. Environ. Microbiol. *80*, 3508–3517. https://doi.org/10.1128/AEM.04092-13

Beresford, T.P., Fitszimons, N.A., Brennan, N.L., and Cogan, T.M. (2001). Recent advances in cheese microbiology. Int. Dairy J. *11*, 259–274. https://doi.org/10.1016/S0958-6946(01)00056-5

Bouchard, J.D., and Moineau, S. (2000). Homologous recombination between a lactococcal bacteriophage and the chromosome of its host strain. Virology *270*, 65–75. https://doi.org/10.1006/viro.2000.0226

Bourand, A., Yebra, M.J., Boël, G., Mazé, A., and Deutscher, J. (2013). Utilization of D-ribitol by *Lactobacillus casei* BL23 requires a mannose-type phosphotransferase system and three catabolic enzymes. J. Bacteriol. *195*, 2652–2661. https://doi.org/10.1128/JB.02276-12

Breitbart, M., and Rohwer, F. (2005). Here a virus, there a virus, everywhere the same virus? Trends Microbiol. *13*, 278–284.

Briggiler Marcó, M., Garneau, J.E., Tremblay, D., Quiberoni, A., and Moineau, S. (2012). Characterization of two virulent phages of *Lactobacillus plantarum*. Appl. Environ. Microbiol. *78*, 8719–8734. https://doi.org/10.1128/AEM.02565-12

Brüssow, H., and Desiere, F. (2001). Comparative phage genomics and the evolution of Siphoviridae: Insights from dairy phages. Mol. Microbiol. *39*, 213–222.

Canchaya, C., Proux, C., Fournous, G., Bruttin, A., and Brüssow, H. (2003). Prophage genomics. Microbiol. Mol. Biol. Rev. *67*, 238–276.

Capra, M.L., Patrignani, F., Quiberoni, A.D., Reinheimer, J.A., Lanciotti, R., and Guerzoni, M.E. (2009). Effect of high pressure homogenization on lactic acid bacteria phages and probiotic bacteria phages. Int. Dairy J. *19*, 336–341. https://doi.org/10.1016/j.idairyj.2008.11.002

Capra, M.L., Mercanti, D.J., Reinhamer, J.A., and Quiberoni, A.L. (2010). Characterization of three temperate phages released from the same *Lactobacillus paracasei* commercial strain. Int. J. Dairy Technol. 63, 396–405. https://doi.org/10.1111/j.1471-0307.2010.00600.x

Capra, M.L., Mercanti, D.J., Rossetti, L.C., Reinheimer, J.A., and Quiberoni, A. (2011). Isolation and phenotypic characterization of *Lactobacillus casei* and *Lactobacillus paracasei* bacteriophage-resistant mutants. J. Appl. Microbiol. *111*, 371–381. https://doi.org/10.1111/j.1365-2672.2011.05056.x

Casey, E., Mahony, J., Neve, H., Noben, J., Bello, F.D., and van Sinderen, D. (2015). Novel phage group infecting *Lactobacillus delbrueckii* subsp. *lactis*, as revealed by genomic and proteomic analysis of bacteriophage Ldl1. Appl. Environ. Microbiol. *81*, 1319–1326. http://doi.org/10.1128/AEM.03413-14

Claesson, M.J., van Sinderen, D., and O'Toole, P.W. (2007). The genus *Lactobacillus* – a genomic basis for understanding its diversity. FEMS Microbiol. Lett. *269*, 22–28.

Coddeville, M., and Ritzenthaler, P. (2010). Control of directionality in bacteriophage mv4 site-specific recombination: Functional analysis of the Xis factor. J. Bacteriol. *192*, 624–635. https://doi.org/10.1128/JB.00986-09

Coddeville, M., Spinella, J.F., Cassart, P., Girault, G., Daveran-Mingot, M.L., Le Bourgeois, P., and Ritzenthaler, P. (2014). Bacteriophage mv4 site-specific recombination: The central role of the P2 mv4Int-binding site. J. Virol. *88*, 1839–1842. https://doi.org/10.1128/JVI.02735-13

Deltcheva, E., Chylinski, K., Sharma, C.M., Gonzales, K., Chao, Y., Pirzada, Z.A., Eckert, M.R., Vogel, J., and Charpentier, E. (2011). CRISPR RNA maturation by trans-encoded small RNA and host factor RNase III. Nature *471*, 602–607. https://doi.org/10.1038/nature09886

Deng, K., Fang, W., Zheng, B., Miao, S., and Huo, G. (2018). Phenotypic, fermentation characterization, and resistance mechanism analysis of bacteriophage-resistant mutants of *Lactobacillus delbrueckii* ssp. *bulgaricus* isolated from traditional Chinese dairy products. J. Dairy Sci. *101*, 1901–1914. https://doi.org/10.3168/jds.2017-13823

Desiere, F., Lucchini, S., Canchaya, C., Ventura, M., and Brüssow, H. (2002). Comparative genomics of phages and prophages in lactic acid bacteria. In Lactic Acid Bacteria: Genetics, Metabolism and Applications. Siezen, R.J., Kok, J., Abee, T., Schasfsma, G., eds. (Springer, Netherlands), pp. 73–91. https://doi.org/10.1007/978-94-017-2029-8_5

Desmyter, A., Farenc, C., Mahony, J., Spinelli, S., Bebeacua, C., Blangy, S., Veesler, D., van Sinderen, D., and Cambillau, C. (2013). Viral infection modulation and neutralization by camelid nanobodies. Proc. Natl. Acad. Sci. U.S.A. *110*, E1371–1379. https://doi.org/10.1073/pnas.1301336110

Dieterle, M.E., Jacobs-Sera, D., Russell, D., Hatfull, G., and Piuri, M. (2014a). Complete genome sequences of *Lactobacillus* Phages J-1 and PL-1. Genome Announc. 2, e00998–13. https://doi.org/10.1128/genomeA.00998-13

Dieterle, M.E., Bowman, C., Batthyany, C., Lanzarotti, E., Turjanski, A., Hatfull, G., and Piuri, M. (2014b). Exposing the secrets of two well-known *Lactobacillus casei* phages, J-1 and PL-1, by genomic and structural analysis. Appl. Environ. Microbiol. *80*, 7107–7121. https://doi.org/10.1128/AEM.02771-14

Dieterle, M.E., Fina Martin, J., Durán, R., Nemirovsky, S.I., Sanchez Rivas, C., Bowman, C., Russell, D., Hatfull, G.F., Cambillau, C., and Piuri, M. (2016). Characterization of prophages containing 'evolved' Dit/Tal modules in the genome of *Lactobacillus casei* BL23. Appl. Microbiol.

Biotechnol. *100*, 9201–9215. https://doi.org/10.1007/s00253-016-7727-x

Dieterle, M.E., Spinelli, S., Sadovskaya, I., Piuri, M., and Cambillau, C. (2017). Evolved distal tail carbohydrate binding modules of *Lactobacillus* phage J-1: A novel type of anti-receptor widespread among lactic acid bacteria phages. Mol. Microbiol. *104*, 608–620. https://doi.org/10.1111/mmi.13649

Díez-Villaseñor, C., Guzmán, N.M., Almendros, C., García-Martínez, J., and Mojica, F.J.M. (2013). CRISPR-spacer integration reporter plasmids reveal distinct genuine acquisition specificities among CRISPR-Cas I-E variants of *Escherichia coli*. RNA Biol. *10*, 792–802. https://doi.org/10.4161/rna.24023

Douglas, L.J., and Wolin, M.J. (1971). Cell wall polymers and phage lysis of *Lactobacillus plantarum*. Biochemistry *10*, 1551–1555.

Douillard, F.P., Ribbera, A., Kant, R., Pietilä, T.E., Järvinen, H.M., Messing, M., Randazzo, C.L., Paulin, L., Laine, P., Ritari, J., *et al.* (2013). Comparative genomic and functional analysis of 100 *Lactobacillus rhamnosus* strains and their comparison with strain GG. PLOS Genet. *9*, e1003683. https://doi.org/10.1371/journal.pgen.1003683

Duplessis, M., and Moineau, S. (2001). Identification of a genetic determinant responsible for host specificity in *Streptococcus thermophilus* bacteriophages. Mol. Microbiol. *41*, 325–336. https://doi.org/10.1046/j.1365-2958.2001.02521.x

Dupont, K., Vogensen, F.K., Neve, H., Bresciani, J., and Josephsen, J. (2004). Identification of the receptor-binding protein in 936-species lactococcal bacteriophages. Appl. Environ. Microbiol. *70*, 5818–5824.

Durmaz, E., Miller, M.J., Azcarate-Peril, M.A., Toon, S.P., and Klaenhammer, T.R. (2008). Genome sequence and characteristics of Lrm1, a prophage from industrial *Lactobacillus rhamnosus* strain M1. Appl. Environ. Microbiol. *74*, 4601–4609. https://doi.org/10.1128/AEM.00010-08

Emond, E., and Moineau, S. (2007). Bacteriophages and food fermentations. In Bacteriophage: Genetics and Molecular Biology, McGrath, S., and van Sinderen D., eds. (Caister Academic Press, UK), pp. 9–124.

FAO/WHO (2006). Probiotics in Food: Health and Nutritional Properties and Guidelines for Evaluation (FAO/WHO, Italy).

Feiner, R., Argov, T., Rabinovich, L., Sigal, N., Borovok, I., and Herskovits, A.A. (2015). A new perspective on lysogeny: Prophages as active regulatory switches of bacteria. Nat. Rev. Microbiol. *13*, 641–650. https://doi.org/10.1038/nrmicro3527

Garneau, J.E., and Moineau, S. (2011). Bacteriophages of lactic acid bacteria and their impact on milk fermentations. Microb. Cell Fact. 10 (Suppl. 1), S20. https://doi.org/10.1186/1475-2859-10-S1-S20

Garneau, J.E., Dupuis, M.È., Villion, M., Romero, D.A., Barrangou, R., Boyaval, P., Fremaux, C., Horvath, P., Magadán, A.H., and Moineau, S. (2010). The CRISPR/Cas bacterial immune system cleaves bacteriophage and plasmid DNA. Nature *468*, 67–71. https://doi.org/10.1038/nature09523

Goulet, A., Lai-Kee-Him, J., Veesler, D., Auzat, I., Robin, G., Shepherd, D.A., Ashcroft, A.E., Richard, E., Lichière, J., Tavares, P., *et al.* (2011). The opening of the SPP1 bacteriophage tail, a prevalent mechanism in Gram-positive-infecting siphophages. J. Biol. Chem. *286*, 25397–25405. https://doi.org/10.1074/jbc.M111.243360

Groth, A.C., and Calos, M.P. (2004). Phage integrases: Biology and applications. J. Mol. Biol. *335*, 667–678.

Guglielmotti, D., Marcó, M.B., Vinderola, C., de Los Reyes Gavilán, C., Reinheimer, J., and Quiberoni, A. (2007). Spontaneous *Lactobacillus delbrueckii* phage-resistant mutants with acquired bile tolerance. Int. J. Food Microbiol. *119*, 236–242.

Guinane, C.M., Kent, R.M., Norberg, S., Hill, C., Fitzgerald, G.F., Stanton, C., and Ross, R.P. (2011). Host specific diversity in *Lactobacillus johnsonii* as evidenced by a major chromosomal inversion and phage resistance mechanisms. PLOS ONE *6*, e18740. https://doi.org/10.1371/journal.pone.0018740

Guo, T., Zhang, C., Xin, Y., Xin, M., and Kong, J. (2016). A novel chimeric prophage vB_LdeS-phiJB from commercial *Lactobacillus delbrueckii* subsp. *bulgaricus*. J. Ind. Microbiol. Biotechnol. *43*, 681–689. https://doi.org/10.1007/s10295-016-1739-5

Hale, C.R., Zhao, P., Olson, S., Duff, M.O., Graveley, B.R., Wells, L., Terns, R.M., and Terns, M.P. (2009). RNA-guided RNA cleavage by a CRISPR RNA-Cas protein complex. Cell *139*, 945–956. https://doi.org/10.1016/j.cell.2009.07.040

Hatfull, G.F. (2015). Dark matter of the biosphere: The amazing world of bacteriophage diversity. J. Virol. *89*, 8107–8110. https://doi:.org/10.1128/JVI.01340-15

Horvath, P., Coûté-Monvoisin, A.C., Romero, D.A., Boyaval, P., Fremaux, C., and Barrangou, R. (2009). Comparative analysis of CRISPR loci in lactic acid bacteria genomes. Int. J. Food Microbiol. *131*, 62–70. https://doi.org/10.1016/j.ijfoodmicro.2008.05.030

Hultberg, A., Tremblay, D.M., de Haard, H., Verrips, T., Moineau, S., Hammarström, L., and Marcotte, H. (2007). *Lactobacillli* expressing llama VHH fragments neutralise *Lactococcus* phages. BMC Biotechnol. 7, 58.

Hyman, P., and Abedon, S.T. (2010). Bacteriophage host range and bacterial resistance. Adv. Appl. Microbiol. *70*, 217–248. https://doi.org/10.1016/S0065-2164(10)70007-1

Kenny, J.G., McGrath, S., Fitzgerald, G.F., and van Sinderen, D. (2004). Bacteriophage Tuc2009 encodes a tail-associated cell wall-degrading activity. J. Bacteriol. *186*, 3480–3491. https://doi.org/10.1128/JB.186.11.3480-3491.2004

Koskella, B., and Brockhurst, M.A. (2014). Bacteria-phage coevolution as a driver of ecological and evolutionary processes in microbial communities. FEMS Microbiol. Rev. *38*, 916–931. https://doi.org/10.1111/1574-6976.12072

Krumsiek, J., Arnold, R., and Rattei, T. (2007). Gepard: A rapid and sensitive tool for creating dotplots on genome scale. Bioinformatics 23, 1026–1028.

Kutter, E., Raya, R.R., and Carlson, K. (2005). Molecular mechanism of phage infection. In Bacteriophages: Biology and Application, Kutter, E., and Sulakvelidze, A., eds. (CRC Press, Boca Raton), pp. 165–222.

Labrie, S.J., and Moineau, S. (2007). Abortive infection mechanisms and prophage sequences significantly influence the genetic makeup of emerging lytic lactococcal phages. J. Bacteriol. *189*, 1482–1487.

Labrie, S.J., Samson, J.E., and Moineau, S. (2010). Bacteriophage resistance mechanisms. Nat. Rev. Microbiol. *8*, 317–327. https://doi.org/10.1038/nrmicro2315

Ledeboer, A.M., Bezemer, S., de Hiaard, J.J., Schaffers, I.M., Verrips, C.T., van Vliet, C., Düsterhöft, E.M., Zoon, P., Moineau, S., and Frenken, L.G. (2002). Preventing phage lysis of *Lactococcus lactis* in cheese production using a neutralizing heavy-chain antibody fragment from llama. J. Dairy Sci. *85*, 1376–1382.

Lepeuple, A.-S., Van Gemert, E., and Chapot-Chartier, M.-P. (1998). Analysis of the bacteriolytic enzymes of the autolytic *Lactococcus lactis* subsp. *cremoris* strain AM2 by renaturing polyacrylamide gel electrophoresis: Identification of a prophage-encoded enzyme. App. Environ. Microbiol., *64*, 4142–4148.

Lima-Mendez, G., Toussaint, A., and Leplae, R. (2011). A modular view of the bacteriophage genomic space: Identification of host and lifestyle marker modules. Res. Microbiol. *162*, 737–746. https://doi.org/10.1016/j.resmic.2011.06.006

Lin, C.F., Lo, T.C., Kuo, Y.C., and Lin, T.H. (2013). Stable integration and expression of heterologous genes in several lactobacilli using an integration vector constructed from the integrase and attP sequences of phage ΦAT3 isolated from *Lactobacillus casei* ATCC 393. Appl. Microbiol. Biotechnol. *97*, 3499–3507. https://doi.org/10.1007/s00253-012-4393-5

Liu, M., Bischoff, K.M., Gill, J.J., Mire-Criscione, M.D., Berry, J.D., Young, R., and Summer, E.J. (2015). Bacteriophage application restores ethanol fermentation characteristics disrupted by *Lactobacillus fermentum*. Biotechnol. Biofuels *8*, 132. https://doi.org/10.1186/s13068-015-0325-9

Lo, T.C., Shih, T.C., Lin, C.F., Chen, H.W., and Lin, T.H. (2005). Complete genomic sequence of the temperate bacteriophage PhiAT3 isolated from *Lactobacillus casei* ATCC 393. Virology *339*, 42–55.

Lortal, S., and Chapot-Chartier, M.P. (2005). Role, mechanisms and control of lactic acid bacteria lysis in cheese. Int. Dairy J. *15*, 857–871. https://doi.org/10.1016/j.idairyj.2004.08.024

Mahony, J., and van Sinderen, D. (2012). Structural aspects of the interaction of dairy phages with their host bacteria. Viruses *4*, 1410–1424. https://doi.org/10.3390/v4091410

Mahony, J., and van Sinderen, D. (2014). Current taxonomy of phages infecting lactic acid bacteria. Front. Microbiol. *5*, 7. https://doi.org/10.3389/fmicb.2014.00007

Mahony, J., and van Sinderen, D. (2015). Novel strategies to prevent or exploit phages in fermentations, insights from phage-host interactions. Curr. Opin. Biotechnol. *32*, 8–13. https://doi.org/10.1016/j.copbio.2014.09.006

Mahony, J., Ainsworth, S., Stockdale, S., and van Sinderen, D. (2012). Phages of lactic acid bacteria: the role of genetics in understanding phage-host interactions and their co-evolutionary processes. Virology *434*, 143–150. https://doi.org/10.1016/j.virol.2012.10.008

Mahony, J., Kot, W., Murphy, J., Ainsworth, S., Neve, H., Hansen, L.H., Heller, K.J., Sørensen, S.J., Hammer, K., Cambillau, C., *et al.* (2013). Investigation of the relationship between lactococcal host cell wall polysaccharide genotype and 936 phage receptor binding protein phylogeny. Appl. Environ. Microbiol. *79*, 4385–4392. https://doi.org/10.1128/AEM.00653-13

Marcó, M.B., Moineau, S., and Quiberoni, A. (2012). Bacteriophages and dairy fermentations. Bacteriophage *2*, 149–158. https://doi.org/10.4161/bact.21868

Mercanti, D.J., Carminati, D., Reinheimer, J.A., and Quiberoni, A. (2011). Widely distributed lysogeny in probiotic lactobacilli represents a potentially high risk for the fermentative dairy industry. Int. J. Food Microbiol. *144*, 503–510. https://doi.org/10.1016/j.ijfoodmicro.2010.11.009

Mercanti, D.J., Rousseau, G.M., Capra, M.L., Quiberoni, A., Tremblay, D.M., Labrie, S.J., and Moineau, S. (2016). Genomic diversity of phages infecting probiotic strains of *Lactobacillus paracasei*. Appl. Environ. Microbiol. *82*, 95–105. https://doi.org/10.1128/AEM.02723-15

Mikkonen, M., Räisänen, L., and Alatossava, T. (1996). The early gene region completes the nucleotide sequence of *Lactobacillus delbrueckii* subsp. *lactis* phage LL-H. Gene *175*, 49–57.

Moineau, S., and Lévesque, C. (2005). Control of bacteriophages in industrial fermentation. In Bacteriophages: Biology and Application, Kutter, E., and Sulakvelidze, A., eds. (CRC Press, Boca Raton), pp. 285–296.

Moineau, S., Pandian, S., and Klaenhammer, T.R. (1994). Evolution of a lytic bacteriophage via DNA acquisition from the *Lactococcus lactis* chromosome. Appl. Environ. Microbiol. *60*, 1832–1841.

Muñoz-Provencio, D., Rodríguez-Díaz, J., Collado, M.C., Langella, P., Bermúdez-Humarán, L.G., and Monedero, V. (2012). Functional analysis of the *Lactobacillus casei* BL23 sortases. Appl. Environ. Microbiol. *78*, 8684–8693. https://doi.org/10.1128/AEM.02287-12

Munsch-Alatossava, P., and Alatossava, T. (2013). The extracellular phage-host interactions involved in the bacteriophage LL-H infection of *Lactobacillus delbrueckii* ssp. *lactis* ATCC 15808. Front. Microbiol. *4*, 408. https://doi.org/10.3389/fmicb.2013.00408

Murphy, J., Bottacini, F., Mahony, J., Kelleher, P., Neve, H., Zomer, A., Nauta, A., and van Sinderen, D. (2016). Comparative genomics and functional analysis of the 936 group of lactococcal Siphoviridae phages. Sci. Rep. *6*, 21345. https://doi.org/10.1038/srep21345

Nes, I.F., Brendehaug, J., and von Husby, K.O. (1988). Characterization of the bacteriophage B2 of *Lactobacillus plantarum* ATCC 8014. Biochimie *70*, 423–427.

Piuri, M., Sanchez-Rivas, C., and Ruzal, S.M. (2003). Adaptation to high salt in *Lactobacillus*: Role of peptides and proteolytic enzymes. J. Appl. Microbiol. *95*, 372–379.

Rabinovich, L., Sigal, N., Borovok, I., Nir-Paz, R., and Herskovits, A.A. (2012). Prophage excision activates *Listeria* competence genes that promote phagosomal escape and virulence. Cell *150*, 792–802. https://doi.org/10.1016/j.cell.2012.06.036

Räisänen, L., Draing, C., Pfitzenmaier, M., Schubert, K., Jaakonsaari, T., Von Aulock, S., Hartung, T., and

Alatossava, T. (2007). Molecular interaction between lipoteichoic acids and *Lactobacillus delbrueckii* phages depends on D-alanyl and α-glucose substitution of poly(glycerophosphate) backbones. J. Bacteriol. *189*, 4135–4140. https://doi.org/10.1128/JB.00078-07

Ravin, V., Räisänen, L., and Alatossava, T. (2002). A conserved C-terminal region in Gp71 of the small isometric-head phage LL-H and ORF474 of the prolate-head phage JCL1032 is implicated in specificity of adsorption of phage to its host, *Lactobacillus delbrueckii*. J. Bacteriol. *184*, 2455–2459.

Raya, R.R., Klaenhammer, T.R., Scienc, F., and Daiuy, S. (1989). Characterization of the temperate bacteriophage phiadh and plasmid transduction in *Lactobacillus acidophilus* ADH. Appl. Environ. Microbiol. *55*, 2206–2213.

Revilla-Guarinos, A., Gebhard, S., Alcántara, C., Staron, A., Mascher, T., and Zúñiga, M. (2013). Characterization of a regulatory network of peptide antibiotic detoxification modules in *Lactobacillus casei* BL23. Appl. Environ. Microbiol. *79*, 3160–3170. https://doi.org/10.1128/AEM.00178-13

Riipinen, K.A., Forsman, P., and Alatossava, T. (2011). The genomes and comparative genomics of *Lactobacillus delbrueckii* phages. Arch. Virol. *156*, 1217–1233. https://doi.org/10.1007/s00705-011-0980-5

Rochat, T., Bermúdez-Humarán, L., Gratadoux, J.J., Fourage, C., Hoebler, C., Corthier, G., and Langella, P. (2007). Anti-inflammatory effects of *Lactobacillus casei* BL23 producing or not a manganese-dependant catalase on DSS-induced colitis in mice. Microb. Cell Fact. *6*, 22.

Samson, J.E., and Moineau, S. (2013). Bacteriophages in food fermentations: New frontiers in a continuous arms race. Annu. Rev. Food Sci. Technol. *4*, 347–368. https://doi.org/10.1146/annurev-food-030212-182541

Sanozky-Dawes, R., Selle, K., O'Flaherty, S., Klaenhammer, T., and Barrangou, R. (2015). Occurrence and activity of a type II CRISPR-Cas system in *Lactobacillus gasseri*. Microbiology *161*, 1752–1761. https://doi.org/10.1099/mic.0.000129

Sciara, G., Bebeacua, C., Bron, P., Tremblay, D., Ortiz-Lombardia, M., Lichière, J., van Heel, M., Campanacci, V., Moineau, S., and Cambillau, C. (2010). Structure of lactococcal phage p2 baseplate and its mechanism of activation. Proc. Natl. Acad. Sci. U.S.A. *107*, 6852–6857. https://doi.org/10.1073/pnas.1000232107

Sechaud, L., Cluzel, P.J., Rousseau, M., Baumgartner, A., and Accolas, J.P. (1988). Bacteriophages of lactobacilli. Biochimie *70*, 401–410.

Shimizu-Kadota, M., Sakurai, T., and Tsuchida, N. (1983). Prophage origin of a virulent phage appearing on fermentations of *Lactobacillus casei* S-1. Appl. Environ. Microbiol. *45*, 669–674.

Stefanovic, E., Fitzgerald, G., and McAuliffe, O. (2017). Advances in the genomics and metabolomics of dairy lactobacilli: A review. Food Microbiol. *61*, 33–49.

Szczepankowska, A.K., Gorecki, R.K., Koakowski, P., and Bardowski, J.K. (2013). Lactic acid bacteria resistance to bacteriophage and prevention techniques to lower phage contamination in dairy fermentation. In Lactic Acid Bacteria: R & D for Food, Health and Livestock Purposes, Kongo, J.M., ed. (InTechOpen), pp. 23–72. http://doi.org/10.5772/51541

Valyasevi, R., Sandine, W.E., and Geller, B.L. (1991). A membrane protein is required for bacteriophage c2 infection of *Lactococcus lactis* subsp. *lactis* C2. J. Bacteriol. *173*, 6095–6100.

Vasu, K., and Nagaraja, V. (2013). Diverse functions of restriction-modification systems in addition to cellular defense. Microbiol. Mol. Biol. Rev. *77*, 53–72. https://doi.org/10.1128/MMBR.00044-12

Veesler, D., and Cambillau, C. (2011). A common evolutionary origin for tailed-bacteriophage functional modules and bacterial machineries. Microbiol. Mol. Biol. Rev. *75*, 423–433. https://doi.org/10.1128/MMBR.00014-11

Veesler, D., Dreier, B., Blangy, S., Lichière, J., Tremblay, D., Moineau, S., Spinelli, S., Tegoni, M., Plückthun, A., Campanacci, V., *et al.* (2009). Crystal structure and function of a DARPin neutralizing inhibitor of lactococcal phage TP901-1: Comparison of DARPin and camelid VHH binding mode. J. Biol. Chem. *284*, 30718–30726. https://doi.org/10.1074/jbc.M109.037812

Veesler, D., Robin, G., Lichière, J., Auzat, I., Tavares, P., Bron, P., Campanacci, V., and Cambillau, C. (2010). Crystal structure of bacteriophage SPP1 distal tail protein (gp19.1): A baseplate hub paradigm in gram-positive infecting phages. J. Biol. Chem. *285*, 36666–36673. https://doi.org/10.1074/jbc.M110.157529

Vegge, C.S., Brøndsted, L., Neve, H., Mc Grath, S., van Sinderen, D., and Vogensen, F.K. (2005). Structural characterization and assembly of the distal tail structure of the temperate lactococcal bacteriophage TP901-1. J. Bacteriol. *187*, 4187–4197.

Vegge, C.S., Vogensen, F.K., Mc Grath, S., Neve, H., van Sinderen, D., and Brøndsted, L. (2006). Identification of the lower baseplate protein as the antireceptor of the temperate lactococcal bacteriophages TP901-1 and Tuc2009. J. Bacteriol. *188*, 55–63.

Ventura, M., Canchaya, C., Kleerebezem, M., de Vos, W.M., Siezen, R.J., and Brüssow, H. (2003). The prophage sequences of *Lactobacillus plantarum* strain WCFS1. Virology *316*, 245–255.

Ventura, M., Canchaya, C., Pridmore, R.D., and Brüssow, H. (2004). The prophages of *Lactobacillus johnsonii* NCC 533: Comparative genomics and transcription analysis. Virology *320*, 229–242. https://doi.org/10.1016/j.virol.2003.11.034

Ventura, M., Canchaya, C., Bernini, V., Altermann, E., Barrangou, R., McGrath, S., Claesson, M.J., Li, Y., Leahy, S., Walker, C.D., *et al.* (2006). Comparative genomics and transcriptional analysis of prophages identified in the genomes of *Lactobacillus gasseri*, *Lactobacillus salivarius*, and *Lactobacillus casei*. Appl. Environ. Microbiol. 72, 3130–3146. https://doi: 10.1128/AEM.72.5.3130-3146.2006

Verreault, D., Gendron, L., Rousseau, G.M., Veillette, M., Massé, D., Lindsley, W.G., Moineau, S., and Duchaine, C. (2011). Detection of airborne lactococcal bacteriophages in cheese manufacturing plants. Appl. Environ. Microbiol. *77*, 491–497. https://doi.org/10.1128/AEM.01391-10

Villion, M., and Moineau, S. (2009). Bacteriophages of L. Front. Biosci. *14*, 1661–1683.

Vinogradov, E., Sadovskaya, I., Grard, T., and Chapot-Chartier, M.P. (2016). Structural studies of the rhamnose-rich cell wall polysaccharide of *Lactobacillus casei* BL23. Carbohydr. Res. *435*, 156–161.

Xin, K.Q., Hoshino, Y., Toda, Y., Igimi, S., Kojima, Y., Jounai, N., Ohba, K., Kushiro, A., Kiwaki, M., Hamajima, K., *et al.* (2003). Immunogenicity and protective efficacy of orally administered recombinant *Lactococcus lactis* expressing surface-bound HIV Env. Blood *102*, 223–228. https://doi.org/10.1182/blood-2003-01-0110

Yokokura, T. (1971). Phage receptor material in *Lactobacillus casei* cell wall. I. Effect of L-rhamnose on phage adsorption to the cell wall. Jpn. J. Microbiol. *15*, 457–463.

Yokokura, T., Kodaira, S., Ishiwa, H., and Sakurai, T. (1974). Lysogeny in lactobacilli. J. Gen. Microbiol. *84*, 277–284. https://doi.org/10.1099/00221287-84-2-277

Zago, M., Orrù, L., Rossetti, L., Lamontanara, A., Fornasari, M.E., Bonvini, B., Meucci, A., Carminati, D., Cattivelli, L., and Giraffa, G. (2017). Survey on the phage resistance mechanisms displayed by a dairy *Lactobacillus helveticus* strain. Food Microbiol. *66*, 110–116. https://doi.org/10.1016/j.fm.2017.04.014

DNA Transfer in *Lactobacillus*: An Overview

8

María Mercedes Palomino[1,2], Joaquina Fina Martin[1,2], Mariana C. Allievi[1,2], María Eugenia Dieterle[1,2], Carmen Sanchez-Rivas[1,2] and Sandra M. Ruzal[1,2*]

[1]Universidad de Buenos Aires, Facultad de Ciencias Exactas y Naturales, Departamento de Química Biológica, Buenos Aires, Argentina.
[2]CONICET – Universidad de Buenos Aires, Instituto de Química Biológica de la Facultad de Ciencias Exactas y Naturales (IQUIBICEN), Buenos Aires, Argentina.

*Correspondence: sandra@qb.fcen.uba.ar

https://doi.org/10.21775/9781910190890.08

Abstract

An overview of the fundamental basis for the introduction of DNA into *Lactobacillus* species is given in this chapter. Natural ways for introducing DNA such as transformation, transduction or conjugation are reported in few strains. No evidence for natural competence is available in particular in probiotic *Lactobacillus* strains, although various components of natural gene transfer systems are present in various genomes. Protoplast fusion and transfection are also described. The most popular method used for introducing foreign DNA is electroporation, mostly based on cell wall weakening prior to electroshock. The composition of the envelope, in particular the presence or not of an S-layer, and the content of endogenous plasmids in the different *Lactobacillus* species result in different transformation efficiencies. The use of ssRNA as donor and CRISPR–Cas as the recombineering factor open the way for obtaining recombinants with the need of high efficiency DNA transfer systems to succeed in the genetic modifications.

Introduction

Since the 1980s molecular biology has evolved and developments of methods for genetic manipulation of strains were necessary to study bacterial genetics and metabolism. New molecular tools like vectors and systems to modify *loci* in the genome had been uncover and used in the definition of gene function. Development of transformation methods to optimize the delivery of plasmid or lineal DNAs into *Lactobacillus* bacterial cells is fundamental in genetic manipulation to engineer these bacteria. Introduction of DNA into cells is necessary for investigations of gene structure and function. However, the behaviour and responses of probiotic and commensal lactic acid bacteria (LAB) to accept exogenous DNA and recombine are different and correlated with their capacity to respond to multi-stress conditions.

Three main types of horizontal gene transfer (HGT) mechanisms recognized in bacteria are: conjugation, transduction and transformation.

Conjugation and phage transduction involve replicative elements (plasmids, phages or phage-like particles). HGT events resulting from conjugation and transduction can be considered as a side effect of the propagative behaviour of these replicative elements. In contrast, natural transformation does not depend on external vehicles and is innate to the species (Blokesch, 2016, 2017).

Different strategies to make *Lactobacillus* competent have been described. In this chapter, we will update the methods described for transduction and conjugation in *Lactobacillus*. We would detail components of the natural gene transfer systems found in various genomes. It would be worthwhile to remark that evidences for natural transformation or natural competence has essentially been observed in pathogenic strains in example *Streptococcus pneumoniae* (Charpentier *et al.*, 2012). Several reports of conjugation and transduction for probiotic lactobacilli have been published, however with low efficiencies, restricting its use for obtaining recombinants. Transfer of a non-conjugative vector by means of conjugative mobilization has been performed in *Lactobacillus* (Jacobsen *et al.*, 2007; Muriana and Klaenhammer, 1987). On the other hand, *in vitro* transformation has been performed with protoplasts although this method yields low transfer or recombinants rates. In addition, intrageneric and intergeneric protoplast fusion and transfection have been successfully applied to lactobacilli. These methods include protoplast fusion or PEG induced transfection. Nevertheless the highest transfer rates were obtained by means of artificial transfer using electroporation. Procedures to modify cells walls have been developed to improve DNA uptake in *Lactobacillus*. Electroporation is the most widely used method for introducing DNA and a variety of techniques have been developed to increase the efficiency of electrotransformation.

Genetic modification of lactobacilli using traditional genetic systems has helped to understand and establish functionalities regarding their probiotic or pathogenic characteristics as well as metabolic functions. Recently, the ability to perform targeted mutagenesis in the chromosome of probiotic *Lactobacillus* strains using ssDNA recombineering was demonstrated (van Pijkeren and Britton, 2012) or was being developed (Xin *et al.*, 2017). Moreover, numerous *Lactobacillus* strains harbour CRISPR–Cas systems in their genomes, with an unusually high frequency of Type II systems. In *L. reuteri* CRISPR–Cas9 selection combined with single-stranded DNA (ssDNA) recombineering has been a successful approach for obtaining mutants at high efficiencies (Briner and Barrangou, 2016; Horvath *et al.*, 2009; Oh and van Pijkeren, 2014). The presence CRISPR–Cas Type II systems in these bacteria perhaps reflects the widespread of phages exposure, creating opportunities for the exploitation of endogenous systems for genome editing (Hidalgo-Cantabrana *et al.*, 2017; Stout *et al.*, 2017). In view of this new technology, highly efficient DNA transfer systems are needed to succeed in the genetic modifications.

In this chapter we describe an overview of the fundaments and methods for transfer or introduction of DNA into *Lactobacillus*.

Natural gene transfer systems: transformation, transduction and conjugation

Transformation

Naturally transformable bacteria use a highly conserved DNA uptake system to internalize DNA and integrate it in their chromosome by homologous recombination (Attaiech and Charpentier, 2017). Bacteria can undergo genetic transformation by actively integrating genetic information from related or unrelated organisms. Transferable antibiotic resistance is a major health concern and should be investigated in *Lactobacillus* since they might experience horizontal gene transfer with the autochthonous microbiota and they might potentially serve as hosts of antibiotic resistance genes, with the risk of transferring these genes to other bacteria, however this fact has never been clearly established. Insight of acquired antibiotic resistance mechanisms in *Lactobacillus* are described in Chapter 10.

When Gram-positive transformable species enter a state of natural competence, exogenous DNA translocation proceeds through the DNA-uptake machinery, comprising the proteins ComEA, ComEC, ComFA, ComFC and a nuclease (EndA in *Streptococcus pneumoniae*) encoded by the late competence (*com*) genes. It was proposed that genetic transformation contributes to pneumococcal plasticity and plays a central role in the adaptation of this species to host defences since although SOS is generally considered as widespread in bacteria, *S. pneumoniae* was reported to lack SOS repair and inducible mutagenesis (Charpentier *et al.*, 2012). Therefore the absence of SOS is compensated by the induction of the *com* functions; a gene dosage effect was observed in arrested bacteria where the location of *com* genes near the origin of replication allow activation of competence functions (Slager *et al.*, 2014).

Other late competence genes encode for proteins that compose pili-like structures (ComGA-GG) or protect internalized DNA against degradation (SsbA, SsbB, DprA and RecA). Competence relies on an alternative σ factor to positively regulate gene expression. The best characterized alternative σ factor, σX of *S. pneumoniae,* is encoded by two identical *comX* genes and is widely distributed among the Firmicutes. ComX of non-sporulating streptococci is the master activator of the genetic competence pathway, participating directly in the transcription of late com genes. In *S. thermophilus,* the quorum sensing ComRS complex regulated expression of *comX,* comprising the pheromone-like peptide ComS and transcriptional regulator ComR, encoded by the *comRS* operon. Addition of a synthetic peptide that resembles the active competence pheromone has proven a successful strategy to induce natural competence in several bacterial species.(Mulder *et al.,* 2017). In *Lactococcus lactis* constitutively or inducible overexpression of *comX* resulted in activation of natural transformation (David *et al.,* 2017). In *Bacillus subtilis,* σH, a characteristic sporulation regulator, starts the process as a final response to starvation that activates a complex response leading to the development of genetic competence and to spore formation as an ultimate outcome.

In attempts to predict a new transformable species, one strategy is to identify a putative central competence regulator although it is important to bear in mind that its identification might not be sufficient to define the conditions that are most favourable for the spontaneous transformation of a species. In Fig. 8.1 we show the genetic context of *sigH* and sigH-like, *comX, comEA* genes in members of *Lactobacillus.*

In *Lactobacillus sakei,* the coding sequence for a factor ComX/σH (sigH), orthologous to *B. subtilis* was found in the genome (Schmid *et al.,* 2012). This observation suggested a link between σH type factors and natural competence in non-sporulators, such as the genus *Lactobacillus.* Through a microarray study, when overexpressing the *sigH* gene of *L. sakei,* activation of genes mainly involved in genetic competence was observed, however they failed to detect natural genetic transformation. Similarly, although *Streptococcus pyogenes* encodes all the genes that are required for natural transformation as well as the alternative sigma factor σX, it has not been possible to transform this species under laboratory conditions, despite attempts to synthetically activate the *com* regulon and the demonstration of natural activation of the *com* regulon by the competence-inducing peptide XIP via σX. It has been suggested that transformation is blocked at the stage of DNA uptake, but this requires further study (Woodbury *et al.,* 2006).

We have investigated the presence of ComX coding in the genome of *Lactobacillus casei* BL23 (Fina Martin *et al.,* 2017). We found YP_001986877.1, Gene ID: 6404650 annotated as alternative sigma factor of the RNA polymerase, *comX/sigH* orthologous. We evaluated if there is a condition that was able to modify this gene expression. For this purpose we checked conditions that resembled those found in the gastrointestinal tract like salt stress or acid pH with bile salts and compared them to those known to promote competence in *Bacillus,* including starvation, UV or heating. Expression of *comX* gene was analysed by qPCR and related to their expression in early stationary phase growth condition and the housekeeping gene 16S rRNA. Results showed an increase in the expression of *comX* in the condition of UV and acid pH with bile salts. We also verified if the transfer of an antibiotic resistance marker (Cm^R) bring by plasmidic DNA was obtained in any of these conditions by quantifying presence of Cm^R transformants with the most probable number (MPN). No Cm^R transformants were obtained in any of the conditions tested compared to the non-induced condition. Although an induction of expression of *comX* was observed but it was not enough to provide competence for plasmid uptake and transformation to *L. casei* (Fina Martin *et al.,* 2017). We speculate that competence functions found in *Lactobacillus* genome might provide this genus with systems for DNA uptake to be used as a nutrient acquisition rather than for obtaining and processing DNA for genetic transformation and recombination as suggested for Gram-negatives (Finkel and Kolter, 2001)

The results in *L. sakei* and *L. casei* argue in favour of the non-natural transformability of lactobacilli although it is not sufficient to ensure the total absence of transfer mechanisms in these bacteria. At least four potential warnings must be considered to disesteem or not their influence in DNA uptake. First, it is possible that there is an alternative

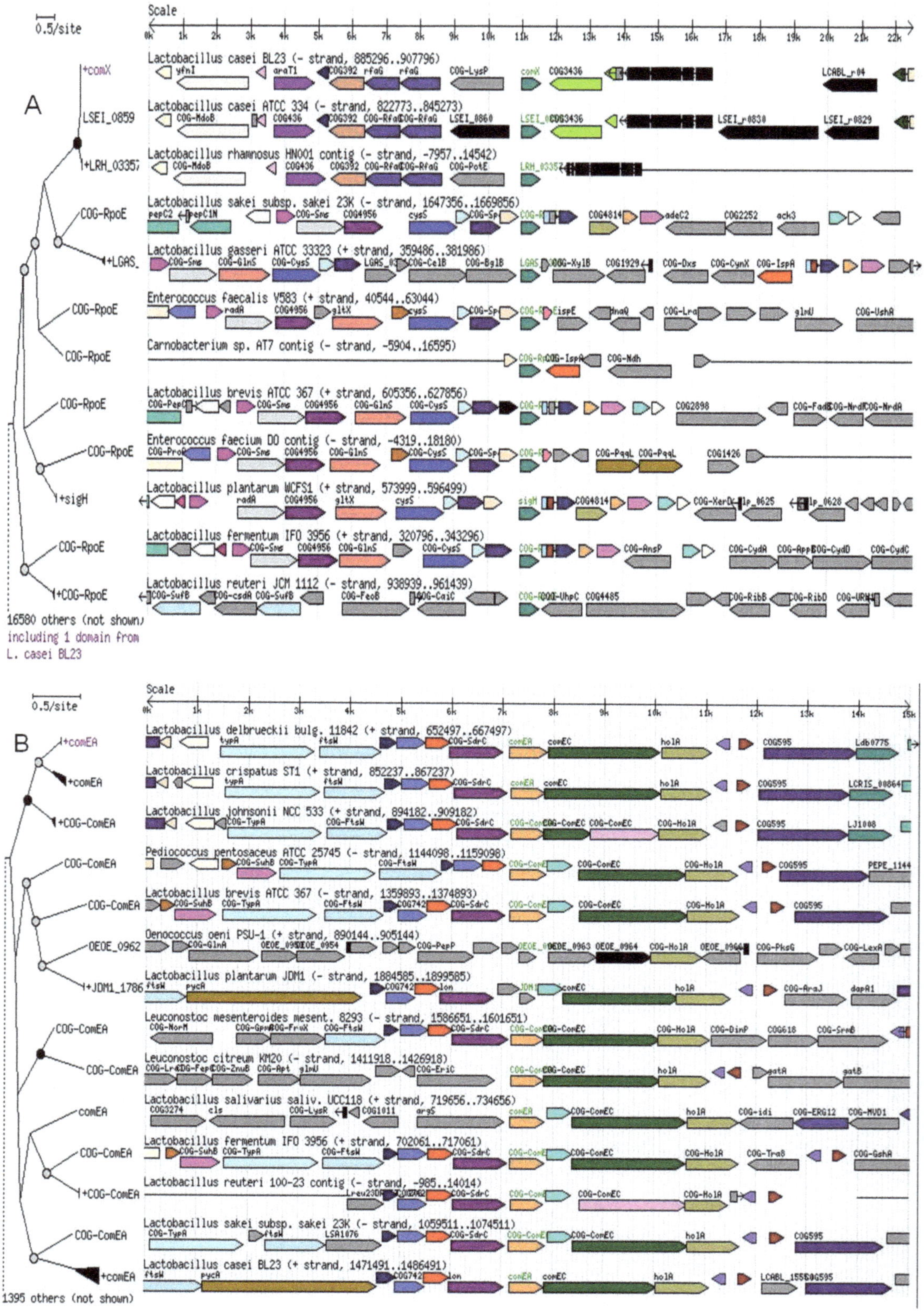

Figure 8.1 The genetic context of *sigH* and sigH-like, *comX* (A) and *comEA* (B) genes in members of *Lactobacillus*. Filled circles represent a well-supported split in the tree. Grey circles represent a moderately-support split in the tree. The figure was generated using the MicrobesOnline facility: http://www.microbesonline.org.

mechanism of regulation. Second, the possible existence of species-competence-inducing signals (Johnston *et al.*, 2014). Third DNA uptake system is directly subjected to silencing by small non-coding RNA which prevents expression (Attaiech and Charpentier, 2016) or CRISPR–Cas processing (Mohanraju *et al.*, 2016). Finally the last step leading to recombinants, the homologous recombination should be proficient (time for induction of the recombinases) or require unknown conditions (Yang *et al.*, 2015).

Universal trends indicate that recombination between species decreases exponentially with sequence divergence (Reenen and Dicks, 2011), however horizontal gene transfer allows an organism to effectively compete in a new environment. In an ever changing environment such as the gastrointestinal tract, one can speculate that the introduction of new organisms such as probiotics may eventually lead to the acquisition or loss of specific functions. Fortunately, major gene loss seems to occur rather than gain in Lactobacillales, which indicates early adaptation to nutritionally rich environments like that in the gastrointestinal tract (Makarova *et al.*, 2006). Furthermore, as we now know that lactic acid bacteria as members of the intestinal microbiota have clustered regularly interspaced short palindromic repeats (CRISPRs) and their associated genes (Hidalgo-Cantabrana *et al.*, 2017; Horvath *et al.*, 2009; Makarova *et al.*, 2015). CRISPR–Cas are significantly represented in Proteobacteria and Firmicutes, especially those of medical importance (Godde and Bickerton, 2006; Horvath *et al.*, 2009; Makarova *et al.*, 2015; Marraffini, 2015; Mohanraju *et al.*, 2016). A high percentage of *Lactobacillus* (62.9%) encodes CRISPR–Cas systems. CRISPR immunity has recently been linked to speciation events in several genera, including, *Lactobacillus* (Sun *et al.*, 2015). CRISPRs memory permits the recognition and neutralization of the invaders on the subsequent infections in spite of the frequent horizontal transference of genes CRISPR–Cas loci (Mohanraju *et al.*, 2016). Hence the CRISPR–Cas system may function as a barrier for gene transfer by blocking non-self-conjugative plasmids. Computational analysis of the composition of CRISPR-based metagenome sequencing data are feasible and it would provides an efficient approach to finding new potential CRISPR matrices and to analyse the ecosystem and history of lactobacilli associated with microbiomes as well as those free-living species (Mangericao *et al*, 2016) to determine whether CRISPR–Cas systems influence the establishment of recombinant DNA.

Transduction

Transduction occurs when bacterial DNA is included in bacteriophages during replication, causing the transfer of foreign DNA to other bacterial genomes in subsequent phage-bacterial infections. Both lytic and lysogenic bacteriophages for *Lactobacillus* species have been described and the genetic analysis of some *Lactobacillus* phages is quite advanced and is the subject of Chapter 7.

While *Lactobacillus*-specific bacteriophages occur in abundance, isolated from diverse habitats, and numerous natural isolates harbour temperate bacteriophage, however horizontally transfer genes by incorporation into host genomes via transduction have a limited host range. Few reports of transduction in *Lactobacillus* have been published. Generalized transduction has been reported in *Lactobacillus salivarius* (Toyama *et al.*, 1971). Luchansky *et al.* (1989) employed a phage lysate of *L. acidophilus* ADH carrying pGK12 by inducing the native temperate phage (Øadh) with mitomycin C, although low frequencies were obtained, 10^{-8} transductants/pfu. In a subsequent study, Raya and Klaenhammer (Raya and Klaenhammer, 1992) showed transfer of vector plasmids by transduction and were able to increase the transduction frequency of plasmid pGK12 in *Lactobacillus gasseri* ADH by 10^{2}- to 10^{5}-fold by inserting restriction fragments of bacteriophage into the vector. Frequency of plasmid transduction increase by three orders when inserting restriction fragments of bacteriophage into the vector. The increase frequency correlated with the degree of homology existing between the phage and plasmid DNAs. In light of today's knowledge about CRISPR it seems clear that avoiding the nuclease system enables plasmid stability.

Strains characterized genomically showed prophage-associated genes. It has been hypothesized that prophages may be involved in lateral gene transfer, by embarking extra-chromosomal elements into their genome. Streptococci, lactobacilli and bifidobacteria are known to be prone to phage infections notably during industrial dairy and other food fermentations as well as in the gut environment (Douillard *et al.*, 2018). Presence of prophages in different *Lactobacillus* strains has been reported (Mercanti *et al.*, 2016). Prophage stability should be considered when using these strains since induction of resident prophages and concomitant cell lysis are an ongoing threat for the dairy industry. High rates of spontaneous induction of prophages and their ability to acquire bacterial genes and transduce them to related strains were described in *Lactobacillus gasseri* ADH suggesting that temperate bacteriophages likely contribute to horizontal gene transfer (HGT) (Baugher *et al.*, 2014). Dieterle *et al.* (2016) identified in the well-studied probiotic strain *Lactobacillus casei* BL23, three complete prophages (PLE1, PLE2, and PLE3) located at different integration sites. All of them showed mosaic architectures with homologous sequences to *Streptococcus, Lactococcus, Lactobacillus,* and *Listeria* phages or strains. Using a combination of quantitative real-time PCR, genomics, and proteomics, they showed that PLE2 and PLE3 can be induced in the presence of mitomycin C but with different kinetics, although PLE1 remains as a prophage due to a small deletion in the integration/immunity region that could account for this different behaviour. However in the absence of mitomycin C spontaneous induction occurred at very low rates for all prophages, indicating a very low frequency of the excision event or that excision only occurs in a small bacterial population. This low induction rate, lack of SOS response and recombination deficiency observed in this strain, would be related to that observed for a prophage mutant repressor in *Bacillus subtilis* (Rubinstein *et al.*, 1993) and should be further investigated.

Conjugation

Conjugative transfer of DNA is mediated by cell-to-cell contact, mating-pair formation and plasmid DNA transfer through conjugative pili. Examples of homologous and heterologous conjugation systems in lactobacilli have been reported. Although plasmids are very common in lactobacilli, literature on the conjugal transfer of native *Lactobacillus* plasmids is limited. Native conjugal transfer systems have been demonstrated for *L. casei* (Chassy and Rokaw, 1981) and *L. acidophilus* (Muriana and Klaenhammer, 1987) for the transfer of lactose fermenting ability and bacteriocin production and immunity, respectively and reviewed by Gasson and Fitzgerald (Gasson and Fitzgerald, 1994). Reniero and co-workers (Reniero *et al.*, 1992) reported that the production of an aggregation-promoting protein stimulated the uptake of pAMβ1 in *L. plantarum* strain with transfer frequencies as high as 10^{-2} using solid matings. Langella and co-workers (Langella *et al.*, 1996) observed that conjugal transfer of pAMβ1 between *L. sakei* strains is also possible on solid surface agar, although at lower frequencies.

Examples of heterologous conjugation systems in laboratory had been shown for *L. plantarum, L. reuteri, L. fermentum* and *L. murinus* that can function as donors of pAMβ1 to other lactic acid bacteria *in vitro* and in the gastrointestinal tract of mice (Gasson and Fitzgerald, 1994). Transfer of a non-conjugative vector by means of conjugative mobilization has been performed with *L. plantarum* (Shrago and Dobrogosz, 1988). Also *Lactobacillus* isolated from fermented dry sausages that harboured the tetracycline resistance gene *tet(M)* previously reported in the pathogenic species *Neisseria meningitidis* and *Staphylococcus aureus,* were able to transfer tetracycline resistance *in vitro* to *Enterococcus faecalis* at frequencies ranging from 10^{-4} to 10^{-6} transconjugants per recipient and erythromycin resistance could be transferred from *L. plantarum* to *E. faecalis* in the gastrointestinal tracts of gnotobiotic rats (Gevers *et al.*, 2003; Jacobsen *et al.*, 2007). In a later study, Mater and co-workers (Mater *et al.*, 2008) showed the transfer of *vanA* resistance from *E. faecium* to a commercially available probiotic strain of *L. acidophilus* was achieved *in vivo.*

Several mobile elements have been found in lactobacilli, including ISL2 in *Lactobacillus helveticus,* ISL3 in *Lactobacillus delbrueckii,* IS1223 in *Lactobacillus johnsonii,* IS1163 and IS1520 in *Lactobacillus sakei* and ISLp11 in *Lactobacillus plantarum.* DNA sequence comparisons of various LAB indicated that during cheese manufacturing it was possible for insertion elements to be horizontally transferred, most likely through conjugation (Nicoloff and Bringel, 2003). IS elements are known to be involved in chromosomal deletions and/or rearrangements, thus playing a role in ecological adaptation and species diversification (Makarova *et al.*, 2006).

Most *Bifidobacterium* strains and *Lactobacillus* strains were predicted to have a relative low number of transposons, when compared to the 27 IS transposons harboured by *Streptococcus thermophilus* (Douillard *et al.*, 2018). Conjugative transposons are found in several species of the genus *Lactobacillus* and are described in Chapter 10.

It was suggested that food *Lactobacillus* spp. can act as reservoir organisms of acquired antibiotic resistance genes that can be disseminated to other bacteria (van Reenen and Dicks, 2011), However, the indigenous gut microbiota severely restricts transfer, thus minimizing the number of detectable events (Feld *et al.*, 2008). Despite the fact that a relatively low transfer frequency between *L. plantarum* and *E. faecalis* was observed *in vitro*, in that study authors showed that a high concentration of transconjugants can be obtained in the gastrointestinal tract of gnotobiotic rats and this concentration can be considerably further raised during treatment with erythromycin (Jacobsen *et al.*, 2007). However, there are a variety of defence mechanisms that protect bacteria against foreign mobile DNA elements including genes encoding for restriction–modification (R/M) systems, i.e. restriction endonucleases and methyltransferases and CRISPR–Cas loci.

Artificial means of DNA transfer: protoplast and electroporation

Protoplast transformation

Protoplast transformations in *Lactobacillus* using protocols and plasmids of other Gram-positive bacteria have been reviewed (Mercenier and Chassy, 1988). Successful transformation of DNA into protoplasts of *L. acidophilus* has been reported. Protocols involved controlled removal of the cell wall using a muralytic enzyme: lysozyme, mutanolysin or a combination of both have been used to remove cell walls. Watanabe *et al.* (1990) obtained protoplasts of *Lactobacillus casei* using phage lysin for removing the cell wall.

The stabilization of protoplasts in an osmotically protective medium is needed to avoid osmosis of water into the protoplasts and lyses. Composition of protoplast buffers is quite complex and includes an osmotic stabilizer (for example, sucrose or succinate) and salts (Gasson and Fitzgerald, 1994). Mixtures of protoplasts and DNA are incubated for few minutes with polyethylene glycol (PEG) and dilutions are platted on media allowing wall-regeneration of protoplasts and further selection of transformants. PEG is an essential element in the protoplast transformation process. Generally PEG with molecular weights between 1000 and 6000 at concentrations ranging between 20% and 40% has been used for transformation of most lactic acid bacteria with treatment times ranging between 2 and 20 minutes. In other Gram-positive bacteria such as *Bacillus subtilis*, for protoplast transformation with plasmids, regeneration of the cell wall on solid media is required before the selection of the plasmid coded functions (antibiotic resistance) are expressed. In fact, regeneration in liquid media can be obtained but plasmidic function cannot be induced until cell wall was regenerated (Sanchez-Rivas, 1988). Also for strains already bearing plasmids, a high incidence of plasmid curing is specially obtained in this condition either in *Bacillus subtilis* (Cofré and Sanchez-Rivas, 1983) and *Staphylococcus aureus* (Novick *et al.*, 1980).

However transformation frequencies are generally very low (Boizet *et al.*, 1988; John *et al.*, 2008; Luchansky *et al.*, 1988; Singhvi *et al.*, 2010; Tanaka and Ohmomo, 2001; Watanabe *et al.*, 1990) and no mayor progress has been made with protoplast mainly due to the difficulties in protoplast regeneration of *Lactobacillus*.

Protoplast regeneration

The procedure of protoplast transformation is time consuming, very sensitive to experimental variation and operator changes and is extremely strain specific. A limiting factor is the protoplast capacity to regenerate the cell wall and to begin cell division. Protoplast regeneration has been successful only for some strains of the *Lactobacillus* genus such as *Lactobacillus reuteri*, *Lactobacillus gasseri* (Connell *et al.*, 1988; Vescovo *et al.*, 1984) and *Lactobacillus casei* (Lee-Wickner and Chassy, 1984).

For bacterial protoplast regeneration, to avoid possible osmosis of water into the protoplasts and their subsequent lyses, an environment of high colloid osmotic pressure may be necessary due to their extreme fragility. Improvement of the regeneration medium was obtained by replacing or increasing colloidal substances such as gelatin, agar, agarose or calcium alginate. Media with extremely high concentrations of colloidal substances enabled efficient protoplast regeneration, as did increasing the $MgCl_2$ concentration in the medium up to 0.1 M. The higher regeneration frequency of *L. curvatus* protoplasts on calcium alginate medium was obtained when added as an overlay on gelatine plate medium (Tanaka and Ohmomo, 2001).

It is difficult to establish a universal method of protoplast regeneration for all lactobacilli, mainly due to differences in their requirement for a colloidal substance in protoplast regeneration.

Protoplast fusion

Protoplast fusion is a common recombination method and allows recombination between virtually any two or more cells. As we described above protoplasts are cells stripped of their cell wall by digestion with cell wall-digesting enzymes. Fusion is promoted by subjecting protoplasts through an electric pulse or by incubating them in the presence of PEG or surfactants that alter membrane fluidity. Recombination can then take place with genetic material from two or more cells enclosed within a single plasma membrane. Fusants are thereafter allowed to regenerate, and viable recombinants can be submitted to screening and selection. However a phenomenon of genetic inactivation has been pointed in *Bacillus subtilis* ex-fusants that reveal again the instability of regenerating protoplasts to express functions (Hotchkiss and Gabor, 1980; Grandjean *et al.*, 1998).

However, protoplast fusion open the possibility of inactivated one parent strain, either isogenic or not (Rubinstein and Sanchez Rivas, 1989) and to transfer functions or plasmids (Baigori *et al.*, 1988). Also in this variant one parent protoplast can be inactivated by exposure to UV light or heat and exposure to PEG fusion allow recover plasmids or functions by recombination to repair fatal lesions.

In *Lactobacillus*, protoplast fusion has been reported by Iwata *et al.* (1986) between isogenic derivatives of *Lactobacillus fermentum* and fusants harbouring tetracycline- and erythromycin-resistance plasmids derived from each of the parental strains were obtained. Kanatani *et al.* (1990) have described protoplast fusion between isogenic *Lactobacillus plantarum* derivatives during which recombination between a range of chromosomal markers (including those for isoleucine, leucine and phenylalanine requirement) occurred.

Intergeneric protoplast fusion has also been demonstrated among lactic acid bacteria. Cocconcelli *et al.* (1986) have described intergeneric protoplast fusion between *L. lactis* and *Lactobacillus reuteri* in which transfer of plasmid occurred at a frequency of approximately 10^{-5}, while recombinants of the *Lactobacillus* strain able of fermenting trehalose were recovered at a frequency of 8.3×10^{-5}.

Classical methods for microbial strain improvement depend on either mutagenesis followed by selection for improved properties, or manipulation of specific genes known to play an important role in the desired phenotype. Genome shuffling of selected strains is an efficient method for evolution of strains of microbes with desirable phenotypes (John *et al.*, 2008). The pooled population is shuffled by homologous recombination using protoplast fusion to allow DNA exchange and recombination. This procedure allows rapid evolution of strains with multiple beneficial mutations, for example acid tolerance in *Lactobacillus* (Patnaik *et al.*, 2002; Wang *et al.*, 2007). An acid-tolerant mutant of *Lactobacillus delbrueckii* was crossed by protoplast fusion with *Bacillus amyloliquefaciens*, a bacterium notable for its efficient starch utilization. In this example, the phenotypes sought were found in two distinct organisms, whose genomes were used as starting diversity to construct an ideal cell for the production of desirable products that would grow rapidly to high density, be sustained at high viability even under no or minimal growth, utilize an inexpensive substrate and secrete the product of interest at high rate, maximum yield with little or no by-products (Wang *et al.*, 2007).

Protoplast transfection with phage DNA

When studying phage and phage mutants, in order to manipulate phage DNA and avoiding *in vitro* encapsidation or even to use phage DNA as a tool for bacteria manipulation, a recombinant phage DNA transfection system is necessary in *Lactobacillus* spp. strains. DNA Electroporation followed by phage particle formation has been described (Chassy and Flickinger, 1987); however, efficiency was low and we were unable to reproduce it in our laboratory. It is believed that big DNA molecules (>45Kbp) along with low recombination efficiencies described for *L. casei* (recombination between different phage DNA particles is needed in order to restore complete phage DNA) and could be part of the problem. Therefore, protoplast transfection seems to be a good alternative. Since there is no need of protoplast regeneration for phages to multiply in the cytoplasm of a protoplast and render bacteriophage infective particles there are several reports of phage DNA transfection. Either naked phage DNA or encapsulated in liposome have been used to transform various *Lactobacillus* species protoplasts. Boizet *et al.* (1988) transfected *Lactobacillus delbrueckii* protoplasts with phage DNA at a high frequency but this was achieved only after incubation of transfected cells for 24 h prior to plating in an overlay agar containing the indicator host. Transfection of *L. casei* protoplasts by naked PL-1 DNA has been achieved only in the presence of a high concentration of PEG (Watanabe *et al.*, 1990).

In our hands protoplast of *L. casei* BL23 were transfected with phage DNA. We developed an improved protoplast transfection system for *L. casei* BL23 and efficiencies up to 1.8×10^4 PFU/µg DNA were obtained. Transfection efficiencies (PFU/µg DNA PL-1) increased with incubation time post-transfection from 10^2 after 4 hours up to 10^4 after 8 hours' incubation. Addition of sucrose increases the initial title by a factor of two however no increase of phage final production was obtained by further incubation. The increase up to 10^9 PFU/µg after overnight incubation would result from new re-infections of bacteria by phage particles generated after new phage cycles. A detailed protocol is described in Box 8.1.

Electroporation

Electroporation is the process of introducing macromolecules such as transforming DNA into bacterial cells. It would probably be more proper to refer to the process as 'electrotransformation' or 'transformation by electropermeabilization' as suggested by Chassy *et al.* (1988). Electroporation has a number of advantages over other more traditional methods for the introduction of DNA into cells; the most significant being the amount of DNA required is reduced. Since electroporation utilizes a physical rather than biological mechanism of DNA transfer, the technique has the power to overcome many of the barriers that are normally found when using more traditional techniques. For example, electroporation can be utilized for the direct movement of nonconjugative plasmids across species barriers, making it a powerful tool in molecular biology. For

Box 8.1 Protoplast transfection protocol

Protoplast formation

L. casei BL23 cells were grown in MRS NaCl 0.8 M until OD 600 nm: 1.
Cells were washed in water and resuspended in protoplast buffer (0.5 M sucrose; 10 mM $MgCl_2$; 0.02 M sodium maleate pH 6.5; 10 mM $CaCl_2$, 0.2% glycine-betaine) to an OD 600 nm: 2.5.
Mutanolysin 25 U/ml and lysozyme 250 µg/ml were added and incubated for 2.5 hours at 37°C. 100× optical microscope was used to follow the protoplast formation. Up to 99% of total cells form protoplasts in this conditions.
Protoplasts were frozen at −70°C until use.

PL-1 phage DNA transfection
1 ml of protoplasts was centrifuged at 3500 g for 5 minutes and the pellet resuspended in 0.3 ml of 10 mM $MgCl_2$; 10 mM $CaCl_2$ and 0.6 M sucrose).
500 ng of phage DNA and 30% of PEG4000 were added and incubated for 5 minutes at room temperature.
1 ml of MRS was then added and cells centrifuged at 1500 g. Pellets were resuspended in 1 ml of MRS. Different incubation times at 37°C were tested and 200 µl was taken for phage recovery. 100 µl was used for phage tittering in MRS and the other 100 µl was used in MRS plus 0.6 M sucrose. For phage titration, *L. casei* BL23 cells were used as indicator cells.

many cell types, electroporation is either the most efficient or the only mean known to provide foreign DNA gene transfer easy and rapid into a variety of different micro-organisms (Nickoloff, 1995).

Although the molecular mechanism of electrotransformation is not completely understood, the electrical pulse at low electric field strengths produces that the membranes of the cells become polarized is thought to result in a rearrangement of cell wall and membrane components to generate transient pores through which DNA can pass into the cell. The efficiency of electro-transformation depends on the level of permeabilization and pretreatment with chemicals that cells are subjected to. Electroporation efficiency is often strongly affected by growth conditions and growth phase at the time of cell harvest. Therefore, detailed information about growth conditions for the particular organism is needed.

Because specific approaches are not always successful, comparisons of procedures used with similar (or even quite different) organisms might provide valuable insight to researchers working to solve a particular problem. Many of the procedures for electrotransformation of different organisms are similar, differences are often important, especially when an experimental design requires optimum transformation. However it is specifically design to each bacterial species and compounds used to alter their cell wall permeability. In this section we will compare methods up to date.

Since the first report on transformation of *Lactobacillus casei* by electroporation (Chassy and Flickinger, 1987), several procedures for the electrotransformation of different *Lactobacillus* strains and species have been published and reviewed (Aukrust *et al.*, 1995a; Beasley *et al.*, 2004; Kim *et al.*, 2005; Landete *et al.*, 2014; Nickoloff, 1995; Palomino *et al.*, 2010; Spath *et al.*, 2012; Welker *et al.*, 2015) although the transformation efficiency is extremely variable, ranging from 10^2 to 10^7 transformants/μg of DNA.

Limited transformability is in part consequence of their thick cell wall, which is highly resistant to mechanical disruption. In particular, a high resistance to cell wall hydrolases like lysozyme or mutanolysin is observed in several lactobacilli species (Piuri *et al.*, 2005).

Many factors can influence the transformation efficiency; some are linked to the plasmid origin, size and replication mechanisms, whereas others are linked to the characteristics of the strains used. Optimization of a number of parameters including the growth phase, cell density, medium compositions and electric conditions has been reported to affect the electrotransformation efficiency:

1. the growth stage at which cells are harvested, which depends on the species or even the strain used;
2. the composition of the wash and electroporation buffers, which has been shown to play an important role in the transformation of several lactobacilli;
3. the parameters of the electrical pulse. Electrical pulse strength and length must be tested for the optimization of electroporation efficiency for a certain species of bacteria.
4. the source of the DNA used to transform, since restriction-modification systems can severely inhibit transformation with foreign DNA.

Another difficulty in the development of a transformation procedure is the choice of the plasmid used. Indeed, the ability of a plasmid to replicate and to express its selectable marker in a given species is not predictable. The use of an endogenous plasmid carrying a native selectable marker often solves this problem. The electroporation efficiencies should be compared with different plasmids of different sizes, selection markers, copy number per cell and replication mechanism. Also the plasmid source should be defined since restriction-modification systems can inhibit transformation with foreign DNA (Alegre *et al.*, 2004). Traditional strategies to overcome these defence systems involved modifying the methylation pattern using different intermediate hosts with varying methylation patterns or incubating the plasmid with a commercially available DNA methyltransferase (Spath *et al.*, 2012). Palomino *et al.* (2010) found no difference when plasmid pNZ273, prepared from a *L. casei* transformant, was used to electroporate the plasmid-free strain compared to the plasmid was isolated from *Escherichia coli*. However electrotransformation of *Lactobacillus plantarum* CECT 220 (ATCC 8014) with plasmid DNA isolated from *Escherichia coli* grown in minimal medium and *in vitro* modification of the DNA by cell-free extracts of the host *L. plantarum* showed improved transformation. These strategies were shown to increase the transformation efficiencies in the human gut- associated *Lactobacillus plantarum* (Spath *et al.*, 2012). The size of the plasmid is also an important factor that should be evaluated. The same replication origin for pRV500 derivatives of different sizes resulted in similar efficiencies (Palomino *et al.*, 2010).

Establishment of plasmid might be controlled by CRISPR systems. The occurrence of clustered regularly interspaced short palindromic repeats (CRISPRs) found in a wide diversity of prokaryotes (Marraffini, 2015). These elements are short palindromic sequences 24–40 bp in length that contain highly conserved inner and terminal inverted repeats of up to 11 bp and are usually adjacent to CRISPR-associated (*Cas*) genes. More than one CRISPR locus may occur in a particular genome, and wide variation occurs in the CRISPR loci of strains of the same species. CRISPRs provide acquired resistance not only to phages but also to plasmids (Godde and Bickerton, 2006; Horvath *et al.*, 2009).

One of the most important factors to be considered to improve the introduction of a foreign DNA into bacteria is the structure and density of cell wall. Especially Gram-positive bacteria with their thick cell wall show lower frequency and efficiency of transformation than Gram-negative bacteria (Trevors *et al.*, 1992). Because the composition of growth medium directly affects the cell wall density and thickness, a number of investigation have been tried to improve the transformation efficiency by weakening the cell walls by the addition of glycine, and treatment with lysozyme or penicillin. Pre-treatments used to increase transformation efficiency have been studied: treatment of cells with penicillin G (Wei *et al.*, 1995), glycine (Holo and Nes 1989; Thompson and Collins, 1996) and application of osmotic stress (Palomino *et al.*, 2010); We have previously observed in *L. casei* BL23 that growth in high salt increases its sensitivity to lyses, probably as a consequence of the decreased peptidoglycan cross-linking (Piuri *et al.*, 2005). The addition of glycine in the growth medium inhibits formation of cross-linkages in the cell wall where L-alanine is replaced by glycine (Hammes *et al.*, 1973), thereby weakening the cell wall. This might also affect the response to electric pulse.

Also, the membrane composition contributes to the success of the electroporation. Strains containing the fatty acid lactobacillic acid cyclopropane ((1R,2S)-2-hexylcyclopropanedecanoic acid) (Broadbent *et al.*, 2014), are transformed with reasonable efficiency, while the group lacking this fatty acid exhibited lower transformation efficiencies and more variable results (Aukrust and Blom, 1992; Aukrust *et al.*, 1995b). Lowering the temperature congeals the lipid membrane by decreasing fluidity and stabilizing the ionic shield. Electric pulse creates an imbalance on either side of the bacterial membrane and the hypothesis is that some of these rearrangements consist of temporary aqueous pathways ('pores'), with the electric field playing the dual role of causing pore formation and providing a local driving force for ionic and molecular transport through the pores and DNA can swept through the generated pore.

Results show that polysaccharides have an adverse effect on transformation efficiency. They may represent an additional barrier to DNA penetration, or, more likely, the effect may be due to less efficient washing and concentration of the cells. When magnesium was included in the washing solution there was less dissociated slime present after washing, giving a positive effect on transformation frequency. Medium shift to reduce slime production had a favourable effect on the transformation results (Aukrust and Blom, 1992; Aukrust *et al.*, 1995b).

Gram-positive cell envelope polymers and membrane lipids have negatively charges which repelled those of DNA phosphates. In chemical transformation cations are added (Ca^{2+}, Mg^{2+}) to interact with the negative charges of the membrane creating an electrostatically neutral situation. This is clearly not possible in an electroporation process were low ionic strength is a requirement to increase resistance and avoid the risk of arcing.

The presence of plasmids in *Lactobacillus* strains may reduce the transformation efficiency (due to incompatibility or rearrangements). Host specificity also plays an important role in the transformation efficiency. Many researchers (Posno *et al.*, 1991; Serror *et al.*, 2002) had proposed that these instabilities may be caused by incompatibility between foreign and endogenous plasmid, differences in restriction-modification systems of host and plasmid size.

A key factor of this procedure was the requirement of high cell density, since, when less than 10^{10}

cfu/ml were used, low levels of transformation were obtained. High cell densities and high voltages in the presence of 10% glycerol caused an increase in time constant and reduced the risk of arcing.

Optimizing the voltage is considered a crucial factor for obtaining successful transformation. The transformation efficiencies increased for some species when the voltage was enhanced. Also resistance between 100 Ω and 800 Ω are needed to be tested to establish optimum transformation frequencies.

Pulse number is another parameter to modify. Kim *et al.* (2005) were able to increase transformation efficiency at a pulse number of 10, whereas transformation efficiency decreased at a pulse number of 15. Electric pulses damage the bacterial cell membrane and allow the foreign DNA enter into the cells. An optimization of the electrical conditions is necessary to have the highest transformation efficiency with minimum damage of the cells. Determination of cell death following electroporation is always a necessary parameter to evaluate the procedure.

When using large DNA molecules, like intact phage DNA, one possible problem concerning low transformation frequencies is that big DNA molecules (>45 kbp) might not be transfer as an intact molecule and would requires recombination between different fragments of phage DNA in order to restore complete molecule. Therefore activation of the recombination machinery might result in increase transformation frequencies. As we have described before, absence of prophage induction in *L. casei* BL23 seems to be related to the presence of a deficient SOS system. To improved efficiency we hypothesized that co-introduction of a damaged DNA will induce SOS system and hence recombination functions. Co-electroporation of phage DNA together with a UV irradiated plasmid from Gram-negative origin was tested. This procedure improved phage particles formation which duplicate PFU/μg phage DNA efficiencies although they were still low.

Optimization of the *in vivo* recombination process is a way to enhance the mutants selection efficiency. Yang and coworkers (Yang *et al.*, 2015) examined the induction time of the recombinases in *Lactobacillus plantarum* after electroporation between 3 and 8 h. They found that allele replacement occurs during recovery cultivation. Therefore, by extending the recovery cultivation time they increased the selection efficiency (Yang *et al.*, 2015)

Another new protocol involving treatment of cells with lithium acetate and dithiothreitol (DTT) was shown to increase the electrotransformation efficiency of *Lactococcus lactis* subsp. *lactis* as well as *L. plantarum*, *L. buchneri* (Spath *et al.*, 2012) and *L. casei* (Welker *et al.*, 2015). In this last work the effect of growth to different cell densities, either with or without glycine, and with or without 0.9 M NaCl, as well as with water or with lithium acetate/DTT pretreatment on the electrotransformation efficiencies of five strains isolated from different environmental origins was evaluated. Using this combined approach and optimized electroporation conditions, they obtained improved transformation efficiencies of 10^6 CFU/μg of transforming DNA

However no unique protocol is best for all strains. For example, strains ATCC 334 and BL23 transformed with the highest efficiency when grown with 0.9 M NaCl and pretreated with lithium acetate/DTT solution but had 10- to 100-fold lower transformation efficiencies when grown with 1% glycine and pretreated with water. In contrast, strain 12A transformed with highest efficiency when grown with 1% glycine and pretreated with water. Strains 32G and A2-362 had lower transformation efficiencies than the other strains. These findings show that a combined approach employing multiple strategies for changing cell membrane fluidity, cell wall integrity and electroporation parameters can improve the transformation efficiency of *L. casei* (Welker *et al.*, 2015).

Table 8.1 summarizes electroporation protocols with all methods reported up to date.

Concluding remarks

Explanations for variations in transformation efficiencies are numerous and include variation in membrane lipid composition, affected by Tween 80, a component of MRS (Broadbent *et al.*, 2014), variation in cell wall composition and integrity, affected by growth in the presence of NaCl (Palomino *et al.*, 2013) or glycine (Holo and Nes, 1989) and by treatment with lithium acetate/DTT (Spath *et al.*, 2012), and the presence or absence of capsules, incompatible endogenous plasmids or restriction endonucleases targeting the transforming DNA.

Table 8.1 Comparison of electroporation protocols

Species	Dilution and growth	Wash solution	Electroporation buffer	Concentration factor from initial growth	Voltage	Recovery	Range of transformants/µg DNA (type)	Reference
L. casei	MRS	EB buffer: 7 mM HEPES pH 7.4, 272 mM sucrose, 1 mM $MgCl_2$	EB buffer on ice	100	5000 V/cm at 25 µF	LCM, glucose 1 g/l	10^4 (PL-1 phage DNA 40 kb)	Chassy and Flickinger (1987)
L. casei, *L. pentosus*, *L. plantarum*, *L. acidophilus*, *L. fermentum*, and *L. brevis*	1/50 in MRS + 1% glycine, 37°C, 34 hours	5 mM NaH_2PO_4 1 mM $MgCl_2$ (ice cold)	0.3 M Sucrose 5 mM NaH_2PO_4 1 mM $MgCl_2$ (ice cold)	100	6250–7000 V/cm, 25 µF 100Ω	90 min. MRS	10^2–10^7 (vectors based on cryptic plasmids or shuttle vectors)	Posno (1991)
L. plantarum and *L. sake*	1/50 in MRS, 1% glycine, 30°C, OD600.=0.6	1 mM $MgCl_2$ (RT°C)	30% PEG 1500	100	7500 V/cm, 400 Ω	2 hours MRS, 0.5 M sucrose, 0.1 M $MgCl_2$	Up to 10^5 (plasmid DNA)	Aukrust (1992)
L. fermentum and *L. casei*	1/50 in MRS, 1% glycine, 37°C	5 mM NaH_2PO_4 1 mM $MgCl_2$ (0°C)	0.9 M sucrose, 3 mM $MgCl_2$	100	12,500 V/cm, 25 µF 400 Ω	2 hours MRS, 0.5 M sucrose, 0.1 M $MgCl_2$	10^4–10^6	Wei (1995)
L. plantarum	1/20 in MRS, 6% glycine, 37°C	2 × water, 1 × 50 mM EDTA, 2 × 0.3 M sucrose (ice cold)	0.3 M sucrose	50	7500 V/cm, 25 µF, 200 Ω	2 hours MRS	10^4 (plasmid DNA)	Thompson (1996)
L. rhamnosus	MRS2% glycine,	7 mM potassium phosphate pH 7.4, 0.5 M sucrose, 1 mM $MgCl_2$	7 mM potassium phosphate pH 7.4, 0.5 M sucrose, 1 mM $MgCl_2$	100	1.5 kV, 0.2 cm, 200Ω, 25 µF	MRS, 20 mM $MgCl_2$, 2 mM $CaCl_2$	Not reported	Varmanen (1998)
L. delbrueckii	MRS, 42°C to OD600 = 1.7	3 × ice cold, 1 mM $MgCl_2$, 5 mM KH_2PO_4	0.4 M sucrose, 1 mM $MgCl_2$, 5 mM KH_2PO_4, pH 6	30 Heat shock 45°C for 20 min., then ice for 10 min.	5000 V/cm, 25 µF, 800 Ω,	3 hours 0.2 M sucrose, 5% skim milk, 0.1% yeast extract, 1% casamino acids, 25 mM $MgCl_2$) 37°C	10^4 (plasmid DNA)	Serror (2002)

Table 8.1 Continued

Species	Dilution and growth	Wash solution	Electroporation buffer	Concentration factor from initial growth	Voltage	Recovery	Range of transformants/µg DNA (type)	Reference
L. plantarum	1/10 in LAB, 37°C, MRS, 30°C	21 × 10 ml 10 mM chilled $MgCl_2$, 1 × 0 ml chilled 0.5 M sucrose, 10% glycerol	0.5 M sucrose, 10% glycerol	30	13,000 V/cm, 200 Ω, 25 µF	2 hours MRS, 80 mM $MgCl_2$	10^3–10^5	Alegre (2004)
L. acidophilus, *L. helveticus* and *L. brevis*	1/50 in 1% glycine, 37°C, OD600 = 0.20.3	Ice 10 min., 21 × w/5 mM NaH_2PO_4, 1 mM $MgCl_2$ (0°C)	1 M sucrose, 3 mM $MgCl_2$	100	12,500 V/cm	2 hours MRS 0.5 M Sucrose, 0.1 M $MgCl_2$	10^4	Kim (2005)
L. casei, *L. paracasei*, *L. delbreuckii*, *L. plantarum* and *L. acidophilus*	Grown with 0.7–0.9 M NaCl	Ice-cold water	Ice-cold water	100	2.5 kV, 200 Ω, 25 µF, 12,500 V/cm	2 hours MRS	10^5–10^6	Palomino (2010)
L. casei, *L. brevis*, *L. rhamnosus* and *L. reuteri*	1/50 MRS, 1% glycine, 0.3 M sucrose	0.3 M sucrose, 10% glycerol, 5 mM KH_2PO_4 pH 7.4, 2 mM $MgCl_2$	0.3 M sucrose, 10% glycerol, 5 mM KH_2PO_4 pH 7.4, 2 mM $MgCl_2$	100	1.7 kV, 0.2 cm, 200Ω, 25 µF	MRS, 0.3 M sucrose, 20 mM $MgCl_2$, 2 mM $CaCl_2$	10^3–10^6	Landete (2014)
L. casei	Grown with 0.9 M NaCl or grown with 1% glycine	Sterile distilled water	100 mM lithium acetate 10 mM dithiothreitol	100	Variable parameters	0.5 M sucrose in MRS broth	Up to10^6	Welker (2015)
L. casei and L. plantarum	2% inoculum overnight cultures in MRS with 0.75 M sorbitol and 1% glycine cultured at 37°C statically	952 mM sucrose, 3.5 mM $MgCl_2$ in ice for 10 min.	952 mM sucrose, 3.5 mM $MgCl_2$	100	2000 V, 25 µF, 400 Ω	SMRS (MRS supplemented with 0.1 M $MgCl_2$ and 0.5 M sucrose) recovered at 37°C for 1 hour	10^2–10^4	Yang (2015), Xin (2017)

Cell stress probably occurs because of relatively non-specific chemical exchange with the extracellular environment. Whether or not the cell survives probably depends on the cell type, the extracellular medium composition, and the ratio of intra- to extracellular volume.

Future trends

Curing of plasmids

Plasmids are extrachromosomal genetic elements, autonomously replicating that are widely present in the genus *Lactobacillus*. About 38% of the species in this genus contain endogenous plasmids (Fang *et al.*, 2008). The functions of these plasmids have traditionally been correlated with phenotypical properties, including drug resistance, carbohydrate metabolism, amino acid metabolism, and bacteriocin production (Pouwels and Leer, 1994). Plasmid free strains are useful for genetic manipulations and the elimination of stable plasmids may help in the construction of new strains with desirable probiotic characteristics. Also in genetic analysis the complementation with plasmids is done by verifying that curing of the plasmid is accompanied by loss of the acquired phenotype.

Usually plasmid curing agents are toxic and mutagenic molecules like acriflavine, ethidium bromide, novobiocin and SDS are used (Karthikeyan and Santosh, 2010; Spengler *et al.*, 2006). Therefore efficient non toxic procedures for plasmid curing in *Lactobacillus* have been sought.

The effectiveness of curing obtained by protoplasting was demonstrated for *Lactobacillus reuteri* (Vescovo *et al.*, 1984) as was previously reported for *Staphylococcus aureus* (Novick *et al.*, 1980) and *Bacillus subtilis* (Cofre et al., 1983).

Electroporation has been used to cure *E. coli* of plasmid DNA. Electrocuring and direct transfer procedures of prokaryotes involve two steps: preparation of cells and the electroporation procedure. Both of these steps can affect the overall efficiency and success of DNA transfer or curing. Heery *et al.* (1989) reported a curing efficiency of 80–90%. However, the level of curing is dependent on both the strain being cured and the plasmid to be cured. However, curing of plasmids by electroporation generates higher frequencies of plasmid-free cells than traditional curing methods. Whether this approach would be applicable to *Lactobacillus* remains to be investigated.

Non-canonical DNA transfer

Besides the classical well-known mechanisms of transformation, conjugation and transduction and their canonical mobile genetic elements (MGE) involved, such as plasmids, transposons and bacteriophages, in the last years, the list of transfer strategies has been expanded beyond these canonical ones (García-Aljaro *et al.*, 2017). Lateral gene transfer frequency in nature plays a major role in prokaryote genome evolution. Recently described structures membrane vesicles (MVs) and nanotubes are considered to influence the lateral acquisition of genes between species (Popa and Dagan, 2011).

Membrane vesicles (MVs) are released from all living cells. The mechanism of MVs biogenesis is not yet fully understood. MVs are lumen-containing nanometric spheres of lipid-bilayers derived from the cell surface. MVs are biologically active and contain various components, including genetic material. Genetic material about 100 Kbp has been detected inside MVs, although smaller fragments are more frequent that can be chromosomal, plasmid and/or of phage origin, as well as different types of RNA have been detected in MVs. MV-mediated transfer of genes coding for antibiotic resistance, virulence and metabolic traits has been reported (Brown *et al.*, 2015; Domingues and Nielsen, 2017). MVs are elements that contribute to HGT and are worth mentioning here. Strategy to communicate between beneficial bacteria and intestinal mucosa cells; and MVs could shuttle mediators, which modulated the host immune and defence responses. Mediated transfer of mRNAs and ssRNAs is a novel mechanism of genetic exchange between cells. MVs have been described in *Lactobacillus* species, *L. reuteri*, *L. plantarum* and *L. casei*, and found to be associated with DNA (Domínguez Rubio *et al.*, 2017; Grande *et al.*, 2017; Li *et al.*, 2017).

An additional transfer mechanism – Nanotubes – was discovered recently. These are tubular protrusions composed of membrane components that can bridge between neighbouring cells and conduct the transfer of DNA and proteins. The exact mechanism of DNA transfer via Nanotubes is yet unknown (Dubey and Ben-Yehuda, 2011).

Examples of integeneric transfer that are proposed to be mediated by nanotuces are: sucrose utilization in *Pediococcus pentosaceus* from *Lactobacillus plantarum*, *Oenococcus oeni* malolactic fermentation in wine is connected with three different *Lactobacillus* species (*L plantarum* as donor and *L. casei* and *L. brevis* as recipients) (Popa and Dagan, 2011). Nanotubes are yet to be found in lactobacilli. The YmdB protein (locus CAB13570), a phosphodiesterase that was reported to function by decreasing the expression of motility genes, was found to be required for nanotube formation and intercellular molecular exchange in *Bacillus subtilis* (Dubey *et al.*, 2016). Remarkably when analysing with microbial protein BLAST search the *Lactobacillus* (taxid 1578) it is possible to find sequences producing significant alignments with over 60% of identity to that of *B. subtilis* in *Lactobacillus* species like *L. sakei*, *L. brevis*, *L. curvatus*, *L. plantarum*, *L. parakefiri*, and *L. buchneri* among others. Therefore it would not be surprising that in the near future we will find results of nanotube transfer between *Lactobacillus* to other genera.

In Fig. 8.2 we show the genetic context of COG1692 coding predicted calcineurin-like phosphoesterase belonging to the YmdB-like protein family of putative phosphoesterases genes in members of *Lactobacillus*.

Web resources

http://www.microbesonline.org
https://openwetware.org/wiki/Lactobacillus_transformation_(Serror_2002)

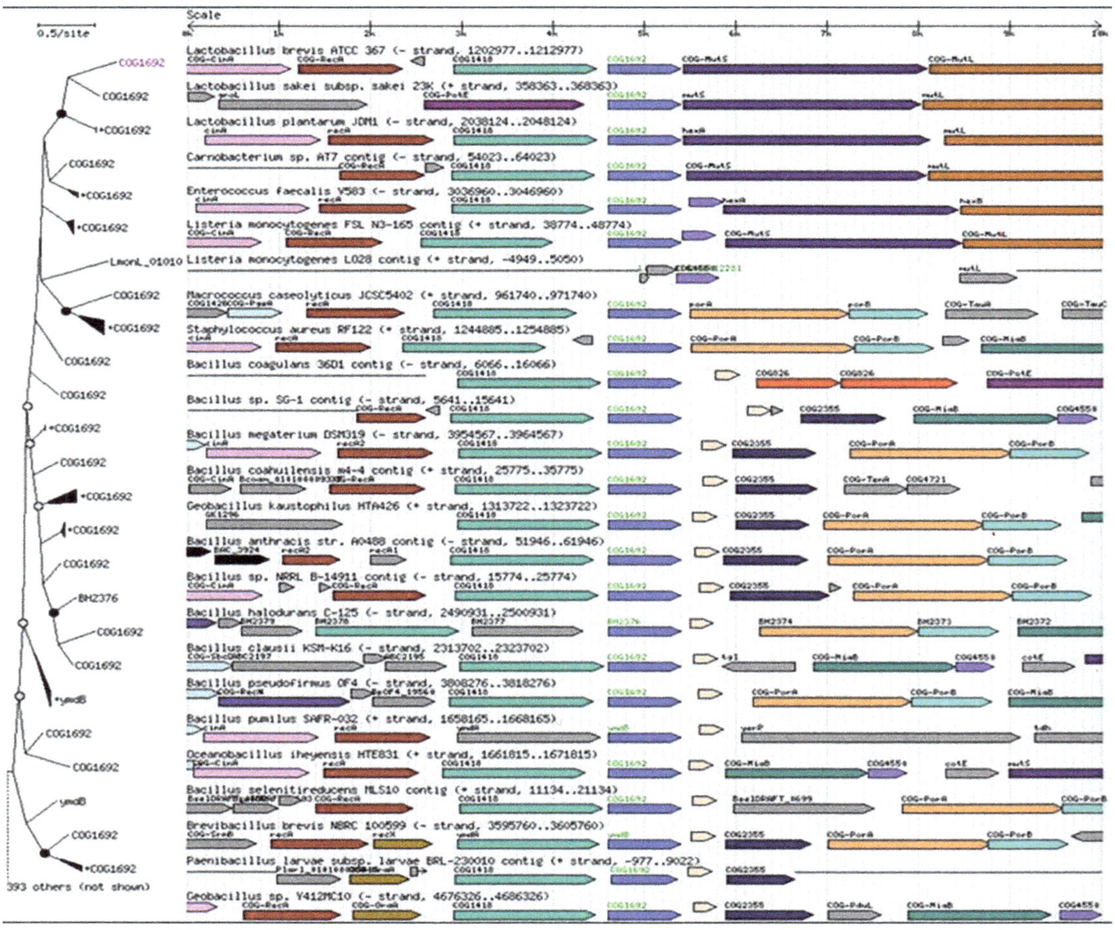

Figure 8.2 The genetic context of COG1692 coding predicted Calcineurin-like phosphoesterase belonging to the YmdB-like in members of *Lactobacillus*. Symbol description same as Figure 8.1. The figure was generated using MicrobesOnline facility: http://www.microbesonline.org.

https://openwetware.org/wiki/Lactobacillus_transformation_(Kim_2005)
https://openwetware.org/wiki/Electro-transformation_of_Lactobacillus_spp.

References

Alegre, M.T., Rodríguez, M.C., and Mesas, J.M. (2004). Transformation of *Lactobacillus plantarum* by electroporation with in vitro modified plasmid DNA. FEMS Microbiol. Lett. *241*, 73–77. https://doi.org/10.1016/j.femsle.2004.10.006

Attaiech, L., and Charpentier, X. (2017). Silently transformable: The many ways bacteria conceal their built-in capacity of genetic exchange. Curr. Genet. *63*, 451–455. https://doi.org/10.1007/s00294-016-0663-6

Aukrust, T., and Blom, H. (1992). Transformation of *Lactobacillus* strains used in meat and vegetable fermentations. Food Res. Int. *25*, 253–261.

Aukrust, T.W., Brurberg, M.B., and Nes, I.F. (1995a). Transformation of *Lactobacillus* by electroporation. Methods Mol. Biol. *47*, 201–208. https://doi.org/10.1385/0-89603-310-4:201

Aukrust, T.W., Brurberg, M.B., and Nes, I.F. (1995b). Transformation of *Lactobacillus* by electroporation. In Electroporation Protocols for Microorganisms, Nickoloff, J.A. ed. (Humana Press, New Jersey), pp. 201–208. https://doi.org/10.1385/0-89603-310-4:201

Baigorí, M., Sesma, F., de Ruiz Holgado, A.P., and de Mendoza, D. (1988). Transfer of plasmids between *Bacillus subtilis* and *Streptococcus lactis*. Appl. Environ. Microbiol. *54*, 1309–1311.

Baugher, J.L., Durmaz, E., and Klaenhammer, T.R. (2014). Spontaneously induced prophages in *Lactobacillus gasseri* contribute to horizontal gene transfer. Appl. Environ. Microbiol. *80*, 3508–3517. https://doi.org/10.1128/AEM.04092-13

Beasley, S.S., Takala, T.M., Reunanen, J., Apajalahti, J., and Saris, P.E. (2004). Characterization and electrotransformation of *Lactobacillus crispatus* isolated from chicken crop and intestine. Poult. Sci. *83*, 45–48. https://doi.org/10.1093/ps/83.1.45

Blokesch, M. (2016). Natural competence for transformation. Curr. Biol. *26*, R1126–R1130. https://doi.org/10.1016/j.mib.2017.04.001

Blokesch, M. (2017). In and out-contribution of natural transformation to the shuffling of large genomic regions. Curr. Opin. Microbiol. *38*, 22–29. https://doi.org/10.1016/J.CUB.2016.08.058

Boizet, B., Flickinger, J.L., and Chassy, B.M. (1988). Transfection of *Lactobacillus bulgaricus* protoplasts by bacteriophage DNA. Appl. Environ. Microbiol. *54*, 3014–3018.

Briner, A.E., and Barrangou, R. (2016). Deciphering and shaping bacterial diversity through CRISPR. Curr. Opin. Microbiol. *31*, 101–108. https://doi.org/10.1016/j.mib.2016.03.006

Broadbent, J., Oberg, T., Hughes, J., Ward, R., Brighton, C., Welker, D., and Steele, J. (2014). Influence of polysorbate 80 and cyclopropane fatty acid synthase activity on lactic acid production by *Lactobacillus casei* ATCC 334 at low pH. J. Ind. Microbiol. Biotechnol. *41*, 545–553. https://doi.org/10.1007/s10295-013-1391-2

Brown, L., Wolf, J.M., Prados-Rosales, R., and Casadevall, A. (2015). Through the wall: extracellular vesicles in Gram-positive bacteria, mycobacteria and fungi. Nat. Rev. Microbiol. *13*, 620–630. https://doi.org/10.1038/nrmicro3480

Charpentier, X., Polard, P., and Claverys, J.P. (2012). Induction of competence for genetic transformation by antibiotics: Convergent evolution of stress responses in distant bacterial species lacking SOS? Curr. Opin. Microbiol. *15*, 570–576. https://doi.org/10.1016/j.mib.2012.08.001.

Chassy, B.M., and Flickinger, J.L. (1987). Transformation of *Lactobacillus casei* by electroporation. FEMS Microbiol. Lett. *44*, 173–177. https://doi.org/10.1111/j.1574-6968.1987.tb02263.x

Chassy, B.M., and Rokaw. E. (1981). Conjugal transfer of plasmid-associated lactose metabolism in *Lactobacillus casei* subsp. *casei*. In Molecular Biology, Pathogenesis, and Ecology of Bacterial Plasmids, Levy, S., Clowes, R., and Koenig, E., eds. (Plenum Press, New York), p. 590.

Chassy, B.M., Mercenier, A., and Flickinger, J. (1988). Transformation of bacteria by electroporation. Trends Biotechnol. *6*, 303–309. https://doi.org/10.1016/0167-7799(88)90025-X

Cocconcelli, P.S., Morelli, L., Vescovo, M., and Bottazzi, V. (1986). Intergeneric protoplast fusion in lactic acid bacteria. FEMS Microbiol. Lett. *35*, 211–214.

Cofré, G., and Sanchez-Rivas, C. (1983). Curing of plasmids during a protoplast-cell wall regeneration cycle in *Bacillus subtilis* strains. FEMS Microbiol Letters *19*, 11–15. https://doi:10.1111/j.1574-6968.1983.tb00501.x

Connell, H., Lemmon, J., and Tannock, G.W. (1988). Formation and regeneration of protoplasts and spheroplasts of gastrointestinal strains of Lactobacilli. Appl. Environ. Microbiol. *54*, 1615–1618.

David, B., Radziejwoski, A., Fontaine, L., Henry de Frahan, M., van Dillen, S., Boyaval, P., Horvath, P., Fremaux, C., and Hols, P. (2017). Natural DNA transformation is functional in *Lactococcus lactis* subsp. *cremoris* KW2. Appl. Environ. Microbiol. *83*, e01074-17. https://doi.org/10.1128/AEM.01074-17.

Dieterle, M.E., Fina Martin, J., Durán, R., Nemirovsky, S.I., Sanchez Rivas, C., Bowman, C., Russell, D., Hatfull, G.F., Cambillau, C., and Piuri, M. (2016). Characterization of prophages containing 'evolved' Dit/Tal modules in the genome of *Lactobacillus casei* BL23. Appl. Microbiol. Biotechnol. *100*, 9201–9215. https://doi.org/10.1007/s00253-016-7727-x

Domingues, S., and Nielsen, K.M. (2017). Membrane vesicles and horizontal gene transfer in prokaryotes. Curr. Opin. Microbiol. *38*, 16–21. https://doi.org/10.1016/j.mib.2017.03.012

Domínguez Rubio, A.P., Martínez, J.H., Martínez Casillas, D.C., Coluccio Leskow, F., Piuri, M., and Pérez, O.E. (2017). *Lactobacillus casei* BL23 produces microvesicles carrying proteins that have been associated with its probiotic effect. Front. Microbiol. *8*, 1783. https://doi.org/10.3389/fmicb.2017.01783

Douillard, F.P., Mora, D., Eijlander, R.T., Wels, M., and de Vos, W.M. (2018). Comparative genomic analysis of the multispecies probiotic-marketed product VSL#3. PLOS ONE *13*, e0192452. https://doi.org/10.1371/journal.pone.0192452

Dubey, G.P., and Ben-Yehuda, S. (2011). Intercellular nanotubes mediate bacterial communication. Cell *144*, 590–600. https://doi.org/10.1016/j.cell.2011.01.015

Dubey, G.P., Malli Mohan, G.B., Dubrovsky, A., Amen, T., Tsipshtein, S., Rouvinski, A., Rosenberg, A., Kaganovich, D., Sherman, E., Medalia, O., *et al.* (2016). Architecture and characteristics of bacterial nanotubes. Dev. Cell *36*, 453–461. https://doi.org/10.1016/j.devcel.2016.01.013

Fang, F., Flynn, S., Li, Y., Claesson, M.J., van Pijkeren, J.P., Collins, J.K., van Sinderen, D., and O'Toole, P.W. (2008). Characterization of endogenous plasmids from *Lactobacillus salivarius* UCC118. Appl. Environ. Microbiol. *74*, 3216–3228. https://doi.org/10.1128/AEM.02631-07

Feld, L., Schjørring, S., Hammer, K., Licht, T.R., Danielsen, M., Krogfelt, K., and Wilcks, A. (2008). Selective pressure affects transfer and establishment of a *Lactobacillus plantarum* resistance plasmid in the gastrointestinal environment. J. Antimicrob. Chemother. *61*, 845–852. https://doi.org/10.1093/jac/dkn033

Fina Martin, J., Palomino, M.M., Rivas, C.S., Ruzal, S.M., and Allievi, M.C. (2017). Transferibilidad de DNA en *Lactobacillus*: ¿ hay riesgo de transferencia horizontal desde la microbiota a probióticos y viceversa? QuímicaViva *16*, 38–55. http://www.redalyc.org/html/863/86351157004/

Finkel, S.E., and Kolter, R. (2001). DNA as a nutrient: Novel role for bacterial competence gene homologs. J. Bacteriol. *183*, 6288–6293. https://doi.org/10.1128/JB.183.21.6288-6293.2001

García-Aljaro, C., Ballesté, E., and Muniesa, M. (2017). Beyond the canonical strategies of horizontal gene transfer in prokaryotes. Curr. Opin. Microbiol. *38*, 95–105. https://doi.org/10.1016/j.mib.2017.04.011

Gasson, M.J., and Fitzgerald, G.F. (1994). Gene transfer systems and transposition. In Genetics and Biotechnology of Lactic Acid Bacteria, Gasson, M.J., De Vos, W.M., eds. (Springer, Dordrecht), pp. 1–51. https://doi.org/10.1007/978-94-011-1340-3_1

Gevers, D., Huys, G., and Swings, J. (2003). In vitro conjugal transfer of tetracycline resistance from *Lactobacillus* isolates to other Gram-positive bacteria. FEMS Microbiol. Lett. *225*, 125–130. https://doi.org/10.1016/S0378-1097(03)00505-6

Godde, J.S., and Bickerton, A. (2006). The repetitive DNA elements called CRISPRs and their associated genes: Evidence of horizontal transfer among prokaryotes. J. Mol. Evol. *62*, 718–729. https://doi.org/10.1007/s00239-005-0223-z

Grande, R., Celia, C., Mincione, G., Stringaro, A., Di Marzio, L., Colone, M., Di Marcantonio, M.C., Savino, L., Puca, V., Santoliquido, R., *et al.* (2017). Detection and physicochemical characterization of membrane vesicles (MVs) of *Lactobacillus reuteri* DSM 17938. Front. Microbiol. *8*, 1040. https://doi.org/10.3389/fmicb.2017.01040

Grandjean, V., Hauck, Y., Beloin, C., Le Hégarat, F., and Hirschbein, L. (1998). Chromosomal inactivation of *Bacillus subtilis* exfusants: A prokaryotic model of epigenetic regulation. Biol. Chem. *379*, 553–557.

Heery, D.M., Powell, R., Gannon, F., and Dunican, L.K. (1989). Curing of a plasmid from *E.coli* using high-voltage electroporation. Nucleic Acids Res. *17*, 10131.

Hidalgo-Cantabrana, C., O'Flaherty, S., and Barrangou, R. (2017). CRISPR-based engineering of next-generation lactic acid bacteria. Curr. Opin. Microbiol. *37*, 79–87. https://doi.org/10.1016/j.mib.2017.05.015

Holo, H., and Nes, I.F. Transformation of *Lactococcus* by electroporation. In Electroporation Protocols for Microorganisms, Nickoloff, J.A., ed. (Humana Press, New Jersey), pp. 195–200. https://doi.org/10.1385/0-89603-310-4:195

Horvath, P., Coûté-Monvoisin, A.C., Romero, D.A., Boyaval, P., Fremaux, C., and Barrangou, R. (2009). Comparative analysis of CRISPR loci in lactic acid bacteria genomes. Int. J. Food Microbiol. *131*, 62–70. https://doi.org/10.1016/j.ijfoodmicro.2008.05.030

Hotchkiss, R.D., and Gabor, M.H. (1980). Biparental products of bacterial protoplast fusion showing unequal parental chromosome expression. Proc. Natl. Acad. Sci. U.S.A. 77, 3553–3557.

Iwata, M., Mada, M., and Ishiwa, H. (1986). Protoplast fusion of *Lactobacillus fermentum*. Appl. Environ. Microbiol. *52*, 392–393.

Jacobsen, L., Wilcks, A., Hammer, K., Huys, G., Gevers, D., and Andersen, S.R. (2007). Horizontal transfer of *tet*(M) and *erm*(B) resistance plasmids from food strains of *Lactobacillus plantarum* to *Enterococcus faecalis* JH2-2 in the gastrointestinal tract of gnotobiotic rats. FEMS Microbiol. Ecol. *59*, 158–166. https://doi.org/10.1111/j.1574-6941.2006.00212.x

John, R.P., Gangadharan, D., and Madhavan Nampoothiri, K. (2008). Genome shuffling of *Lactobacillus delbrueckii* mutant and *Bacillus amyloliquefaciens* through protoplasmic fusion for L-lactic acid production from starchy wastes. Bioresour. Technol. *99*, 8008–8015. https://doi.org/10.1016/j.biortech.2008.03.058

Johnston, C., Martin, B., Fichant, G., Polard, P., and Claverys, J.P. (2014). Bacterial transformation: Distribution, shared mechanisms and divergent control. Nat. Rev. Microbiol. *12*, 181–196. https://doi.org/10.1038/nrmicro3199

Kanatani, K., Yoshida, K., Tahara, T., Sakamoto, M., and Oshimura, M. (1990). Intraspecific protoplast fusion of *Lactobacillus plantarum*. Agric. Biol. Chem. *54*, 225–227.

Karthikeyan, V., and Santosh, S.W. (2010). Comparing the efficacy of plasmid curing agents in *Lactobacillus acidophilus*. Benef. Microbes *1*, 155–158. https://doi.org/10.3920/BM2009.0038

Kim, Y.H., Han, K.S., Oh, S., You, S., and Kim, S.H. (2005). Optimization of technical conditions for the transformation of *Lactobacillus acidophilus* strains by electroporation. J. Appl. Microbiol. *99*, 167–174. https://doi.org/10.1111/j.1365-2672.2005.02563.x

Landete, J.M., Arqués, J.L., Peirotén, A., Langa, S., and Medina, M. (2014). An improved method for the electrotransformation of lactic acid bacteria: A comparative survey. J. Microbiol. Methods *105*, 130–133. https://doi.org/10.1016/j.mimet.2014.07.022

Langella, P., Zagorec, M., Ehrlich, S.D., and Morel-Deville, F. (1996). Intergeneric and intrageneric conjugal transfer of plasmids pAMBeta1, pIL205 and pIP501 in *Lactobacillus sake*. FEMS Microbiol. Lett. *139*, 51–56. https://doi.org/10.1111/j.1574-6968.1996.tb08178.x

Lee-Wickner, L.J., and Chassy, B.M. (1984). Production and regeneration of *Lactobacillus casei* protoplasts. Appl. Environ. Microbiol. *48*, 994–1000.

Li, M., Lee, K., Hsu, M., Nau, G., Mylonakis, E., and Ramratnam, B. (2017). *Lactobacillus*-derived extracellular vesicles enhance host immune responses against vancomycin-resistant enterococci. BMC Microbiol. *17*, 66. https://doi.org/10.1186/s12866-017-0977-7

Luchansky, J.B., Muriana, P.M., and Klaenhammer, T.R. (1988). Application of electroporation for transfer of plasmid DNA to *Lactobacillus, Lactococcus, Leuconostoc, Listeria, Pediococcus, Bacillus, Staphylococcus, Enterococcus and Propionibacterium*. Mol. Microbiol. *2*, 637–646.

Luchansky, J.B., Kleeman, E.G., Raya, R.R., and Klaenhammer, T.R. (1989). Genetic transfer systems for delivery of plasmid deoxyribonucleic acid to *Lactobacillus acidophilus* ADH: Conjugation, electroporation, and transduction. J. Dairy Sci. 72, 1408–1417. https://doi.org/10.3168/jds.S0022-0302(89)79248-1

Makarova, K., Slesarev, A., Wolf, Y., Sorokin, A., Mirkin, B., Koonin, E., Pavlov, A., Pavlova, N., Karamychev, V., Polouchine, N., *et al.* (2006). Comparative genomics of the lactic acid bacteria. Proc. Natl. Acad. Sci. U.S.A. *103*, 15611–15616. https://doi.org/10.1073/pnas.0607117103

Makarova, K.S., Wolf, Y.I., Alkhnbashi, O.S., Costa, F., Shah, S.A., Saunders, S.J., Barrangou, R., Brouns, S.J., Charpentier, E., Haft, D.H., *et al.* (2015). An updated evolutionary classification of CRISPR-Cas systems. Nat. Rev. Microbiol. *13*, 722–736. https://doi.org/10.1038/nrmicro3569

Mangericao, T.C., Peng, Z., and Zhang, X. (2016). Computational prediction of CRISPR cassettes in gut metagenome samples from Chinese type-2 diabetic patients and healthy controls. BMC Syst. Biol. 10 (Suppl. 1), 5. https://doi.org/10.1186/s12918-015-0248-x

Marraffini, L.A. (2015). CRISPR-Cas immunity in prokaryotes. Nature *526*, 55–61. https://doi.org/10.1038/nature15386

Mater, D.D., Langella, P., Corthier, G., and Flores, M.J. (2008). A probiotic *Lactobacillus* strain can acquire vancomycin resistance during digestive transit in mice. J. Mol. Microbiol. Biotechnol. *14*, 123–127. https://doi.org/10.1159/000106091

Mercanti, D.J., Rousseau, G.M., Capra, M.L., Quiberoni, A., Tremblay, D.M., Labrie, S.J., and Moineau, S. (2016). Genomic diversity of phages infecting probiotic strains of *Lactobacillus paracasei*. Appl. Environ. Microbiol. *82*, 95–105. https://doi.org/10.1128/AEM.02723-15

Mercenier, A., and Chassy, B.M. (1988). Strategies for the development of bacterial transformation systems. Biochimie *70*, 503–517.

Mohanraju, P., Makarova, K.S., Zetsche, B., Zhang, F., Koonin, E.V., and van der Oost, J. (2016). Diverse evolutionary roots and mechanistic variations of the CRISPR-Cas systems. Science 353, aad5147. https://doi.org/10.1126/science.aad5147

Mulder, J., Wels, M., Kuipers, O.P., Kleerebezem, M., and Bron, P.A. (2017). Unleashing natural competence in *Lactococcus lactis* by induction of the competence regulator ComX. Appl. Environ. Microbiol. *83*, e01320-17. https://doi.org/10.1128/AEM.01320-17

Muriana, P.M., and Klaenhammer, T.R. (1987). Conjugal transfer of plasmid-encoded determinants for bacteriocin production and immunity in *Lactobacillus acidophilus* 88. Appl. Environ. Microbiol. *53*, 553–560.

Nickoloff, J.A. (1995). Electroporation Protocols for Microorganisms (Humana Press, New Jersey). https://doi.org/10.1385/0896033104

Nicoloff, H., and Bringel, F. (2003). ISLpl1 is a functional IS30-related insertion element in *Lactobacillus plantarum* that is also found in other lactic acid bacteria. Appl. Environ. Microbiol. *69*, 6032–6040.

Novick, R., Sanchez-Rivas, C., Gruss, A., and Edelman, I. (1980). Involvement of the cell envelope in plasmid maintenance: Plasmid curing during the regeneration of protoplasts. Plasmid 3, 348–358.

Oh, J.H., and van Pijkeren, J.P. (2014). CRISPR-Cas9-assisted recombineering in *Lactobacillus reuteri*. Nucleic Acids Res. *42*, e131. https://doi.org/10.1093/nar/gku623

Palomino, M.M., Allievi, M.C., Prado-Acosta, M., Sanchez-Rivas, C., and Ruzal, S.M. (2010). New method for electroporation of *Lactobacillus* species grown in high salt. J. Microbiol. Methods *83*, 164–167. https://doi.org/10.1016/j.mimet.2010.08.017

Patnaik, R., Louie, S., Gavrilovic, V., Perry, K., Stemmer, W.P., Ryan, C.M., and del Cardayré, S. (2002). Genome shuffling of *Lactobacillus* for improved acid tolerance. Nat. Biotechnol. *20*, 707–712. https://doi.org/10.1038/nbt0702-707

Piuri, M., Sanchez-Rivas, C., and Ruzal, S.M. (2005). Cell wall modifications during osmotic stress in *Lactobacillus casei*. J. Appl. Microbiol. *98*, 84–95.

Popa, O., and Dagan, T. (2011). Trends and barriers to lateral gene transfer in prokaryotes. Curr. Opin. Microbiol. *14*, 615–623. https://doi.org/10.1016/j.mib.2011.07.027

Posno, M., Leer, R.J., van Luijk, N., van Giezen, M.J., Heuvelmans, P.T., Lokman, B.C., and Pouwels, P.H. (1991). Incompatibility of *Lactobacillus* vectors with replicons derived from small cryptic *Lactobacillus plasmids* and segregational instability of the introduced vectors. Appl. Environ. Microbiol. *57*, 1822–1828.

Pouwels, P.H., and Leer, R.J. (1993). Genetics of lactobacilli: Plasmids and gene expression. Antonie Van Leeuwenhoek *64*, 85–107. https://doi.org/10.1007/BF00873020

Raya, R.R., and Klaenhammer, T.R. (1992). High-frequency plasmid transduction by *Lactobacillus gasseri* bacteriophage phiadh. Appl. Environ. Microbiol. *58*, 187–193.

Reniero, R., Cocconcelli, P., Bottazzi, V., and Morelli, L. (1992). High frequency of conjugation in *Lactobacillus* mediated by an aggregation-promoting factor. J. Gen. Microbiol. *138*, 763–768.

Rubinstein, C.P., and Sanchez-Rivas, C. (1989). Genetic manipulations in auto-induced protoplasts of *Bacillus thuringiensis*. Mem. Inst. Oswaldo Cruz *84* (Suppl. 3), 35–37.

Rubinstein, C.P., Coso, O.A., Ruzal, S., and Sanchez-Rivas, C. (1993). Anti-SOS effects induced in *Bacillus subtilis* by a phi 105 mutant prophage. Arch. Microbiol. *160*, 486–491.

Sanchez-Rivas, C. (1988). Expression and maintenance of plasmid resistance in regenerating protoplasts of *Bacillus subtilis*. Ann. Inst. Pasteur Microbiol. *139*, 403–409. https://doi.org/10.1016/0769-2609(88)90103-2

Schmid, S., Bevilacqua, C., and Crutz-Le Coq, A.M. (2012). Alternative sigma factor σH activates competence gene expression in *Lactobacillus sakei*. BMC Microbiol. *12*, 32. https://doi.org/10.1186/1471-2180-12-32

Serror, P., Sasaki, T., Ehrlich, S.D., and Maguin, E. (2002). Electrotransformation of *Lactobacillus delbrueckii* subsp. *bulgaricus* and *L. delbrueckii* subsp. *lactis* with various plasmids. Appl. Environ. Microbiol. *68*, 46–52.

Shrago, A.W., and Dobrogosz, W.J. (1988). Conjugal transfer of group B streptococcal plasmids and comobilization of *Escherichia coli-Streptococcus* shuttle plasmids to *Lactobacillus plantarum*. Appl. Environ. Microbiol. *54*, 824–826.

Singhvi, M., Joshi, D., Gaikaiwari, S., and Gokhale, D.V. (2010). Protoplast formation and regeneration in *Lactobacillus delbrueckii*. Indian J. Microbiol. *50*, 97–100. https://doi.org/10.1007/s12088-010-0008-y

Slager, J., Kjos, M., Attaiech, L., and Veening, J.W. (2014). Antibiotic-induced replication stress triggers bacterial competence by increasing gene dosage near the origin. Cell *157*, 395–406. https://doi.org/10.1016/J.CELL.2014.01.068

Spath, K., Hein, S., and Grabherr, R. (2012). Direct cloning in *Lactobacillus plantarum*: Electroporation with non-methylated plasmid DNA enhances transformation efficiency and makes shuttle vectors obsolete. Microb. Cell Fact. *11*, 141. https://doi.org/10.1186/1475-2859-11-141

Spengler, G., Molnár, A., Schelz, Z., Amaral, L., Sharples, D., and Molnár, J. (2006). The mechanism of plasmid curing in bacteria. Curr. Drug Targets *7*, 823–841. https://doi.org/10.2174/138945006777709601

Stout, E., Klaenhammer, T., and Barrangou, R. (2017). CRISPR-Cas technologies and applications in food bacteria. Annu. Rev. Food Sci. Technol. *8*, 413–437. https://doi.org/10.1146/annurev-food-072816-024723

Sun, Z., Harris, H.M., McCann, A., Guo, C., Argimón, S., Zhang, W., Yang, X., Jeffery, I.B., Cooney, J.C., Kagawa, T.F., *et al.* (2015). Expanding the biotechnology potential of Lactobacilli through comparative genomics of 213 strains and associated genera. Nat. Commun. *6*, 8322. https://doi.org/10.1038/ncomms9322

Tanaka, O., and Ohmomo, S. (2001). Efficient protoplast regeneration for some homofermentative Lactobacilli and *Pediococci*. Arch. Microbiol. *177*, 36–40. https://doi.org/10.1007/s00203-001-0357-3

Thompson, K., and Collins, M.A. (1996). Improvement in electroporation efficiency for *Lactobacillus plantarum* by the inclusion of high concentrations of glycine in the growth medium. J. Microbiol. Methods *26*, 73–79. https://doi.org/10.1016/0167-7012(96)00845-7

Toyama, K., Sakurai, T., and Ari, H. (1971). [Transduction by temperate phage PLS-1 in *Lactobacillus salivarius*.] Nippon. Saikingaku Zasshi *26*, 482–487.

van Pijkeren, J.P., and Britton, R.A. (2012). High efficiency recombineering in lactic acid bacteria. Nucleic Acids Res. *40*, e76. https://doi.org/10.1093/nar/gks147

van Reenen, C.A., and Dicks, L.M. (2011). Horizontal gene transfer amongst probiotic lactic acid bacteria and other intestinal microbiota: What are the possibilities? A review. Arch. Microbiol. *193*, 157–168. https://doi.org/10.1007/s00203-010-0668-3

Varmanen, P., Rantanen, T., Palva, A., and Tynkkynen, S. (1998). Cloning and characterization of a prolinase gene (pepR) from *Lactobacillus rhamnosus*. Appl. Environ. Microbiol. *64*, 1831–1836.

Vescovo, M., Morelli, L., Cocconcelli, P.S., and Bottazzi, V. (1984). Protoplast formation, regeneration and plasmid curing in *Lactobacillus reuteri*. FEMS Microbiol. Lett. *23*, 333–336. https://doi.org/10.1016/0378-1097(84)90235-0

Wang, Y., Li, Y., Pei, X., Yu, L., and Feng, Y. (2007). Genome-shuffling improved acid tolerance and L-lactic acid volumetric productivity in *Lactobacillus rhamnosus*. J. Biotechnol. *129*, 510–515. https://doi.org/10.1016/j.jbiotec.2007.01.011

Watanabe, K., Kakita, Y., Nakashima, Y., and Sasaki, T. (1990). Protoplast transfection of *Lactobacillus casei* by phage PL-1 DNA. Agric. Biol. Chem. *54*, 937–941.

Wei, M.Q., Rush, C.M., Norman, J.M., Hafner, L.M., Epping, R.J., and Timms, P. (1995). An improved method for the transformation of *Lactobacillus* strains using electroporation. J. Microbiol. Methods *21*, 97–109. https://doi.org/10.1016/0167-7012(94)00038-9

Welker, D.L., Hughes, J.E., Steele, J.L., and Broadbent, J.R. (2015). High efficiency electrotransformation of *Lactobacillus casei*. FEMS Microbiol. Lett. *362*, 1–6. https://doi.org/10.1093/femsle/fnu033

Woodbury, R.L., Wang, X., and Moran, C.P. (2006). Sigma X induces competence gene expression in *Streptococcus pyogenes*. Res. Microbiol. *157*, 851–856. https://doi.org/10.1016/j.resmic.2006.07.002

Xin, Y., Guo, T., Mu, Y., and Kong, J. (2017). Identification and functional analysis of potential prophage-derived recombinases for genome editing in *Lactobacillus casei*. FEMS Microbiol. Lett. *364*,. https://doi.org/10.1093/femsle/fnx243

Yang, P., Wang, J., and Qi, Q. (2015). Prophage recombinases-mediated genome engineering in *Lactobacillus plantarum*. Microb. Cell Fact. *14*, 154. https://doi.org/10.1186/s12934-015-0344-z

Recombinant Gene Expression in Lactobacilli: Strategies and Applications

9

Clemens Peterbauer[1], Stefan Heinl[2], Aleš Berlec[3] and Reingard Grabherr[2]*

[1]Department of Food Sciences and Technology, University of Natural Resources and Life Sciences Vienna, Vienna, Austria.
[2]Department of Biotechnology, University of Natural Resources and Life Sciences Vienna, Vienna, Austria.
[3]Department of Biotechnology, Jožef Stefan Institut, Ljubljana, Slovenia.

*Correspondence: reingard.grabherr@boku.ac.at

https://doi.org/10.21775/9781910190890.09

Abstract

Bacteria-based expression systems play a substantial role in industrial and medical biotechnology for the production of various proteins. In contrast to Gram-negative bacteria such as *Escherichia coli*, Gram-positive lactic acid bacteria possess a specific cell wall structure that allows for secretion of proteins directly into their environment. In spite of some limitations, this advantage and the fact that many of these bacteria have GRAS (generally recognized as safe) status made them into an attractive tool in biotechnology. In this chapter we give an overview of the potential of lactobacilli to be used for recombinant protein expression and their feasibility in various applications. In the past years lactobacilli have been used for the production of therapeutics, as vaccines, and as a platform for protein surface display. Many tools for genetic modification of lactobacilli have been developed and optimized. In addition, a great variety of gene regulatory elements have been isolated, characterized and established in the context of recombinant protein expression. We discuss the properties of different plasmids that may be used for the expression of intracellular or secreted target proteins. We describe the mechanisms of inducible and constitutive promoters, the influence of codon usage and ribosomal binding sites on protein expression levels, and strategies for stable integration of target genes into the genome of lactobacilli using modern technologies. Overall, there exists a high variety of different lactobacilli species and strains that are highly specialized and have adapted to specific ecological niches. Depending on the application, the desired product or the required process, genetic tools and regulatory elements often have to be tested, adapted and optimized individually.

Introduction

Lactobacilli are members of the heterogeneous group of lactic acid bacteria (LAB), Gram-positive, microaerophilic, non-spore-forming bacteria with a complex taxonomic relationship and very variable phenotypic traits. LAB are divided into six families, Lactobacillaceae, Streptococcaceae, Enterococcaceae, Leuconostocaceae, Carnobacteriaceae and Aerococcaceae. The best-studied model organism for LAB is *Lactococcus lactis*, a member of the Streptococcaceae, most species with biotechnological relevance are found among the Lactobacillaceae, namely the genus *Pediococcus* as well as the largest genus, *Lactobacillus*, with over 200 species (Bosma *et al.*, 2017). These species are found in a variety of, mostly nutrient-rich, ecological niches such as decomposing plant materials, wine, meat and raw milk (O'Sullivan *et al.*, 2009; Topisirovic *et al.*, 2006) and are often commensals to plants and animals including humans (O'Callaghan and O'Toole, 2013). Due to their multifarious physiological and metabolic abilities, lactobacilli are commercially highly relevant in the feed, food and beverages industries and play a role as fermentation starter cultures for the manufacturing of a wide range of fermented food products with improved shelf-life, taste and nutritional properties (Guerzoni *et al.*, 2007) as well as for ensiled fodder (Ávila *et al.*, 2014; Eikmeyer *et al.*, 2013). Lactobacilli are also important producers of industrial platform chemicals such as lactic acid (Kim *et al.*, 2010), gamma-aminobutyric acid (Hasegawa *et al.*, 2017; Li *et al.*, 2010), 1,3-propanediol (Pflugl *et al.*, 2014) or exopolysaccharides (Rani *et al.*, 2017; Ryan *et al.*, 2015; Zannini *et al.*, 2016). They are implied to have health-promoting effects as so-called probiotics, beneficial microbes that form transient or stable populations in animals and humans, mostly in the gastrointestinal tract, by a number of interactions with the epithelial surface and the immune system, beneficial enzymatic activities etc. (Lebeer *et al.*, 2008; Salvetti and O'Toole, 2017; van den Nieuwboer *et al.*, 2016). Due to their nature as part of the human microbiome, their capability of establishing themselves as transient or permanent colonizers of the gastrointestinal tract and their alleged health-promoting properties as probiotics, lactobacilli have attracted increased interest for applications as delivery vehicles for therapeutic proteins and oral and mucosal vaccinations (Rosales-Mendoza *et al.*, 2016; Schwarzer *et al.*, 2011; Seegers, 2002; Wyszynska *et al.*, 2015). Previously recognized weaknesses of *Lactobacillus*-based cell factories, namely limited production of biomass and lower maximum yield particularly for produced proteins, do not play a major role for such applications due to the direct, local and often long-term application. *Lactobacillus* spp. that are capable of persistent colonization enable a long term-release of the vaccine or therapeutic protein (Berlec *et al.*, 2012; Berlec *et al.*, 2015; Nishiyama *et al.*, 2016). This chapter aims to survey the most relevant tools, methods and applications for recombinant gene expression in lactobacilli described in recent studies.

Plasmid based protein expression in lactobacilli

Plasmids are autonomously replicating, naturally occurring, DNA elements. They are widespread in bacteria, especially among lactobacilli, where they act as mobile genetic elements allowing genetic exchange between strains and species as well as intragenomic reorganization, thereby contributing to adaptation to environmental changes (Fernández *et al.*, 1999; Heinl *et al.*, 2012; Wang and Lee, 1997). Some plasmids found in lactobacilli confer antibiotic resistance (Ahn *et al.*, 1992; Egervärn *et al.*, 2009; Fons *et al.*, 1997), resistance to phage infection (Eguchi *et al.*, 2000), the ability to transport or metabolize certain substrates (Chen *et al.*, 2012; Fernández *et al.*, 1999; Mayo *et al.*, 1994; Shimizu-Kadota, 1987) or for the synthesis of bacteriocins (Miller *et al.*, 2005; Vaughan *et al.*, 2003) to their host. Others are so called cryptic plasmid, which means that no obvious function for their host can be assigned to them. Plasmids are, due to their relatively easy manipulability and the availability of transformation protocols for many species, important tools for genetic engineering of bacterial hosts. Cryptic plasmids from *L. plantarum* (Bouia *et al.*, 1989; Bringel *et al.*, 1989), *L. hilgardii* (Josson *et al.*, 1990), *L. pentosus* (Posno *et al.*, 1991a), *L. casei* (An *et al.*, 2007; Chen *et al.*, 2014; Shimizu-Kadota *et al.*, 1991; Suebwongsa *et al.*, 2016), *L. curvatus* (Klein *et al.*, 1993), *L. helveticus* (Yamamoto and Takano, 1996), *L. fermentum* (Aleshin *et al.*, 1999; Pavlova *et al.*, 2002), *L. reuteri* (Lin *et al.*, 2001), *L. sakei* (Alpert *et al.*, 2003), *L. delbrueckii* (Lee *et al.*, 2007), *L. salivarius* (Fang *et al.*, 2008) and *L. buchneri* (Heinl *et al.*, 2011b) have been adapted as *E. coli*/*Lactobacillus* shuttle vectors for cloning or gene expression purposes.

Lactobacillus plasmids differ in size from a few kilobases (kb) up to a few hundred kb. Small plasmids can easily be isolated and genetically modified; as such they are often the tool of choice for recombinant protein expression in bacterial hosts (Shareck *et al.*, 2004; Wang and Lee, 1997). Besides size, plasmids also differ in terms of how replication takes place. Several replication mechanisms have been described (del Solar *et al.*, 1998). The most widespread replication modes in lactobacillal plasmids are the theta-replication and rolling circle replication (del Solar *et al.*, 1998; Khan, 2005; Ruiz-Maso *et al.*, 2015). Theta replicating plasmids depend on host factors for replication and are therefore often quite species or even strain specific, while rolling-circle-replicating ones have a broader host range. Not all plasmids are compatible with each other (plasmid incompatibility), as plasmids with a very similar replication mechanism compete for cellular replication factors or plasmid encoded proteins (e.g. Rep proteins) that are essential for replication, with the consequence that only one representative of one type of plasmid will be stably maintained in the cell (Posno *et al.*, 1991b). Beside replication type, plasmids can also be categorized according to the number of copies that exist in a single cell at the same time, the so called plasmid copy number (PCN). Low PCN plasmids are present at a number of 1–3 copies per cell, whereas high PCN plasmids can reach a copy number of 50–200 copies per cell or even higher. It has been shown for several plasmid based expression systems that gene copy number which corresponds to PCN has a strong influence on overall expression of heterologous proteins. While normally an increased PCN leads to higher expression rates, sometimes a too high replication rate can be detrimental for cell growth (Grabherr *et al.*, 2002). Today, many cloning and expression vectors used for recombinant gene expression in lactobacilli are derivatives of the low copy plasmids p256 (Sorvig *et al.*, 2005b) and pWV01 (Kok *et al.*, 1984), or the high PCN pSH71 (David *et al.*, 1989). PCNs were determined and estimated to be around three for p256 and 200 for pSH71. Another high copy number plasmid (pCD034-1) was isolated from a *L. buchneri* strain CD034. The plasmid was shown to be suitable for replication in *L. plantarum* (Heinl *et al.*, 2011b), and copy numbers were estimated to exceed 200 (Spath *et al.*, 2011). In order to investigate the influence of PCN on the recombinant expression level (Tauer *et al.*, 2014) compared several plasmids, either containing the low copy number ori from p256, designated pCD256, or the high copy number ori from pCD034-1, isolated from *L. buchneri* CD034 (Heinl *et al.*, 2011a) resulting in the plasmid pCDLbu-1. Strains carrying the high copy number plasmids were shown to produce less biomass during fermentation. This might be because the overall metabolic load due to plasmid replication and overproduction of the recombinant protein hampers the growth rate. However, when comparing overall transcriptional activities the strong P_{11} promoter (Rud *et al.*, 2006) in combination with a high copy number plasmid backbone turned out to be the strongest.

Inducible gene expression

In many cases expression of a gene of interest is only desired at a certain time point or interval, e.g. to improve biomass formation, minimize stress as result of the metabolic burden for the host cell, or to control the expression time span. In such cases, inducible expression systems are preferred over constitutive systems.

NICE system in lactobacilli

Many LAB synthesize antimicrobial peptides, bacteriocins, that help them establish themselves in nutrient-rich habitats among strong competition from ecologically and physiologically related bacteria (Boris and Barbés, 2000; Eijsink *et al.*, 2002). Biosynthesis of such bacteriocins is often subject to autoinduction and is characterized by strongly inducible and tightly controlled promoters. Kuipers *et al.* (1995) were the first to use induction of the nisin promoter, *PnisA*, in *Lactococcus lactis* by sub-inhibitory concentrations of nisin for the production of a heterologous enzyme, *E. coli* glucuronidase. This was the birth of the NIsin Controlled Expression, or NICE, system, one of the most widespread and successful expression system for Gram-positive bacteria. Custom strains were constructed that carry the regulatory genes (*nisK* and *nisR*) under control of their own constitutive promoters integrated in the genome, freeing up space on the plasmid for the gene(s) of interest under control of *PnisA*. The system has since been used for the production of a large number of heterologous enzymes and therapeutic proteins as well as, because of its inducer-concentration-dependent transcriptional response, for metabolic engineering studies (reviewed by (Mierau and Kleerebezem, 2005). It has been optimized and upscaled for industrial conditions (Mierau *et al.*, 2005), and variants were constructed, together with the required deletion strains, to allow the use of food-grade, non-antibiotics-based selection (de Ruyter *et al.*, 1996). It has also been transferred to other gram-positive bacteria such as *Leuconostoc lactis, Enterococcus faecalis,* various species of *Streptococcus* and *Lactobacillus* as well as *Bacillus subtilis,* both as single-plasmid- or double-plasmid systems and in some cases by integrating required genes into the genome (Mierau and Kleerebezem, 2005).

pSIP system in lactobacilli

Based on similar principles, namely promoters controlling the biosynthesis of class II bacteriocins, a series of expression systems were developed mostly for use in species of the genus *Lactobacillus*. The promoters of the genes encoding the precursors of sakacin A and sakacin P, originally derived from *Lactobacillus sakei*, are induced not by the bacteriocin itself, but by an inducing pheromone, IP (Axelsson *et al.*, 2003). Various vectors (the pSIP series) were constructed for use in *L. sakei* and *L. plantarum*, generally containing origins of replication that are functional in *E. coli* and LAB, in order to permit vector construction in the easier-to-handle *E. coli* (Mathiesen *et al.*, 2004; Sorvig *et al.*, 2003; Sorvig *et al.*, 2005a). These vectors contain either the promoter for sakacin A (*PsppA*) or the promoter of the uncharacterized ORF X from the same regulon (*PorfX*), which led to higher GusA activities. Such results are, however, dependent on the target gene and the host strain, as, e.g. the performance of the two promoters was equal for the production of the aminopeptidase PepN, or later the yield of LacLM proteins from different *Lactobacillus* spp. More refined vectors, omitting unnecessary DNA regions, such as pSIP412, generally resulted in the highest production of the reporter proteins GusA and PepN (Sorvig *et al.*, 2005a). The NICE system has previously been observed to be poorly regulated in *Lactobacillus* spp., and integration of the regulatory genes *nisK* and *nisR* was suggested to avoid the strong background expression and maintained induction (Pavan *et al.*, 2000). The pSIP vectors derived from pSIP401 still contain the required genes for regulation, *sppK* and *sppR*, and are thus able to function as 'stand alone' systems without requiring a customized host strain, but do not show the regulation-related issues of the NICE system in *Lactobacillus* spp.

Further experiments using β-galactosidase genes (*lacLM*) from four different *Lactobacillus* spp. revealed a significant spread in the achieved protein yields, despite the use of otherwise entirely identical vectors and cultivation and induction conditions (Halbmayr *et al.*, 2008). This observation was traced back to strong variations in mRNA stability of the respective *lacLM* genes, which in turn was likely caused by a higher frequency of unfavourable codons in the 5′-portion of the encoding genes (Nguyen *et al.*, 2011).

Vectors like the pSIP series essentially conform to 'self-cloning' criteria, as they do not contain foreign DNA, i.e. DNA that is derived from phylogenetically distant non-LAB organisms. Food-grade expression additionally requires that no antibiotic resistance genes are used (de Vos, 1999). A food-grade version of pSIP was constructed by exchanging the erythromycin-resistance gene against the alanine-racemase gene *alr* from *L. plantarum* WCFS1, and a customized recipient strain by knocking out the genomic *alr* gene in this strain, followed by Cre-mediated marker removal. This strain was observed to give slightly higher yields of recombinantly produced protein, probably due to an increased plasmid stability (Nguyen *et al.*, 2011).

Other inducible systems

Besides the well-established peptide inducible expression systems pSIP and NICE, there is a number of inducible systems based on the regulation of carbohydrate utilization pathways. Lokman *et al.* (1994) tested two different promoters connected with xylose utilization in *L. pentosus*. They used the chloramphenicol acetyltransferase (CAT) gene as a reporter on a multicopy plasmid. With the *xylA* promoter, they observed 60- to 80-fold induction by xylose, but only 3- to 10-fold induction in the presence of glucose and xylose. Expression of CAT under control of the *xylR* promoter was shown to be constitutive (Lokman *et al.*, 1994). Lactose inducible expression of the reporter gene *gusA* was achieved by stable integration of the gene into the chromosomal lactose operon of *L. casei* CECT 5276. Expression was induced by lactose and repressed by glucose (Gosalbes *et al.*, 2000). With expression vectors based on the plasmid pWV01 replicon (Vosman and Venema, 1983), Duong *et al.* tested promoter and repressor elements associated with transport and utilization of carbohydrates like fructooligosaccharides (FOS), lactose (lac), trehalose (tre) and genes directing glycolysis from *L. acidophilus*. To characterize expression from each promoter, a β-glucuronidase reporter gene was used. The tested FOS, lac and tre based vectors were shown to be highly inducible by their specific carbohydrate and repressed by glucose (Duong *et al.*, 2011). Recently, Heiss *et al.* engineered several inducible promoter/repressor systems and compared them using the red fluorescent protein mCherry as a reporter: *PlacA* a promoter/repressor pair derived from a β-galactosidase from *L. plantarum* 3NSH, *PxylA* from a putative *Bacillus megaterium* DSMZ 319 xylose utilization cluster and *PlacSynth*, a synthetic promoter based on the *Escherichia coli lac* operon, controlled by a codon optimized version of the LacI repressor. They found that *L. plantarum PlacA* was inducible only by lactose, but not by the non-metabolizable allolactose analogue isopropyl β-D-1-thiogalactopyranoside (IPTG) and methyl β-D-thiogalactopyranoside (TMG). *PxylA* was inducible by xylose but showed significant basal expression under non-induced conditions. *PlacSynth* was inducible with TMG and IPTG, but also showed basal expression under non-inducing conditions. The promoter *PlacSynth* was, moreover, used for establishment of an orthogonal, dual plasmid expression system, based on phage T7 RNA polymerase. All systems showed moderate expression levels when compared to the very strong constitutive promoter P_{11} (Heiss *et al.*, 2016).

Yet other inducible expression systems are based on temperature responsive elements. Binishofer *et al.* (2002) developed such an inducible expression system based on the promoter phiFSW-TI obtained from a temperate *L. casei* phage. The promoter was repressed at 28°C but was induced at 42°C in *L. casei* (Binishofer *et al.*, 2002).

Crutz-Le Coq and Zagorec (2008) developed a shuttle vector based on the replicon of the theta replicating low copy plasmid pRV500 from *L. sakei* and a copper-inducible promoter. Expression of the downstream reporter gene lacZ was shown to be induced at a concentration of 30 μM $CuSO_4$ (Crutz-Le Coq and Zagorec, 2008).

Due to the lack of a superoxide dismutase, lactobacilli require and accumulate intracellularly relatively high amounts of manganese to scavenge the toxic product superoxide when growing under aerobic conditions. A manganese starvation-inducible expression system for *L. plantarum* was developed by Bohmer *et al.* (2013), based on the novel promoter from a specific manganese transporter of *L. plantarum* NC8. The expression system does not need an external inducing agent, cultivation on low manganese medium is sufficient for induction.

Constitutive gene expression

Rud *et al.* (2006) have constructed a constitutive promoter library for use in, primarily, *L. plantarum*, by aligning rRNA promoters and establishing consensus −10 and −35 sequences. The intervening spacer sequence was then randomized and used to construct a promoter library, which was tested using the reporter enzymes GusA and PepN in both *L. plantarum* and *L. sakei*. Consistent and stable constitutive expression ranging from rather low levels to levels as high or even higher than induced levels using *PorfX* were obtained with both species, making this promoter library particularly suitable for metabolic engineering approaches (Rud *et al.*, 2006).

The homologous *PldhL* promoter was used to express heterologous *B. subtilis* oxalate decarboxylase in *L. plantarum* WCFS1 (Sasikumar *et al.*, 2014).

McCracken and colleagues cloned *Sau*3A fragments of genomic DNA from the probiotic *L. rhamnosus* strain GG and a promoter fragment from *L. fermentums* BR11 into pNZ272, a pSH71 derivative, identified promoter active fragments by expression of reporter gene *gusA*, mapped the transcription start sites by primer-extension analysis and tested promoter activity in three different *Lactobacillus* strains (*L. rhamnosus* GG, *L. fermentum* BR11 and *L. plantarum* BR3). All promoters were active in all tested strains and it could be shown that in most cases transcription starts at a purine-base (McCracken *et al.*, 2000).

Lizier *et al.* (2010) compared constitutive promoters of three different genes (*L. acidophilus* surface layer protein gene (*slp*), *L. acidophilus* lactate dehydrogenase gene (*ldhL*) and enterococcal rRNA adenine N-6-methyltransferase (*ermB*) regarding their ability to drive enhanced green fluorescent protein (eGFP) expression in six *L. reuteri* strains whereof five strains were isolated from chicken crop. Expression was quantified by measuring the eGFP fluorescence in a Qubit™ fluorometer (Invitrogen, Milan, Italy) and also tested by epifluorescence microscopy, and Western blot analysis. The *ermB*-promoter turned out to be the strongest in *L. reuteri*, followed by the *ldhL*-promoter, whereas the promoter of the *L. acidophilus* surface layer protein gene, although it was shown to be a very strong promoter in its original organism (Boot *et al.*, 1996), was active only at very low levels in all tested *L. reuteri* strains (Lizier *et al.*, 2010).

Kahala and Palva (1999) used the expression signals identified upstream of the *slpA* gene encoding a surface layer protein in *L. brevis*. Three reporter genes, β-glucuronidase (*gusA*), luciferase (*luc*) and aminopeptidase N (*pepN*) were cloned into the low-copy-number vector pKTH2095, a vector derived from pGK12. Expression was observed in all tested strains (*L. lactis*, *L. plantarum* and *L. gasseri*), however, GusA, Luc and PepN activities varied considerably among the studied strains (Kahala and Palva, 1999). Further studies of the sequence upstream of the slpA gene by (Hynönen *et al.*, 2010) showed that there actually were two consecutive promoter sequences (P1 and P2) which were tested separately with the *Lactobacillus helveticus* proline iminopeptidase (PepI) gene as a reporter. Both promoters were shown to be non-regulated (under the conditions tested) and highly active in *L. brevis.*

Duong *et al.* showed that a reporter encoding β-glucuronidase as well as two genes from an oxalate utilization pathway were constitutively highly expressed in *L. gasseri* and *L. acidophilus* when under control of the phosphoglycerate mutase promoter from *L. acidophilus* (Duong *et al.*, 2011).

Two other strong promoters of the genes encoding the putative translation elongation factors TU in *L. plantarum* CD033 (*Ptuf33*) and *L. buchneri* CD034 (*Ptuf34*) promoters were identified and characterized by BioLector measurement by Tauer *et al.* using the red fluorescent protein mCherry as reporter (Tauer *et al.*, 2014).

Chromosome based heterologous gene expression

Traditional double crossover integration

While plasmid based expression system are quite versatile, for industrial or medical applications, segregational plasmid instability as well as the metabolic costs for plasmid maintenance might be a disadvantage. Stable plasmid maintenance, especially of low copy plasmids, usually needs some sort of selection pressure, which is a drawback that can be overcome by targeted integration into the *Lactobacillus* chromosome.

Classically, integration of expression cassettes into the chromosome of lactobacilli is obtained by two step homologous recombination (Fitzsimons *et al.*, 1994; Hols *et al.*, 1994; Scheirlinck *et al.*, 1989). The first recombination event, integration of the entire plasmid, can be forced by employing integration vectors that are unstable, e.g. because of weak replication in a specific host or because of incubation under elevated temperatures (Maguin *et al.*, 1992). The second recombination event, eliminating the plasmid backbone and the selection marker, but leaving the expression cassette for the gene of interest in place, poses the bottleneck of the method. Colonies that have undergone this second step are not easy to detect. One method yet to identify colonies exposing the desired phenotype with justifiable screening effort is replica plating. Fitzsimons *et al.* used the unstable integration vector pGIP73 for integration of an expression cassette expressing the *Lactobacillus amylovorus* α-amylase gene at the *cbh*-locus. Transformants were identified by erythromycin resistance. After growth for 30 generations under selective pressure to allow the first integration event to take place, cells were plated on culture plates containing starch and amylase positive colonies, identified by visualization of starch degradation halos by staining with iodine vapours, were grown for 10 generations without selection. Amylase positive cells were again identified and grown for further 30 generations without selective pressure to allow excision of the vector backbone and selection marker. Screening for the second recombination event was performed by replica plating on plates without and with selective pressure. About 10% of the screened colonies showed the desired amylase positive/erythromycin sensitive phenotype (Fitzsimons *et al.*, 1994). A similar strategy was employed by Rossi *et al.* in order to generate genetically modified lactobacilli suitable as microbial silage additive that improves silage quality by enhanced acidification. Therefore, the *celA* gene, encoding an alkaline endo-1,4-β-glucanase from *Bacillus* sp. was integrated in the chromosome of two strains of *L. plantarum* (Lp80 and B41) by two-step homologous recombination at the *cbh* locus. Presence of the endo-1,4-β-glucanase was detected by growth on plates containing methylcellulose and staining with Congo red (degradation of methylcellulose by endo-1,4-β-glucanase results in yellow halos); absence of the resistance marker gene was screened by replica plating (Rossi *et al.*, 2001).

Another strategy to identify colonies that underwent the second recombination event is the use of counterselectable markers. Goh *et al.* (2009) described a *upp*-counterselectable gene replacement system, based on deletion of the *upp*-encoded uracil phosphoribosyltransferase (UPRTase) in *L. acidophilus* NCFM. The deletion mutant is resistant to 5-fluorouracil (5-FU), a toxic uracil analogue that is also a substrate for UPRTase. The first crossover event, integration of the recombinant plasmid, makes the host sensitive to 5-FU due to expression of the plasmid-borne PRTase gene. Excision of the integrated plasmid following a second recombination event restores the 5-FU resistant phenotype of the host, and allows rapid identification of plasmid-free recombinants (Goh *et al.*, 2009). Employing this system a reporter gene encoding a β-glucuronidase (GusA3) was integrated into four intergenic locations on the *L. acidophilus* NCFM chromosome. The integrants were tested for genetic stability and expression of *gusA3* at each locus was tested by measuring GusA3 activity on the substrate 4-methyl-umbelliferyl-β-D-glucuronide. The significant differences found in expression levels between the distinct integration loci underline the importance of a rational targeting of chromosomal expression cassettes (Douglas and Klaenhammer, 2011). The same strategy was adapted to *L. casei* by Yin *et al.* (2016) to stably integrate and express the porcine rotavirus outer capsid protein in *L. casei* (Yin *et al.*, 2016).

Recombineering: genomic integration of expression cassettes or point mutations using phage recombinases

Traditional strategies for obtaining gene deletion or integration variants in bacteria are, as mentioned above, mainly plasmid vector-based double-crossover methods, which are inefficient and laborious (Leenhouts *et al.*, 1989, 1991; Rossi *et al.*, 2001; Scheirlinck *et al.*, 1989).

Recombineering, an alternative genetic engineering method, utilizes linear DNA substrates that are either double-stranded (dsDNA) or single-stranded (ssDNA). In *E. coli*, dsDNA recombineering has been used successfully to create gene deletions, insertions, and inversions, but also integration of heterologous genes was shown to be possible (Murphy, 1998). In 2015, Yang and colleagues showed that prophage recombinases-mediated genome engineering can be used in *L. plantarum* as a smart integration technology that allows junk free integration of target expression cassettes into the host chromosome and thereby provide stable improved strains (Yang *et al.*, 2015). Recently, Xin and colleagues identified and analysed the function of potential prophage recombinases for genome editing in *L. casei*. A λ Red-like recombinase operon LCABL_13040-50-60 was identified on prophage PLE3 in *L. casei* BL23. In their study they could show that LCABL_13040 and LCABL_13060 were analogues to the host nuclease inhibitor (Redγ) and 5′–3′ exonuclease (Redα/RecE), respectively. They optimized recombineering conditions such as induction, homology length, recovery time and double-strand DNA (dsDNA) substrates quantity, the recombineering efficiency reached ~2.2×10^{-7}. They also tested ssDNA recombineering and introduced a precise point mutation into the *rpoB* gene in *L. casei* BL23, suggesting LCABL_13050 encoding a RecT protein, a single strand annealing protein that promotes annealing of complementary DNA strands, strand exchange and strand invasion and thereby allows incorporation of an oligonucleotide into the chromosome. It is thought that the oligonucleotide serves as a primer or as an Okazaki fragment for lagging strand replication. Up to now, ssDNA recombineering in LAB was only possible at sufficient efficiencies in *L. reuteri* and *Lactococcus lactis* (van Pijkeren and Britton, 2012). The LCABL_13040-50-60 proteins were shown to exhibit recombinase activity in six other *L. casei* strains, *L. paracasei* OY and *L. plantarum* WCSF1 (Xin *et al.*, 2017).

CRISPR/Cas9 assisted genetic engineering

The major bottleneck that hampers recombineering techniques are the sometimes quite low recombineering efficiencies. Below certain recombineering efficiency it is no longer possible to identify recombinant colonies by PCR screening methods. To solve this problem, recombineering can be combined with CRISPR/Cas9 systems which eliminate cells whose genomes have not been successfully modified. This method, called CRISPR/Cas9 assisted recombineering, allows reduction of the screening effort after a recombineering experiment to a practicable scale. In lactobacilli, CRISPR/Cas9 assisted recombineering was for the first time successfully employed in the probiotic *Lactobacillus reuteri* strain ATCC PTA 6475 by Oh and van Pijkeren. They introduced small consecutive changes into three independent genes, the *lacL* gene (locus tag HMPREF0536_0317), the *srtA* gene (locus tag HMPREF0536_0973, encoding a sortase) and the *sdp6* gene (locus tag HMPREF0536_0710, encoding a surface protein). In two independent experiments where 20 tetracycline resistant colonies at each time were PCR screened, they obtained an average transformation efficiency between 92 and 100% for the three loci, whereas no recombinants were identified within the pool of cells derived from the control plasmid (Oh and van Pijkeren, 2014).

Beside CRISPR/Cas9 assisted recombineering, CRISPR/Cas9 technology can also be used for the mediation of genomic integration of whole expression cassettes. In 2015, Xu *et al.* developed a single plasmid strategy for CRISPR/Cas9 assisted genome integration for *Clostridium cellulolyticum.* To attenuate the lethality of Cas9 for the bacterial host cell, its function was changed from an endonuclease to a nickase by introducing a single point mutation (D10A). Therefore, CRISPR RNA (crRNA) and trans activating crRNA (tracrRNA) were joined into a so called single guide RNA (sgRNA) which could direct the Cas9-nickase to any sequence defined by the 20 bp protospacer followed by a protospacer-adjacent motif (PAM). The protospacer targets the genomic region that is flanked by homologous regions that are also present in the donor plasmid, carrying the gene that has to be integrated, but lacking the target region. This strategy allows both, gene deletions and integration of whole expression cassettes (Xu *et al.*, 2015). The same strategy was used for *Lactobacillus casei* providing the Cas9-nickase on plasmid pNCNICK. With their strategy, Song *et al.* (2017) achieved deletion of four independent genes as well as chromosomal insertion of an enhanced green fluorescent protein (eGFP) expression cassette at the *L. casei LC2W_1628* locus (Song *et al.*, 2017). Expanding the system to other lactobacilli seems to be feasible.

Additional factors influencing expression strength: 5′-untranslated region, Shine–Dalgarno sequence and codon usage

The 5′-untranslated leader sequence

Besides PCN/gene dosage and promoter strength a few other factors have been identified, influencing recombinant protein production. Since the 1980s we know that the untranslated region upstream of the ribosomal binding site of mRNA has an influence on translation efficiency (Stanssens *et al.*, 1985). Boot *et al.* (1996) showed that the 5′-leader of the untranslated region of the surface layer protein mRNA can form a stable stem–loop structure which stabilizes the transcript and additionally leads to an exposure of the ribosomal binding site. Truncation of the 5′-untranslated leader sequence (UTLS) sequence results in a reduction of expression efficiency in *L. casei* (Boot *et al.*, 1996). Reduced in S-layer protein production was also observed when the UTLS of *L. crispatus slpB*-mRNA was truncated, indicating a similar mechanism (Sun *et al.*, 2013). The mRNA stabilizing properties of UTLSs by stem-lop formation of the *slpA*-mRNA from *L. acidophilus* was exploited by Narita *et al.* (2006a) to develop a high-level expression system for lactic acid bacteria. Therefore, the *slpA*-core promoter sequence was integrated upstream of an α-amylase gene. Plasmids containing the whole UTLS as well as with partially deleted versions of the UTLS were constructed. Expression levels of the different constructs in *L. lactis* and *L. casei* were evaluated by RT-PCR and by α-amylase activity assays. The results of the study showed that exploitation of the mRNA stabilizing effect of the *slpA* UTLS is able to improve protein production in expression systems for lactic acid bacteria (Narita *et al.*, 2006a).

Influence of the relative position of the Shine–Dalgarno sequence on recombinant gene expression

The specific base pairing between the 3′-end of the rRNA and the sequence preceding an initiator AUG provides a mechanism by which the cell can distinguish between initiator AUGs and internal and/or out-of-phase AUG sequences. The degree of base pairing also plays a role in determining the rate of initiation at different AUG initiator codons in polycistronic mRNAs (Shine and Dalgarno, 1974). For fine-tuning translational efficiency, Tauer *et al.* (2014) analysed the effect of varying the distance between the anti-SDS and the translational start-site, in the range between 5 and 12 nucleotides. All constructs were based on the low copy p256 origin of replication, and mCherry expression was under control of the P_{11} promoter (Rud *et al.*, 2006). Spacer sequences shorter than 7 nucleotides turned out to considerably hamper translation efficiency, while 8 nucleotides seemed to be optimal. A slight decrease could be observed when the spacer was designed to be as long as 12 nucleotides.

Codon usage optimization

Since synthetic genes have become affordable for every laboratory and an increasing amount of codon usage data have become available because of the high number of completed genome sequencing projects (Nakamura *et al.*, 1997) (http://www.kazusa.or.jp/codon/), optimization of codon usage for efficient gene expression has gained more and more interest. Companies supplying gene synthesis often offer their inhouse-developed codon optimization services and a lot of codon optimization tools are available on the internet (Chin *et al.*, 2014; Grote *et al.*, 2005; Puigbo *et al.*, 2007; Tuan-Anh *et al.*, 2017; Yu *et al.*, 2017), but it is not always clear if the used approaches and algorithms were only tested with *E. coli*, *B. subtilis* and yeasts, or also with more non-mainstream organisms like lactobacilli. Only a few papers specifically cover the topic of codon usage optimization in lactobacilli.

Pouwels and Leunissen (1994) compared codon usage of 70 *Lactobacillus* species. They observed inter-species as well as intra-species heterogeneity of codon usage and found that codon usage in lactobacilli is highly biased. Moreover they found a positive correlation between codon usage bias and expression level. Highly expressed genes showed highly biased codon usage, whereas genes expressed at a low level showed much less biased codon usage (Pouwels and Leunissen, 1994).

Fuglsang (2003) showed that especially in species with such a strong correlation of expression level and codon bias, such as. *L. plantarum* and *L. lactis*, recombinant gene expression can be optimized by adapting codons of the gene of interest towards the set of codons that are naturally used in the set of highly expressed genes of the expression host organism (Fuglsang, 2003). A comparative study by Nayak *et al.* (2012) including *L. sakei* as well as 13 other *Lactobacillus* species revealed that the major trend of synonymous codon usage was highly correlated with gene expression level, whereas translational selection for highly expressed genes, where more frequent synonymous codons correspond to more abundant cognate tRNAs, was not found to be similar in all tested *Lactobacillus* species (Nayak, 2012). Based on the tRNA-pool and the codon usage preference of *L. casei* BL23 the influenza A virus nucleocapsid protein (NP) gene was re-designed and successfully expressed using the constitutive *ldh* promoter as well as the inducible *nisA* promoter (Suebwongsa *et al.*, 2013). A frequently occurring problem is the over-expression of genes derived from G+C rich coding sequences in an A+T rich *Lactobacillus* host. Johnston *et al.* (2014) demonstrated that codon optimization by synonymous substitutions improves expression levels of *Mycobacterium avium* subsp. *paratuberculosis* antigens in *L. salivarius* (Johnston *et al.*, 2014).

Secretory expression of target proteins

Bacteria export a significant part of their proteome beyond the cellular membrane. Various mechanisms have evolved to facilitate this secretion through the cell membrane, the most important and best investigated of which is the Sec-dependent translocation system. This system is also most commonly exploited for the secretion of heterologously expressed proteins (Mathiesen *et al.*, 2008). The Sec-based system is characterized by N-terminal signal peptides (SP), which direct designated proteins towards the translocon and facilitate translocation, either during or after translation, and which are cleaved off the mature protein by signal peptidases (Tsirigotaki *et al.*, 2017). Heterologously expressed proteins that lack a native SP, or whose SP is not recognized by the respective receptors in the cell with sufficient specificity, but which have to be exported for functionality, can be translationally fused to a native lactococcal or lactobacillal signal peptide. The most widely exploited one in genetic engineering of LAB is the signal peptide of the major lactococcal secretory protein, Usp45 (Pusch *et al.*, 2006; van Asseldonk *et al.*, 1990). In *L. plantarum*, the SP from the *Streptococcus pyogenes* M6 protein, a *Lactobacillus amylovorus* amylase (Hols *et al.*, 1994;, 1997) and the *Lactobacillus brevis* S-layer protein SlpA (Savijoki *et al.*, 1997), among others, have been used to direct heterologous proteins towards secretion. Mathiesen *et al.* (2008) have explored seven signal peptides selected from the genome sequence of *L. plantarum* WCFS1 to facilitate secretion of two reporter proteins, *Staphylococcus aureus* NucA and *L. amylovorus* AmyA, and compared their efficiency to the SPs from Usp45 and the M6 protein, using the pSIP system for inducible expression. While two SPs, Lp_0373 and Lp_0600, gave high yields with both reporter proteins, the overall picture of this and previous studies was less clear: several SPs worked very well with one reporter protein, but were inefficient with another. In a related study on *Bacillus subtilis*, Brockmeier *et al.* (2006) also could not correlate SP D-values with secretion yields, and concluded that a rationalization of secretion efficiency and secretion yields is difficult (Brockmeier *et al.*, 2006). Clearly the SP is not the only determinant of secretion efficiency and secretion yield, and the combinations of host strain, signal peptide and target protein appear to require optimization on an individual basis, as well as a balancing of transcriptional strength and the capacity of the secretory system and other factors (Mathiesen *et al.*, 2008). In an expansion of this study (Mathiesen *et al.*, 2009) investigated almost all predicted (based on presence of a signal peptidase cleavage site) SPs from the genome of *L. plantarum*, altogether 76. The majority of these SPs facilitated secretion of the reporter protein NucA, but again with a wide spread in achievable secretion yields as well as differing efficiencies with other reporter proteins, and no clear correlation of secretion efficiency and various properties such as length of the SP, length of the N-terminal (positively charged) domain or the central, hydrophobic helical domain, overall hydrophobicity, charge or pI (Mathiesen *et al.*, 2009). The set of modular pSIP vectors developed in those studies, that allow the quick exchange of components in order to efficiently generate the optimal combination of an SP with a given target protein, however, are a highly useful tool for the development of heterologous cell factories.

Surface display in lactobacilli

Bacterial surface display is defined as the presentation of recombinant proteins or peptides on the surface of bacterial cells. Surface display of a protein of interest is usually achieved by genetic fusion with a protein that is intrinsically located at the bacterial surface (carrier protein), or a fragment thereof. The carrier protein serves as surface anchor and, so far, five different types of surface anchors have been described for LAB, including transmembrane domains, lipoprotein anchor domains, LPXTG-type domains, lysin motif (LysM) domains and surface layer protein anchor domains (Leenhouts *et al.*, 1999; Michon *et al.*, 2016). Among the most frequently applied surface anchoring domains are LPXTG sequences (such as that of M6 protein of *Streptococcus pyogenes*) (Wieczorek and Martin, 2010), which are recognized by sortases and covalently attached to the cell wall. LysM domains (such as C-terminal part of the AcmA protein of *Lactococcus lactis* or from *L. fermentum* bacteriophage) enable non-covalent binding to the peptidoglycan via N-acetylglucosamine residues (Hu *et al.*, 2010; Lim *et al.*, 2010; Okano *et al.*, 2008). Surface layer proteins are particularly attractive carriers for surface display because they form a two-dimensional lattice that covers the entire bacterial surface (Hynonen and Palva, 2013). The displayed protein should not sterically hinder the formation of the lattice in order for the display to be effective. This was achieved by using surface layer proteins SlpB from *L. crispatus*, displaying the green fluorescent protein; (Hu *et al.*, 2011) and SlpA from *L. brevis*, displaying the c-Myc epitope (Avall-Jaaskelainen *et al.*, 2002). Examples of carrier proteins that anchor via transmembrane domains include PgsA protein from *B. subtilis* (part of poly-γ-glutamic acid synthase complex; (Lee *et al.*, 2006; Narita *et al.*, 2006b; Wei *et al.*, 2010), and mucin-binding protein (Mub) from *Listeria monocytogenes* (Hsueh *et al.*, 2014). It has been shown that different surface anchoring domains localize to different parts of cells (poles, septum, whole cell) and that localization is species dependent (Hu *et al.*, 2010; Steen *et al.*, 2003). Improved surface display approaches are still being sought.

A majority of proteins for surface display are plasmid encoded. Attempts have been made to increase the acceptance of such systems by introducing food-grade or non-GMO approaches. An example of a food-grade system was developed for

L. casei. It consists of a lactose-deficient derivative of *L. casei* ATCC 334 and of a plasmid that contains lactose metabolism genes and surface layer protein of *L. acidophilus* as a surface anchor. A non-GMO approach was devised by using heterologously produced fusion proteins, containing LysM domains, for coating of live non-modified bacteria or inactivated, trichloroacetic acid-treated bacteria-like particles (also known as GEM) (Kobierecka *et al.*, 2015; Ribelles *et al.*, 2013; Zadravec *et al.*, 2015). These particles contain no recombinant DNA and require less rigorous regulatory procedure.

Various biotechnological applications of surface display of recombinant proteins in bacteria have been suggested. These include the display of enzymes as whole cell biocatalysts, the display of antigens for vaccines and the display of protein binders as biosensors in diagnostics, as pathogen or toxin binders in therapy and as heavy metal binders in bioremediation (Wu *et al.*, 2008). Another field of use is the display of proteins for protein engineering by using combinatorial libraries and *in vitro* selection (Daugherty, 2007). *L. plantarum* and *L. casei* are the most common hosts for surface display, while for *L. paracasei*, *L. reuteri*, *L. jensenii*, *L. acidophilus* and *L. salivarius* only a few applications have been developed. Therapeutic applications are intended for human or veterinary medicine, with delivery of antigens for the purpose of vaccination being the most thoroughly studied (see separate section). Display of enzymes on the surface of bacteria enables biotransformation of various substrates into valuable products. The applications of lactobacilli have focused mostly on the transformation of carbohydrates. Carbohydrate-degrading enzymes, such as β-mannanase, chitosanase (Nguyen *et al.*, 2016), xylanase (Hsueh *et al.*, 2014), β-gluconase (Huang *et al.*, 2011), α-amylase (Narita *et al.*, 2006b), and others, have been effectively displayed, and their catalytic potential demonstrated. Enzyme-displaying lactobacilli can be used as whole-cell biocatalysts for the production of oligosaccharides or as feed additives.

Therapeutic applications of lactobacilli

Lactobacilli are generally regarded as safe (GRAS) for human consumption and as food grade organisms they are not only attractive candidates for applications relevant for food and feed processing, but also for applications for therapeutic purposes. In previous studies it has been shown that the immune response of animals to different lactobacilli can vary widely. While some lactobacilli seem to induce oral tolerance, others induce an adaptive immune response as well as dendritic cell maturation and cytokine production (Stoeker *et al.*, 2011). Genetic engineering can be used to recombinantly express therapeutic polypeptides or antigens intracellularly, in a secreted form or to display them on the bacterial cell surface.

An example of intracellular gene expression of an immune response activating protein was given by Stoeker *et al.* (2011). They engineered *L. gasseri* to express *Salmonella enterica* serovar Typhimurium flagellin C (FliC), which resulted in the release of proinflammatory cytokines in dendritic cells. Also a clear interaction with the mammalian immune system could be demonstrated as it was shown that lactobacilli activate the toll-like receptor 2.

Secreted expression of recombinant genes for therapeutic use in lactobacilli was performed with the antigens from enterohaemorrhagic or enterotoxigenic *E. coli*, e.g. expressed in *L. acidophilus* or *L. plantarum* (Lin *et al.*, 2017; Yang *et al.*, 2017a), or antigens from *Helicobacter pylori*, expressed in *L. acidophilus* (Hongying *et al.*, 2014).

For the presentation of recombinant antigens, surface display has been demonstrated in animal trials to be superior to other expression loci (Bermudez-Humaran *et al.*, 2004; Norton *et al.*, 1996), forming the rationale for further development of the approach. *L. plantarum* was used to display HA-2 influenza antigen (Yang *et al.*, 2017c) and mycobacterium antigens (Kuczkowska *et al.*, 2017). Further, Yang *et al.* (2017b) showed that *L. plantarum* can serve as a vector vaccine against the protozoic parasite *Eimeria tenella* in animals. Reveneau *et al.* (2002) engineered *L. plantarum* to serve as a mucosal vaccine delivery vehicle against tetanus toxin. They tested three cellular locations, intracellular, secreted to the supernatant and displayed at the cell-surface. All tested constructs were shown to be able to induce strong specific immune responses against the non-toxic C fragment of tetanus toxin (TTFC). Cell-surface presentation revealed to require lower antigen doses to be immunogenic, however, the highest IgG serum antibody titres were obtained with the *L. plantarum* strain producing TTFC intracellularly (Reveneau *et al.*, 2002). Scheppler *et al.* (2002) engineered a *Lactobacillus johnsonii* strain displaying a cell wall-anchored proteinase PrtB/tetanus toxin mimotope fusion but they did not succeed to get sufficient immune response against the tetanus toxin (Scheppler *et al.*, 2002). The β-toxoid from *Clostridium perfringens* was shown to be protective in animals when expressed in *L. casei* and given by an oral route (Alimolaei *et al.*, 2017). Moreover, *L. casei* was used to display influenza antigens (Li *et al.*, 2015; Tan *et al.*, 2017), E7 antigen of human papillomavirus (Ribelles *et al.*, 2013), antigen of enterotoxigenic *E. coli* (Wei *et al.*, 2010) or SARS coronavirus (Lee *et al.*, 2006). Interestingly, some applications include co-display of additional protein that serves as molecular adjuvant and strengthens the immune response (Kajikawa *et al.*, 2012; Li *et al.*, 2015).

Other therapeutic applications of surface display include display of binding proteins for the neutralization of pathogens. Nanobody against rotavirus was displayed on the surface of exopolysaccharide-deficient *L. rhamnosus* (Alvarez *et al.*, 2015). Similarly, *L. rhamnosus* was applied for the surface display of immunoglobulin-binding proteins and, subsequently, coated with immunoglobulins from bovine colostrum, again targeting rotavirus (Gunaydin *et al.*, 2014). *L. paracasei* was used for the display of single-chain variable fragments against Anthrax oedema toxin (Andersen *et al.*, 2011). Surface display on lactobacilli was also exploited for the delivery of chemokine-binding proteins (Škrlec *et al.*, 2017) or delivery of cytokines (IL-10; Cai *et al.*, 2016), with the intention of modulating cytokine signalling in inflammatory disorders.

Another promising approach for antigen display is virus like particles (VLPs). So far, in lactobacilli VLPs have been only produced in *L. casei*. A lactose-inducible system based on the *Lactobacillus casei* lactose operon promoter (*Plac*) was used for the expression of the HPV-16 L1 protein in *L. casei*. The authors showed that the recombinant protein was able to self-assemble into VLPs intracellularly (Aires *et al.*, 2006).

References

Ahn, C., Collins-Thompson, D., Duncan, C., and Stiles, M.E. (1992). Mobilization and location of the genetic determinant of chloramphenicol resistance from *Lactobacillus plantarum* caTC2R. Plasmid 27, 169–176.

Aleshin, V.V., Semenova, E.V., Doroshenko, V.G., Jomantas, Y.V., Tarakanov, B.V., and Livshits, V.A. (1999). The broad host range plasmid pLF1311 from *Lactobacillus fermentum* VKM1311. FEMS Microbiol. Lett. *178*, 47–53.

Alimolaei, M., Golchin,M., and Ezatkhah, M. (2017). Orally administered recombinant *Lactobacillus casei* vector vaccine expressing beta-toxoid of *Clostridium perfringens* that induced protective immunity responses. Res. Vet. Sci. *115*, 332–339.

Alpert, C.A., Crutz-Le Coq, A.M., Malleret, C., and Zagorec, M. (2003). Characterization of a theta-type plasmid from *Lactobacillus sakei*: a potential basis for low-copy-number vectors in lactobacilli. Appl. Environ. Microbiol. *69*, 5574–5584.

Alvarez, B., Krogh-Andersen, K., Tellgren-Roth, C., Martinez, N., Gunaydin, Lin, G.Y., Martin, M.C., Alvarez, M.A., Hammarstrom, L., and Marcotte, H. (2015). An exopolysaccharide-deficient mutant of *Lactobacillus rhamnosus* GG efficiently displays a protective llama antibody fragment against rotavirus on its surface. Appl. Environ. Microbiol. *81*, 5784–5793.

An, H.Y., Tsuda, H., and Miyamoto, T. (2007). Expression of citrate permease gene of plasmid pCM1 isolated from *Lactococcus lactis* subsp. *lactis* biovar diacetylactis NIAI N-7 in *Lactobacillus casei* L-49-4. Appl. Microbiol. Biotechnol., *74*, 609–616.

Andersen, K.K., Marcotte, H., Alvarez, B., Boyaka, P.N., and Hammarstrom, L. (2011). In situ gastrointestinal protection against anthrax edema toxin by single-chain antibody fragment producing lactobacilli. BMC Biotechnol. *11*, 126.

Avall-Jääskeläinen, S., Kylä-Nikkilä, K., Kahala, M., Miikkulainen-Lahti, T., and Palva, A. (2002). Surface display of foreign epitopes on the *Lactobacillus brevis* S-layer. Appl. Environ. Microbiol. *68*, 5943–5951.

Ávila, C.L., Carvalho, B.F., Pinto, J.C., Duarte, W.F., and Schwan, R.F. (2014). The use of *Lactobacillus* species as starter cultures for enhancing the quality of sugar cane silage. J. Dairy Sci. *97*, 940–951. https://doi.org/10.3168/jds.2013-6987

Axelsson, L., Lindstad, G., and Naterstad, K. (2003). Development of an inducible gene expression system for *Lactobacillus sakei*. Lett. Appl. Microbiol. 37, 115–120.

Berlec, A., Ravnikar, M., and Strukelj, B. (2012). Lactic acid bacteria as oral delivery systems for biomolecules. Pharmazie *67*, 891–898.

Berlec, A., Zavrsnik, J., Butinar, M., Turk, B., and Strukelj, B. (2015). In vivo imaging of *Lactococcus lactis*, *Lactobacillus plantarum* and *Escherichia coli* expressing infrared fluorescent protein in mice. Microb. Cell. Fact. *14*, 181.

Bermudez-Humaran, L.G., Cortes-Perez, N.G., Le Loir, Y., Alcocer-Gonzalez, J.M., Tamez-Guerra, R.S., de Oca-Luna, R.M., and Langella, P. (2004). An inducible surface presentation system improves cellular immunity against human papillomavirus type 16 E7 antigen in mice after nasal administration with recombinant *lactococci*. J. Med. Microbiol. 53, 427–433.

Binishofer, B., Moll, I., Henrich, B., and Blasi, U. (2002). Inducible promoter-repressor system from the *Lactobacillus casei* phage phiFSW. Appl. Environ. Microbiol. *68*, 4132–4135.

Bohmer, N., Konig, S., and Fischer, L. (2013). A novel manganese starvation-inducible expression system for *Lactobacillus plantarum*. FEMS Microbiol. Lett. *342*, 37–44.

Boot, H.J., Kolen, C.P., Andreadaki, F.J., Leer, R.J., and Pouwels, P.H. (1996). The *Lactobacillus acidophilus* S-layer protein gene expression site comprises two consensus promoter sequences, one of which directs transcription of stable mRNA. J. Bacteriol. *178*, 5388–5394.

Boris, S., and Barbés, C. (2000). Role played by lactobacilli in controlling the population of vaginal pathogens. Microbes Infect. *2*, 543–546.

Bosma, E.F., Forster, J., and Nielsen, A.T. (2017). Lactobacilli and *pediococci* as versatile cell factories: Evaluation of strain properties and genetic tools. Biotechnol. Adv. *35*, 419–442.

Bouia, A., Bringel, F., Frey, L., Kammerer, B., Belarbi, A., Guyonvarch, A., and Hubert, J.C. (1989). Structural organization of pLP1, a cryptic plasmid from *Lactobacillus plantarum* CCM 1904. Plasmid *22*, 185–192.

Bringel, F., Frey, L., and Hubert, J.C. (1989). Characterization, cloning, curing, and distribution in lactic acid bacteria of pLP1, a plasmid from *Lactobacillus plantarum* CCM 1904 and its use in shuttle vector construction. Plasmid *22*, 193–202.

Brockmeier, U., Caspers, M., Jockwer, R.A., Noll, T., and Eggert, T. (2006). Systematic screening of all signal peptides from *Bacillus subtilis*: A powerful strategy in optimizing heterologous protein secretion in Gram-positive bacteria. J. Mol. Biol. *362*, 393–402.

Cai, R., Jiang, Y., Yang, W., Yang, W., Shi, S., Shi, C., Hu, J., Gu, W., Ye, L., Zhou, F., *et al.* (2016). Surface-displayed IL-10 by recombinant *Lactobacillus plantarum* reduces Th1 responses of RAW264.7 cells stimulated with Poly(I:C) or LPS. J. Microbiol. Biotechnol. *26*, 421–431. https://doi.org/10.4014/jmb.1509.09030

Chen, C., Ai, L., Zhou, F., Ren, J., Sun, K., Zhang, H., Chen, W., and Guo, B. (2012). Complete nucleotide sequence of plasmid pST-III from *Lactobacillus plantarum* ST-III. Plasmid *67*, 236–244. https://doi.org/10.1016/j.plasmid.2011.12.005

Chen, Z., Lin, J., Ma, C., Zhao, S., She, Q., and Liang, Y. (2014). Characterization of pMC11, a plasmid with dual origins of replication isolated from *Lactobacillus casei* MCJ and construction of shuttle vectors with each replicon. Appl. Microbiol. Biotechnol. *98*, 5977–5989.

Chin, J.X., Chung, B.K., and Lee, D.Y. (2014). Codon Optimization OnLine (COOL): A web-based multi-objective optimization platform for synthetic gene design. Bioinformatics *30*, 2210–2212.

Crutz-Le Coq, A.M., and Zagorec, M. (2008). Vectors for Lactobacilli and other Gram-positive bacteria based on the minimal replicon of pRV500 from *Lactobacillus sakei*. Plasmid *60*, 212–220. https://doi.org/10.1016/j.plasmid.2008.08.002

Daugherty, P.S. (2007). Protein engineering with bacterial display. Curr. Opin. Struct. Biol. *17*, 474–480.

David, S., Simons, G., and De Vos, W.M. (1989). Plasmid transformation by electroporation of Leuconostoc paramesenteroides and its use in molecular cloning. Appl. Environ. Microbiol. *55*, 1483–1489.

de Ruyter, P.G., Kuipers, O.P., and de Vos, W.M. (1996). Controlled gene expression systems for *Lactococcus lactis* with the food-grade inducer nisin. Appl. Environ. Microbiol. *62*, 3662–3667.

de Vos, W.M. (1999). Gene expression systems for lactic acid bacteria. Curr. Opin. Microbiol. *2*, 289–295.

del Solar, G., Giraldo, R., Ruiz-Echevarría, M.J., Espinosa, M., and Díaz-Orejas, R. (1998). Replication and control of circular bacterial plasmids. Microbiol. Mol. Biol. Rev. *62*, 434–464.

Douglas, G.L., and Klaenhammer, T.R. (2011). Directed chromosomal integration and expression of the reporter gene gusA3 in *Lactobacillus acidophilus* NCFM. App. Environ. Microbiol. *77*, 7365–7371.

Duong, T., Miller, M.J., Barrangou, R., Azcarate-Peril, M.A., and Klaenhammer, T.R. (2011). Construction of vectors for inducible and constitutive gene expression in *Lactobacillus*. Microb. Biotechnol. *4*, 357–367. https://doi.org/10.1111/j.1751-7915.2010.00200.x

Egervärn, M., Roos, S., and Lindmark, H. (2009). Identification and characterization of antibiotic resistance genes in *Lactobacillus reuteri* and *Lactobacillus plantarum*. J. Appl. Microbiol. *107*, 1658–1668.

Eguchi, T., Doi, K., Nishiyama, K., Ohmomo, S., and Ogata, S. (2000). Characterization of a phage resistance plasmid, pLKS, of silage-making *Lactobacillus plantarum* NGRI0101. Biosci. Biotechnol. Biochem. *64*, 751–756.

Eijsink, V.G., Axelsson, L., Diep, D.B., Håvarstein, L.S., v. Holo, H., and Nes, I.F. (2002). Production of class II bacteriocins by lactic acid bacteria; an example of biological warfare and communication. Antonie Van Leeuwenhoek *81*, 639–654.

Eikmeyer, F.G., Kofinger, P., Poschenel, A., Junemann, S., Zakrzewski, M., Heinl, S., Mayrhuber, E., Grabherr, R., Puhler, A., Schwab, H., and Schluter, A. (2013). Metagenome analyses reveal the influence of the inoculant *Lactobacillus buchneri* CD034 on the microbial community involved in grass ensiling. J. Biotechnol. *167*, 334–343.

Fang, F., Flynn, S., Li, Y., Claesson, M.J., van Pijkeren, J.P., Collins, J.K., van Sinderen, D., and O'Toole, P.W. (2008). Characterization of endogenous plasmids from *Lactobacillus salivarius* UCC118. Appl. Environ. Microbiol. *74*, 3216–3228.

Fernández, M., Margolles, A., Suárez, J.E., and Mayo, B. (1999). Duplication of the beta-galactosidase gene in some *Lactobacillus plantarum* strains. Int. J. Food Microbiol. *48*, 113–123.

Fitzsimons, A., Hols, P., Jore, J., Leer, R.J., O'Connell, M., and Delcour, J. (1994). Development of an amylolytic *Lactobacillus plantarum* silage strain expressing the *Lactobacillus amylovorus* alpha-amylase gene. Appl. Environ. Microbiol. *60*, 3529–3535.

Fons, M., Hégé,. M., Ladiré, M., Raibaud, P., Ducluzeau, R., and Maguin, E. (1997). Isolation and characterization of a plasmid from *Lactobacillus fermentum* conferring erythromycin resistance. Plasmid *37*, 199–203.

Fuglsang, A. (2003). Lactic acid bacteria as prime candidates for codon optimization. Biochem. Biophys. Res. Commun. *312*, 285–291.

Goh, Y.J., Azcarate-Peril, M.A., O'Flaherty, S., Durmaz, E., Valence, F., Jardin, J., Lortal, S., and Klaenhammer, T.R. (2009). Development and application of a upp-based counterselective gene replacement system for the study of the S-layer protein SlpX of *Lactobacillus acidophilus* NCFM. Appl. Environ. Microbiol. *75*, 3093–3105.

Gosalbes, M.J., Esteban, C.D., Galan, J.L., and Perez-Martinez, G. (2000). Integrative food-grade expression system based on the lactose regulon of *Lactobacillus casei*. Appl. Environ. Microbiol. *66*, 4822–4828.

Grabherr, R., Nilsson, E., Striedner, G., and Bayer, K. (2002). Stabilizing plasmid copy number to improve recombinant protein production. Biotechnol. Bioeng. *77*, 142–147.

Grote, A., Hiller, K., Scheer, M., Münch, R., Nörtemann, B., Hempel, D.C., and Jahn, D. (2005). JCat: a novel tool to adapt codon usage of a target gene to its potential expression host. Nucleic Acids Res. *33*, 526–531.

Guerzoni, M.E., Vernocchi, P., Ndagijimana, M., Gianotti, A., and Lanciotti, R. (2007). Generation of aroma compounds in sourdough: Effects of stress exposure and lactobacilli-yeasts interactions. Food Microbiol. *24*, 139–148.

Gunaydin, G., Zhang, R., Hammarstrom, L., and Marcotte, H. (2014). Engineered *Lactobacillus rhamnosus* GG expressing IgG-binding domains of protein G: Capture of hyperimmune bovine colostrum antibodies and protection against diarrhea in a mouse pup rotavirus infection model. Vaccine *32*, 470–477.

Halbmayr, E., Mathiesen, G., Nguyen, T.H., Maischberger, T., Peterbauer, C.K., Eijsink, V.G., and Haltrich, D. (2008). High-level expression of recombinant beta-galactosidases in *Lactobacillus plantarum* and *Lactobacillus sakei* using a Sakacin P-based expression system. J. Agric. Food Chem. *56*, 4710–4719.

Hasegawa, M., Yamane, D., Funato, K., Yoshida, A., and Sambongi, Y. (2017). Gamma-aminobutyric acid fermentation with date residue by a lactic acid bacterium, *Lactobacillus brevis*. J. Biosci. Bioeng. *125*, 346–352.

Heinl, S., Spath, K., Egger, E., and Grabherr, R. (2011a). Sequence analysis and characterization of two cryptic plasmids derived from *Lactobacillus buchneri* CD034. Plasmid *66*, 159–168.

Heinl, S., Wibberg, D., Eikmeyer, F., Szczepanowski, R., Blom, J., Linke, B., Goesmann, A., Grabherr, R., Schwab, H., Pühler, A., and Schlüter, A. (2012). Insights into the completely annotated genome of *Lactobacillus buchneri* CD034, a strain isolated from stable grass silage. J. Biotechnol. *161*, 153–166.

Heiss, S., Hormann, A., Tauer, C., Sonnleitner, M., Egger, E., Grabherr, R., and Heinl, S. (2016). Evaluation of novel inducible promoter/repressor systems for recombinant protein expression in *Lactobacillus plantarum*. Microb. Cell. Fact. *15*, 50.

Hols, P., Ferain, T., Garmyn, D., Bernard, N., and Delcour, J. (1994). Use of homologous expression-secretion signals and vector-free stable chromosomal integration in engineering of *Lactobacillus plantarum* for alpha-amylase and levanase expression. Appl. Environ. Microbiol. *60*, 1401–1413.

Hols, P., Slos, P., Dutot, P., Reymund, J., Chabot, P., Delplace, B., Delcour, J., and Mercenier, A. (1997). Efficient secretion of the model antigen M6-gp41E in *Lactobacillus plantarum* NCIMB 8826. Microbiology *143*, 2733–2741.

Hongying, F., Xianbo, W., Fang, Y., Yang, B., and Beiguo, L. (2014). Oral immunization with recombinant *Lactobacillus acidophilus* expressing the adhesin Hp0410 of Helicobacter pylori induces mucosal and systemic immune responses. Clin. Vaccine Immunol. *21*, 126–132.

Hsueh, H.Y., Yu, B., Liu, C.T., and Liu, J.R. (2014) Increase of the adhesion ability and display of a rumen fungal xylanase on the cell surface of *Lactobacillus casei* by using a listerial cell wall -anchoring protein. J. Sci. Food Agric. *94*, 576–584.

Hu, S., Kong, J., Kong, W., Guo, T., and Ji, M. (2010). Characterization of a novel LysM domain from *Lactobacillus fermentum* bacteriophage endolysin and its use as an anchor to display heterologous proteins on the surfaces of lactic acid bacteria. Appl. Environ. Microbiol. *76*, 2410–2418.

Hu, S., Kong, J., Sun, Z., Han, L., Kong, W., and Yang, P. (2011). Heterologous protein display on the cell surface of lactic acid bacteria mediated by the s-layer protein. Microb. Cell Fact. *10*, 86.

Huang, S.J., Chen, M.J., Yueh, P.Y., Yu, B., Zhao, X., and Liu, J.R. (2011). Display of Fibrobacter succinogenes β-glucanase on the cell surface of *Lactobacillus reuteri*. J. Agric. Food Chem. *59*, 1744–1751. https://doi.org/10.1021/jf104266x

Hynonen, U., and Palva, A. (2013). *Lactobacillus* surface layer proteins: structure, function and applications. Appl. Microbiol. Biotechnol. *97*, 5225–5243.

Hynönen, U., Avall-Jääskeläinen, S., and Palva, A. (2010). Characterization and separate activities of the two promoters of the *Lactobacillus brevis* S-layer protein gene. Appl. Microbiol. Biotechnol. *87*, 657–668. https://doi.org/10.1007/s00253-010-2500-z

Johnston, C.D., Bannantine, J.P., Govender, R., Endersen, L., Pletzer, D., Weingart, H., Coffey, A., O'Mahony, J., and Sleator, R.D. (2014). Enhanced expression of codon optimized *Mycobacterium avium* subsp. *paratuberculosis* antigens in *Lactobacillus salivarius*: Front. Cell Infect. Microbiol. *4*, 120.

Josson, K., Soetaert, P., Michiels, F., Joos, H., and Mahillon, J. (1990). *Lactobacillus hilgardii* plasmid pLAB1000 consists of two functional cassettes commonly found in other gram-positive organisms. J. Bacteriol. *172*, 3089–3099.

Kahala, M., and Palva, A. (1999). The expression signals of the *Lactobacillus brevis* slpA gene direct efficient heterologous protein production in lactic acid bacteria. Appl. Microbiol. Biotechnol. *51*, 71–78.

Kajikawa, A., Zhang, L., Long, J., Nordone, S., Stoeker, L., LaVoy, A., Bumgardner, S., Klaenhammer, T., and Dean, G. (2012). Construction and immunological evaluation of dual cell surface display of HIV-1 gag and *Salmonella enterica* serovar Typhimurium FliC in *Lactobacillus acidophilus* for vaccine delivery. Clin. Vaccine Immunol. *19*, 1374–1381.

Khan, S.A. (2005). Plasmid rolling-circle replication: highlights of two decades of research. Plasmid, *53*, 126–136.

Klein, J.R., Ulrich, C., and Plapp, R. (1993). Characterization and sequence analysis of a small cryptic plasmid from *Lactobacillus curvatus* LTH683 and its use for construction of new *Lactobacillus* cloning vectors. Plasmid. *30*, 14–29.

Kobierecka, P., Wyszynska, A., Maruszewska, M., Wojtania, A., Zylinska, J., Bardowski, J.,. and Jagusztyn-Krynicka, E.K. (2015). Lactic acid bacteria as a surface display platform for *Campylobacter jejuni* antigens. J. Mol. Microbiol. Biotechnol. *25*, 1–10.

Kok, J., van der Vossen, J.M., and Venema, G. (1984). Construction of plasmid cloning vectors for lactic *streptococci* which also replicate in *Bacillus subtilis* and *Escherichia coli*. Appl. Environ. Microbiol. *48*, 726–731.

Kuczkowska, K., Kleiveland, C.R., Minic, R., Moen, L.F., Øverland, L., Tjaland, R., Carlsen, H., Lea, T., Mathiesen, G., and Eijsink,V.G. (2017). Immunogenic properties of *Lactobacillus plantarum* producing surface-displayed *Mycobacterium tuberculosis* antigens. Appl. Environ. Microbiol. *83*, e02782-16.

Kuipers, O.P., Beerthuyzen, M.M., de Ruyter, P.G., Luesink, E.J., and de Vos, W.M. (1995). Autoregulation of nisin biosynthesis in *Lactococcus lactis* by signal transduction. J. Biol. Chem. *270*, 27299–27304.

Lebeer, S., Vanderleyden, J., and De Keersmaecker, S.C. (2008). Genes and molecules of lactobacilli supporting probiotic action. Microbiol. Mol. Biol. Rev. *72*, 728–764.

Lee, J.H., Halgerson, J.S., Kim, J.H., and O'Sullivan, D.J. (2007). Comparative sequence analysis of plasmids from *Lactobacillus delbrueckii* and construction of a shuttle cloning vector. Appl. Environ. Microbiol. *73*, 4417–4424.

Lee, J.S., Poo, H., Han, D.P., Hong, S.P., Kim, K., Cho, M.W., Kim, E., Sung, M.H., and Kim, C.J. (2006). Mucosal immunization with surface-displayed severe acute respiratory syndrome coronavirus spike protein on *Lactobacillus casei* induces neutralizing antibodies in mice. J. Virol. *80*, 4079–4087.

Leenhouts, K., Buist, G., and Kok, J. (1999). Anchoring of proteins to lactic acid bacteria. Antonie Van Leeuwenhoek 76, 367–376.

Leenhouts, K.J., Kok, J., and Venema, G. (1989). Campbell-like integration of heterologous plasmid DNA into the chromosome of *Lactococcus lactis* subsp. *lactis*. Appl. Environ. Microbiol. *55*, 394–400.

Leenhouts, K.J., Tolner, B., Bron, S., Kok, J., Venema, G., and Seegers, J.F. (1991). Nucleotide sequence and characterization of the broad-host-range lactococcal plasmid pWVO1. Plasmid *26*, 55–66.

Li, H., Qiu, T., Huang, G., and Cao,Y. (2010). Production of gamma-aminobutyric acid by *Lactobacillus brevis* NCL912 using fed-batch fermentation. Microb. Cell. Fact. *9*, 85.

Li, R., Chowdhury, M.Y., Kim, J.H., Kim, T.H., Pathinayake, P., Koo, W.S., Park, M.E., Yoon, J.E., Roh, J.B., Hong, S.P., Sung, M.H., Lee, J.S., and Kim, C.J. (2015). Mucosally administered *Lactobacillus* surface-displayed influenza antigens (sM2 and HA2) with cholera toxin subunit A1 (CTA1) induce broadly protective immune responses against divergent influenza subtypes. Vet. Microbiol. *179*, 250–263.

Lim, S.H., Jahanshiri, F., Rahim, R.A., Sekawi, Z., and Yusoff, K. (2010). Surface display of respiratory syncytial virus glycoproteins in *Lactococcus lactis* NZ9000. Lett. Appl. Microbiol. *51*, 658–664. https://doi.org/10.1111/j.1472-765X.2010.02950.x

Lin, C.F., Ho, J.L., and Chung, T.C. (2001). Characterization of the replication region of the *Lactobacillus reuteri* plasmid pTC82 potentially used in the construction of cloning vector. Biosci. Biotechnol. Biochem., *65*, 1495–1503.

Lin, R., Zhang, Y., Long, B., Li, Y., Wu, Y., Duan, S., Zhu, B., Wu, X., and Fan, H. (2017). Oral immunization with recombinant *Lactobacillus acidophilus* expressing espA-Tir-M confers protection against enterohemorrhagic *Escherichia coli* O157:H7 challenge in mice. Front. Microbiol., *8*, 417.

Lizier, M., Sarra, P.G., Cauda, R., and Lucchini, F. (2010). Comparison of expression vectors in *Lactobacillus reuteri* strains. FEMS Microbiol. Lett. *308*, 8–15.

Lokman, B.C., Leer, R.J., van Sorge, R., and Pouwels, P.H. (1994). Promoter analysis and transcriptional regulation of *Lactobacillus pentosus* genes involved in xylose catabolism. Mol. Gen. Genet. *245*, 117–125.

Maguin, E., Duwat, P., Hege, T., Ehrlich, D., and Gruss, A. (1992). New thermosensitive plasmid for gram-positive bacteria. J. Bacteriol. *174*, 5633–5638.

Mathiesen, G., Sveen, A., Brurberg, M.B., Fredriksen, L., Axelsson, L., and Eijsink, V.G. (2009). Genome-wide analysis of signal peptide functionality in *Lactobacillus plantarum* WCFS1. BMC Genomics *10*, 425. https://doi.org/10.1186/1471-2164-10-425

Mathiesen, G., Sveen, A., Piard, J.C., Axelsson, L., and Eijsink, V.G. (2008). Heterologous protein secretion by *Lactobacillus plantarum* using homologous signal peptides. J. Appl. Microbiol. *105*, 215–226.

Mathiesen, G., Sørvig, E., Blatny, J., Naterstad, K., Axelsson, L., and Eijsink, V.G. (2004). High-level gene expression in *Lactobacillus plantarum* using a pheromone-regulated bacteriocin promoter. Lett. Appl., Microbiol. *39*, 137–143.

Mayo, B., Gonzalez, B., Arca, P., and Suarez, J.E. (1994). Cloning and expression of the plasmid encoded beta-D-galactosidase gene from a *Lactobacillus plantarum* strain of dairy origin. FEMS Microbiol. Lett. *122*, 145–151.

McCracken, A., Turner, M.S., Giffard, P., Hafner, L.M., and Timms, P. (2000). Analysis of promoter sequences from *Lactobacillus* and *Lactococcus* and their activity in several *Lactobacillus* species. Arch. Microbiol. *173*, 383–389.

Michon, C., Langella, P., Eijsink, V.G., Mathiesen, G., and Chatel, J.M. (2016). Display of recombinant proteins at the surface of lactic acid bacteria: strategies and applications. Microb. Cell. Fact. *15*, 70.

Mierau, I., and Kleerebezem, M. (2005). 10 years of the nisin-controlled gene expression system (NICE) in *Lactococcus lactis*: Appl. Microbiol. Biotechnol., *68*, 705–717.

Mierau, I., Olieman, K., Mond, J., and Smid, E.J. (2005). Optimization of the *Lactococcus lactis* nisin-controlled gene expression system NICE for industrial applications. Microb. Cell. Fact. *4*, 16.

Miller, K.W., Ray, P., Steinmetz, T., Hanekamp, T., and Ray, B. (2005). Gene organization and sequences of pediocin AcH/PA-1 production operons in *Pediococcus* and *Lactobacillus* plasmids. Lett. Appl. Microbiol. *40*, 56–62.

Murphy, K.C. (1998). Use of bacteriophage lambda recombination functions to promote gene replacement in *Escherichia coli*. J. Bacteriol. *180*, 2063–2071.

Nakamura, Y., Gojobori, T., and Ikemura, T. (1997). Codon usage tabulated from the international DNA sequence databases. Nucleic Acids Res., 25, 244–245.

Narita, J., Ishida, S., Okano, K., Kimura, S., Fukuda, H., and Kondo, A. (2006a). Improvement of protein production in lactic acid bacteria using 5'-untranslated leader sequence of slpA from *Lactobacillus acidophilus*. Improvement in protein production using UTLS. Appl. Microbiol. Biotechnol. 73, 366–373.

Narita, J., Okano, K., Kitao,T., Ishida, S., Sewaki, T., Sung, M.H., Fukuda, H., and Kondo, A. (2006b). Display of alpha-amylase on the surface of *Lactobacillus casei* cells by use of the PgsA anchor protein, and production of lactic acid from starch. Appl. Environ. Microbiol. *72*, 269–275.

Nayak, K.C. (2012). Comparative study on factors influencing the codon and amino acid usage in *Lactobacillus sakei* 23K and 13 other lactobacilli. Mol. Biol. Rep. *39*, 535–545. https://doi.org/10.1007/s11033-011-0768-4

Nguyen, H.M., Mathiesen, G., Stelzer, E.M., Pham, M.L., Kuczkowska, K., Mackenzie, A., Agger, J.W., Eijsink,V.G., Yamabhai, M., Peterbauer, C.K., Haltrich, D., and Nguyen,T.H. (2016). Display of a beta-mannanase and a chitosanase on the cell surface of *Lactobacillus plantarum* towards the development of whole-cell biocatalysts. Microb. Cell. Fact. *15*, 169.

Nguyen, T.T., Nguyen, T.H., Maischberger, T., Schmelzer, P., Mathiesen, G., Eijsink, V.G., Haltrich, D., and Peterbauer, C.K. (2011). Quantitative transcript analysis of the inducible expression system pSIP: Comparison of the overexpression of *Lactobacillus* spp. *beta-galactosidases* in *Lactobacillus plantarum*. Microb. Cell. Fact. *10*, 46.

Nishiyama, K., Sugiyama, M., and Mukai, T. (2016). Adhesion properties of lactic acid bacteria on intestinal mucin. Microorganisms *4*, 34. https://doi.org/10.3390/microorganisms4030034

Norton, P.M., Brown, H.W., Wells, J.M., Macpherson, A.M., Wilson, P.W., and Le Page, R.W. (1996). Factors affecting the immunogenicity of tetanus toxin fragment C expressed in *Lactococcus lactis*. FEMS Immunol. Med. Microbiol. *14*, 167–177.

O'Callaghan, J., and O'Toole, P.W. (2013). *Lactobacillus*: host-microbe relationships. Curr. Top. Microbiol. Immunol. *358*, 119–154. https://doi.org/10.1007/82_2011_187

O'Sullivan, O., O'Callaghan, J., Sangrador-Vegas, A., McAuliffe, O., Slattery, L., Kaleta, P., Callanan, M., Fitzgerald, G.F., Ross, R.P., and Beresford, T. (2009). Comparative genomics of lactic acid bacteria reveals a niche-specific gene set. BMC Microbiol. *9*, 50.

Oh,J.H., and van Pijkeren,J.P. (2014). CRISPR-Cas9-assisted recombineering in *Lactobacillus reuteri*. Nucleic Acids Res. *42*, e131.

Okano, K., Zhang, Q., Kimura, S., Narita, J., Tanaka, T., Fukuda, H., and Kondo, A. (2008). System using tandem repeats of the cA peptidoglycan-binding domain from *Lactococcus lactis* for display of both N- and C-terminal fusions on cell surfaces of lactic acid bacteria. Appl. Environ. Microbiol. *74*, 1117–11123.

Pavan, S., Hols, P., Delcour, J., Geoffroy, M.C., Grangette, C., Kleerebezem, M., and Mercenier, A. (2000). Adaptation of the nisin-controlled expression system in *Lactobacillus plantarum*: A tool to study in vivo biological effects. Appl. Environ. Microbiol. *66*, 4427–4432.

Pavlova, S.I., Kilic, A.O., Topisirovic, L., Miladinov, N., Hatzos, C., and Tao, L. (2002). Characterization of a cryptic plasmid from *Lactobacillus fermentum* KC5b and its use for constructing a stable *Lactobacillus* cloning vector. Plasmid *47*, 182–192.

Pflugl, S., Marx, H., Mattanovich, D., and Sauer, M. (2014). Heading for an economic industrial upgrading of crude glycerol from biodiesel production to 1,3-propanediol by *Lactobacillus diolivorans*. Bioresour. Technol. *152*, 499–504.

Posno, M., Heuvelmans, P.T., van Giezen, M.J., Lokman, B.C., Leer, R.J., and Pouwels, P.H. (1991a). Complementation of the inability of *Lactobacillus* strains to utilize D-xylose with D-xylose catabolism-encoding genes of *Lactobacillus pentosus*. Appl. Environ. Microbiol. *57*, 2764–2766.

Posno, M., Leer, R.J., van Luijk, N., van Giezen, M.J., Heuvelmans, P.T., Lokman, B.C., and Pouwels, P.H. (1991b). Incompatibility of *Lactobacillus* vectors with replicons derived from small cryptic *Lactobacillus* plasmids and segregational instability of the introduced vectors. Appl. Environ. Microbiol. *57*, 1822–1828.

Pouwels, P.H., and Leunissen, J.A. (1994). Divergence in codon usage of *Lactobacillus* species. Nucleic Acids Res. *22*, 929–936.

Puigbo, P., Guzman, E., Romeu, A., and Garcia-Vallve, S. (2007). OPTIMIZER: A web server for optimizing the codon usage of DNA sequences. Nucleic Acids Res. *35*, W126–131.

Pusch, O., Kalyanaraman, R., Tucker, L.D., Wells, J.M., Ramratnam, B., and Boden, D. (2006). An anti-HIV microbicide engineered in commensal bacteria: Secretion of HIV-1 fusion inhibitors by lactobacilli. Aids *20*, 1917–1922.

Rani, R.P., Anandharaj, M., and Ravindran, D.A. (2017). Characterization of a novel exopolysaccharide produced by *Lactobacillus gasseri* FR4 and demonstration of its in vitro biological properties. Int. J. Biol. Macromol. *109*, 772–783.

Reveneau, N., Geoffroy, M.C., Locht, C., Chagnaud, P., and Mercenier, A. (2002). Comparison of the immune responses induced by local immunizations with recombinant *Lactobacillus plantarum* producing tetanus toxin fragment C in different cellular locations. Vaccine *20*, 1769–1777.

Ribelles, P., Benbouziane, B., Langella, P., Suarez, J.E., and Bermudez-Humaran, L.G. (2013). Protection against human papillomavirus type 16-induced tumors in mice using non-genetically modified lactic acid bacteria displaying E7 antigen at its surface. Appl. Microbiol. Biotechnol. *97*, 1231–1239.

Rosales-Mendoza, S., Angulo, C., and Meza, B. (2016). Food-grade organisms as vaccine biofactories and oral delivery vehicles. Trends Biotechnol. *34*, 124–136.

Rossi, F., Rudella, A., Marzotto, M., and Dellaglio, F. (2001). Vector-free cloning of a bacterial endo-1,4-beta-glucanase in *Lactobacillus plantarum* and its effect on the acidifying activity in silage: Use of recombinant cellulolytic *Lactobacillus plantarum* as silage inoculant. Antonie Van Leeuwenhoek *80*, 139–147.

Rud, I., Jensen, P.R., Naterstad, K., and Axelsson, L. (2006). A synthetic promoter library for constitutive gene expression in *Lactobacillus plantarum*. Microbiology *152*, 1011–1019.

Ruiz-Maso, J.A., Macho, N.C., Bordanaba-Ruiseco, L., Espinosa, M., Coll, M., and Del Solar, G. (2015). Plasmid rolling-circle replication. Microbiol. Spectr. *3*, Plas-0035-2014.

Ryan, P.M., Ross, R.P., Fitzgerald, G.F., Caplice, N.M., and Stanton, C. (2015). Sugar-coated: Exopolysaccharide producing lactic acid bacteria for food and human health applications. Food Funct. *6*, 679–693.

Salvetti, E., and O'Toole, P.W. (2017). The genomic basis of Lactobacilli as health-promoting organisms. Microbiol. Spectr. *5*,. https://doi.org/10.1128/microbiolspec.BAD-0011-2016

Sasikumar, P., Gomathi, S., Anbazhagan, K., Baby, A.E., Sangeetha, J., and Selvam, G.S. (2014). Genetically engineered *Lactobacillus plantarum* WCFS1 constitutively secreting heterologous oxalate decarboxylase and degrading oxalate under in vitro. Curr. Microbiol. *69*, 708–715.

Savijoki, K., Kahala, M., and Palva, A. (1997). High level heterologous protein production in *Lactococcus* and *Lactobacillus* using a new secretion system based on the *Lactobacillus brevis* S-layer signals. Gene *186*, 255–262.

Scheirlinck, T., Mahillon, J., Joos, H., Dhaese, P., and Michiels, F. (1989). Integration and expression of alpha-amylase and endoglucanase genes in the *Lactobacillus plantarum* chromosome. Appl. Environ. Microbiol. *55*, 2130–2137.

Scheppler, L., Vogel, M., Zuercher, A.W., Zuercher, M., Germond, J.E., Miescher, S.M., and Stadler, B.M. (2002). Recombinant *Lactobacillus johnsonii* as a mucosal vaccine delivery vehicle. Vaccine *20*, 2913–2920.

Schwarzer, M., Repa, A., Daniel, C., Schabussova, I., Hrncir, T., Pot, B., Stepankova, R., Hudcovic, T., Pollak, A., Tlaskalova-Hogenova, H., Wiedermann, U., and Kozakova, H. (2011). Neonatal colonization of mice with *Lactobacillus plantarum* producing the aeroallergen Bet v 1 biases towards Th1 and T-regulatory responses upon systemic sensitization. Allergy, *66*, 368–375.

Seegers, J.F. (2002). Lactobacilli as live vaccine delivery vectors: Progress and prospects. Trends Biotechnol. *20*, 508–515.

Shareck, J., Choi, Y., Lee, B., and Miguez, C.B. (2004). Cloning vectors based on cryptic plasmids isolated from lactic acid bacteria: Their characteristics and potential applications in biotechnology. Crit. Rev. Biotechnol. *24*, 155–208.

Shimizu-Kadota, M. (1987). Properties of lactose plasmid pLY101 in *Lactobacillus casei*. Appl. Environ. Microbiol. *53*, 2987–2991.

Shimizu-Kadota, M., Shibahara-Sone, H., and Ishiwa, H. (1991). Shuttle plasmid vectors for *Lactobacillus casei* and *Escherichia coli* with a minus origin. Appl. Environ. Microbiol. *57*, 3292–3300.

Shine, J., and Dalgarno, L. (1974). The 3'-terminal sequence of *Escherichia coli* 16S ribosomal RNA: Complementarity to nonsense triplets and ribosome binding sites. Proc. Natl. Acad. Sci. U S A *71*, 1342–1346.

Škrlec, K., Pucer Janež, A., Rogelj, B., Štrukelj, B., and Berlec, A. (2017). Evasin-displaying lactic acid bacteria bind different chemokines and neutralize CXCL8 production in Caco-2 cells. Microb. Biotechnol. *10*, 1732–1743. https://doi.org/10.1111/1751-7915.12781

Song, X., Huang, H., Xiong, Z., Ai, L., and Yang, S. (2017). CRISPR-Cas9(D10A) Nickase-Assisted Genome Editing in *Lactobacillus casei*. Appl. Environ. Microbiol. *83*, e01259–17.

Sorvig, E., Grönqvist, S., Naterstad, K., Mathiesen, G., Eijsink, V.G., and Axelsson, L. (2003). Construction of vectors for inducible gene expression in *Lactobacillus sakei* and L *plantarum*. FEMS Microbiol. Lett. *229*, 119–126.

Sorvig, E., Mathiesen, G., Naterstad, K., Eijsink, V.G., and Axelsson, L. (2005a). High-level, inducible gene expression in *Lactobacillus sakei* and *Lactobacillus plantarum* using versatile expression vectors. Microbiology *151*, 2439–2449.

Sorvig, E., Skaugen, M., Naterstad, K., Eijsink, V.G., and Axelsson, L. (2005b). Plasmid p256 from *Lactobacillus plantarum* represents a new type of replicon in lactic acid bacteria, and contains a toxin-antitoxin-like plasmid maintenance system. Microbiology *151*, 421–431.

Spath, K., Heinl, S., Egger, E., and Grabherr, R. (2011). *Lactobacillus plantarum* and *Lactobacillus buchneri* as expression systems: Evaluation of different origins of replication for the design of suitable shuttle vectors. Mol. Biotechnol. *52*, 40–48.

Stanssens, P., Remaut, E., and Fiers, W. (1985). Alterations upstream from the Shine-Dalgarno region and their effect on bacterial gene expression. Gene *36*, 211–223.

Steen, A., Buist, G.K., Leenhouts, J., El Khattabi, M., Grijpstra, F., Zomer, A.L., Venema, G., Kuipers, O.P., and Kok, J. (2003). Cell wall attachment of a widely distributed peptidoglycan binding domain is hindered by cell wall constituents. J. Biol. Chem. *278*, 23874–23881.

Stoeker, L., Nordone, S., Gunderson, S., Zhang, L., Kajikawa, A., LaVoy, A., Miller, M., Klaenhammer, T.R., and Dean, G.A. (2011). Assessment of *Lactobacillus gasseri* as a candidate oral vaccine vector. Clin. Vaccine Immunol. *18*, 1834–1844.

Suebwongsa, N., Lulitanond, V., Mayo, B., Yotpanya, P., and Panya, M. (2016). Development of an *Escherichia coli-Lactobacillus* casei shuttle vector for heterologous protein expression in *Lactobacillus casei*. Springerplus 5, 169.

Suebwongsa, N., Panya, M., Namwat, W., Sookprasert, S., Redruello, B., Mayo, B., Alvarez, M.A., and Lulitanond, V. (2013). Cloning and expression of a codon-optimized gene encoding the influenza A virus nucleocapsid protein in *Lactobacillus casei*. Int. Microbiol. *16*, 93–101.

Sun, Z., Kong, J., Hu, S., Kong, W., Lu, W., and Liu, W. (2013). Characterization of a S-layer protein from *Lactobacillus crispatus* K313 and the domains responsible for binding to cell wall and adherence to collagen. Appl. Microbiol. Biotechnol. *97*, 1941–1952. https://doi.org/10.1007/s00253-012-4044-x

Tan, T.S., Syed Hassan, S., and Yap, W.B. (2017). Expression of surface-bound nonstructural 1 (NS1) protein of influenza virus A H5N1 on *Lactobacillus casei* strain C1. Lett. Appl. Microbiol. *64*, 446–451.

Tauer, C., Heinl, S., Egger, E., Heiss, S., and Grabherr, R. (2014). Tuning constitutive recombinant gene expression in *Lactobacillus plantarum*. Microb. Cell. Fact. *13*, 150. https://doi.org/10.1186/s12934-014-0150-z.

Topisirovic, L., Kojic, M., Fira, D., Golic, N., Strahinic, I., and Lozo, J. (2006). Potential of lactic acid bacteria isolated from specific natural niches in food production and preservation. Int. J. Food Microbiol. *112*, 230–235.

Tsirigotaki, A., De Geyter, J., Šoštaric, N., Economou, A., and Karamanou, S. (2017). Protein export through the bacterial Sec pathway. Nat. Rev. Microbiol. *15*, 21–36. https://doi.org/10.1038/nrmicro.2016.161

Tuan-Anh, T., Ly, L.T., Viet, N.Q., and Bao, P.T. (2017). Novel methods to optimize gene and statistic test for evaluation - an application for *Escherichia coli*. BMC Bioinformatics *18*, 100.

van Asseldonk, M., Rutten, G., Oteman, M., Siezen, R.J., de Vos, W.M., and Simons, G. (1990). Cloning of usp45, a gene encoding a secreted protein from *Lactococcus lactis* subsp. *lactis* MG1363. Gene *95*, 155–160.

van den Nieuwboer, M., van Hemert, S., Claassen, E., and de Vos, W.M. (2016). *Lactobacillus plantarum* WCFS1 and its host interaction: A dozen years after the genome. Microb. Biotechnol. *9*, 452–465.

van Pijkeren, J.P., and Britton, R.A. (2012). High efficiency recombineering in lactic acid bacteria. Nucleic Acids Res. *40*, e76. https://doi.org/10.1093/nar/gks147

Vaughan, A., Eijsink,V.G., and Van Sinderen, D. (2003). Functional characterization of a composite bacteriocin locus from malt isolate *Lactobacillus sakei* 5. Appl. Environ. Microbiol. *69*, 7194–7203.

Vosman, B., and Venema, G. (1983). Introduction of a *Streptococcus cremoris* plasmid in *Bacillus subtilis*. J. Bacteriol. *156*, 920–921.

Wang, T.T., and Lee, B.H. (1997). Plasmids in *Lactobacillus*. Crit. Rev. Biotechnol. *17*, 227–272. https://doi.org/10.3109/07388559709146615

Wei, C.H., Liu, J.K., Hou, X.L., Yu, L.Y., Lee, J.S., and Kim, C.J. (2010). Immunogenicity and protective efficacy of orally or intranasally administered recombinant *Lactobacillus casei* expressing ETEC K99. Vaccine *28*, 4113–4118. https://doi.org/10.1016/j.vaccine.2009.05.088

Wieczorek, A.S., and Martin, V.J. (2010). Engineering the cell surface display of cohesins for assembly of cellulosome-inspired enzyme complexes on *Lactococcus lactis*. Microb. Cell. Fact. *9*, 69.

Wu, C.H., Mulchandani, A., and Chen, W. (2008). Versatile microbial surface-display for environmental remediation and biofuels production. Trends Microbiol. *16*, 181–188.

Wyszynska, A., Kobierecka, P., Bardowski, J., and Jagusztyn-Krynicka, E.K. (2015). Lactic acid bacteria: 20 years exploring their potential as live vectors for mucosal vaccination. Appl. Microbiol. Biotechnol. *99*, 2967–2977.

Xin, Y., Guo, T., Mu, Y., and Kong, J. (2017). Development of a counterselectable seamless mutagenesis system in lactic acid bacteria. Microb. Cell Fact. *16*, 116. https://doi.org/10.1186/s12934-017-0731-8

Xu, T., Li, Y., Shi, Z., Hemme, C.L., Zhu, Y., Van Nostrand, J.D., He, Z., and Zhou, J. (2015). Efficient genome editing in *Clostridium cellulolyticum* via CRISPR-Cas9 Nickase. Appl. Environ. Microbiol. *81*, 4423–4431.

Yamamoto, N., and Takano, T. (1996). Isolation and characterization of a plasmid from *Lactobacillus helveticus* CP53. Biosci. Biotechnol. Biochem. *60*, 2069–2070.

Yang, G., Jiang, Y., Tong, P., Li, C., Yang, W., Hu, J., Ye, L., Gu, W., Shi, C., Shan, B., and Wang, C. (2017a). Alleviation of enterotoxigenic *Escherichia coli* challenge by recombinant *Lactobacillus plantarum* expressing a FaeG- and DC-targeting peptide fusion protein. Benef. Microbes *8*, 379–391.

Yang, G., Yao, J., Yang, W., Jiang, Y., Du, J., Huang, H., Gu, W., Hu, J., Ye, L., Shi, C., Shan, B., and Wang, C. (2017b). Construction and immunological evaluation of recombinant *Lactobacillus plantarum* expressing SO7 of *Eimeria tenella* fusion DC-targeting peptide. Vet. Parasitol. *236*, 7–13.

Yang, P., Wang, J., and Qi, Q. (2015). Prophage recombinases-mediated genome engineering in *Lactobacillus plantarum*. Microb. Cell. Fact. *14*, 154.

Yang, W.T., Yang, G.L., Yang, X., Shonyela, S.M., Zhao, L., Jiang, Y.L., Huang, H.B., Shi, C.W., Wang, J.Z., Wang, G., Zhao, J.H., and Wang, C.F. (2017c). Recombinant *Lactobacillus plantarum* expressing HA2 antigen elicits protective immunity against H9N2 avian influenza virus in chickens. Appl. Microbiol. Biotechnol. *101*, 8475–8484.

Yin, J.Y., Guo, C.Q., Wang, Z., Yu, M.L., Gao, S., Bukhari, S.M.L., Tang, J., Xu, Y.G., and Li, Y.J. (2016). Directed chromosomal integration and expression of porcine rotavirus outer capsid protein VP4 in *Lactobacillus casei* ATCC393. Appl. Microbiol. Biotechnol. *100*, 9593–9604.

Yu, K., Ang, K.S., and Lee, D.Y. (2017). Synthetic gene design using Codon Optimization On-Line (COOL). Methods Mol. Biol. *1472*, 13–34. https://doi.org/10.1007/978-1-4939-6343-0_2

Zadravec, P., Strukelj, B., and Berlec, A. (2015). Improvement of LysM-mediated surface display of designed ankyrin repeat proteins (DARPins) in recombinant and nonrecombinant strains of *Lactococcus lactis* and *Lactobacillus* species. Appl. Environ. Microbiol. *81*, 2098–2106.

Zannini, E., Waters, D.M., Coffey, A., and Arendt, E.K. (2016). Production, properties, and industrial food application of lactic acid bacteria-derived exopolysaccharides. Appl. Microbiol. Biotechnol. *100*, 1121–1135.

Genomic Overview of Acquired Antibiotic Resistance Mechanisms in *Lactobacillus*

10

Cecilia Rodríguez[1], Lucía Petrelli[1], María Soledad Ramírez[2], Daniela Centrón[3], Elvira María Hebert[1*] and Lucila Saavedra[1*]

[1]Laboratory of Genetics and Molecular Biology, Reference Centre for Lactobacilli (CERELA-CONICET), San Miguel de Tucumán, Argentina.
[2]Department of Biological Sciences, California State University, Fullerton, CA, USA.
[3]Institute of Microbiology and Medical Parasitology, Faculty of Medicine, University of Buenos Aires (IMPaM, UBA/CONICET), Buenos Aires, Argentina.

*Correspondence: ehebert@cerela.org.ar; lucila@cerela.org.ar

https://doi.org/10.21775/9781910190890.10

Abstract

In recent years, the number of antibiotic-resistant bacteria prevalent in clinical settings has risen, alarming both scientists and government agencies. Studies on the emergence and spread of antibiotic resistance have focused mainly on bacteria of clinical importance. However, commensal and environmental bacteria appear as a reservoir of the determinants of resistance to antibiotics found in bacteria of clinical origin. In particular, the food chain was proposed as the main route for the introduction of resistant bacteria associated with animals and the environment. The emergence of antibiotic resistance can be attributed to several factors, such as the overuse of antibiotics both within and outside of a clinical setting, random bacterial mutations leading to increased resistance, not completing the course of an antibiotic prescription, and the use of antibiotics in farming and agricultural. A summary of acquired antibiotic resistance genes to tetracycline, erythromycin and aminoglycosides and a genomic overview of these resistance genes in *Lactobacillus* is described in this chapter

Introduction

In recent years, the number of antibiotic-resistant bacteria prevalent in clinical settings has risen, alarming both scientists and government agencies (CDC, 2013, Spellberg *et al.*, 2008, WHO, 2017). The Centers for Disease Control and Prevention (CDC) released a report outlining the threat of antibiotic resistance (AR) and identified critical pathogens as urgent, serious, or concerning. They showed that over two million individuals are infected with resistant pathogens each year and that over 23,000 of these infections are mortal (CDC, 2013). However, we have been dealing with this problem since the first antibiotic – penicillin – was discovered in 1928 and was beginning to be used indiscriminately. In 1945, with the World War II going on, the mass production and use of antibiotics began. At that moment, Alexander Fleming cautioned us about the chance of antibiotic resistance occurring if penicillin was used inaccurately (Aminov, 2010).

Considering the history of antibiotics, many examples illustrating the commercial released of an antibiotic and the subsequent occurrence of antibiotic resistance against it, are found (Fig. 10.1). One example of resistance development includes the introduction of methicillin in 1960 and the appearance of methicillin-resistant *Staphylococcus aureus* 2 years later (WHO, 2017). As shown in Fig. 10.1, despite the varying length of time between the introduction of an antibiotic in the market and identification of resistance, the occurrence of resistant strains inevitably follows the extensive use of an antibiotic.

In February 2017, the World Health Organization published a list of 12 bacterial pathogens for antibiotic research and development that are currently extensively or multidrug resistant and are common pathogens in the clinical setting (WHO, 2017).

The emergence of antibiotic resistance can be attributed to several factors, such as the overuse of antibiotics both within and outside of a clinical setting, random bacterial mutations leading to increased resistance, not completing the course of an antibiotic prescription, and the use of antibiotics in farming and agriculture (Davies and Davies, 2010). It was estimated that the global consumption of antimicrobial in animal agriculture industry is bigger than the consumption in human medicine, and it is one of the causes that is contributing to the rise of antibiotic resistance observed in zoonotic pathogens (Van Boeckel *et al.*, 2017). Antibiotic resistance emerged in farming and agriculture could be spread to human by exposition to contaminated food, occupational contact, and spread by air, water and/or manure in areas closed to farming facilities. Many classes of antibiotics, such as β-lactams, tetracyclines, macrolides, and sulfonamides, valuable to treat bacterial infections in human are used in farms and agriculture with the consequent development of antibiotic resistance (Van Boeckel *et al.*, 2015).

Antibiotic resistance mechanisms in bacteria

Several mechanisms of antibiotic resistance have been described and characterized (van Hoek *et al.*, 2011). Resistance to a particular antimicrobial agent can be inherent in a bacterial species, called intrinsic resistance. Some bacterial cells can possess intrinsic mechanisms of resistance, such as efflux pumps, chromosomally encoded-enzymes, or selective permeability of the membrane (Thomas and Nielsen, 2005). In this case, the resistance is characteristic of species. Lactobacilli have intrinsic resistance to vancomycin (Tynkkynen et al., 1998), β-lactams (oxacillin, cefoxitin and ceftriaxone), bacitracin and inhibitors of nucleic acid synthesis

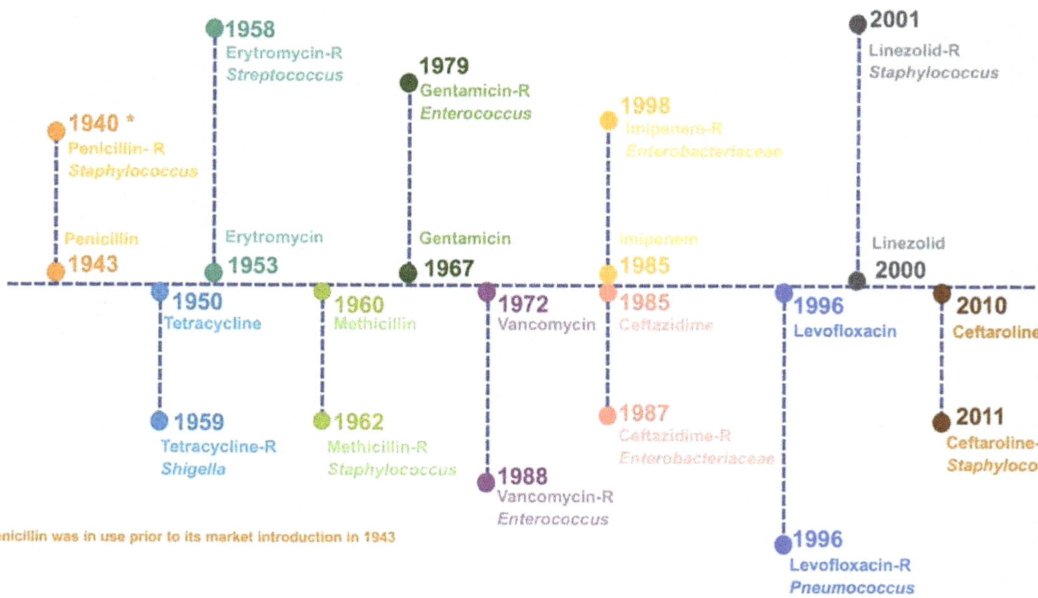

Figure 10.1 First reported cases of bacterial resistance against key antibiotics. Data source: Antibiotic resistance threats in the United States, 2013. US Centers for Disease Control and Prevention (CDC).

such as nalidixic acid, certain quinolones as ciprofloxacin, trimethoprim and sulfamethoxazole (Danielsen and Wind, 2003).

The vancomycin-resistant phenotype of some lactobacilli is one of the best-characterized intrinsic resistance in *Lactobacillus*. On the contrary, the resistance is considered as acquired, when a strain of a normally susceptible species becomes resistant to an antimicrobial drug (van Hoek *et al*., 2011). Some bacteria can also modify the target of an antibiotic through mutation or the addition of a chemical group, such as methylation. (Chatterjee *et al*., 2017; Thomas and Nielsen, 2005). Some bacterial cells produce enzymes, such as β-lactamases or aminoglycoside modifying enzymes that modify or degrade the drug (Klimentová and Stulík, 2015). These enzymes can be either intrinsic or acquired by horizontal gene transfer (HGT).

HGT plays a key role in the evolution and adaptability of members of the bacterial kingdom. Bacteria's capability to acquire and incorporate DNA into their genomes allows them to evolve and survive in hostile environments, with the acquisition of different AR determinants and virulence traits being critical for bacterial pathogens (Boucher *et al*., 2003; Kwon *et al*., 2003; Lorenz and Wackernagel, 1994; Trzciński *et al*., 2004). As bacteria evolve to become more resistant to antibiotics, there is a possibility for a single bacterium to transfer its resistance to another bacterium via HGT, thus increasing the ability of the bacterium to survive in harsh environments. Traditionally, this process occurred by one of three mechanisms: conjugation, transduction and transformation (Thomas and Nielsen, 2005). More recently, a fourth mechanism of HGT, called outer membrane vesicle (OMV) transport, has been identified (Chatterjee *et al*., 2017). This mechanism involves the formation of a vesicle from the outer membrane that acts as a secretion system and is especially common in Gram-negative species (Klimentová and Stulík, 2015). The mechanisms of conjugation and transduction, however, involve a direct or nearly direct transfer of genetic information from one bacterium to another while transformation occurs when a bacterium takes up free DNA from its environment (Thomas and Nielsen, 2005). Many bacterial species are capable of taking up, integrating and expressing DNA from their environment and are therefore considered competent for natural transformation (Thomas and Nielsen, 2005).

Mobile genetic elements and their contribution to the emergence of antimicrobial resistant

In the 1970s the first examples of 'jumping' DNA were discovered in bacteria, genetic elements that presented the novel property of moving from one side to another of the genomes, going from plasmids to chromosomes or *vice versa*. These mobile elements turned out to be genetic units with a wide diversity both in structure and in the mobilization mechanisms they use. They are composed of one or several genes that have the ability to mobilize either within a bacterial cell from the chromosome to a plasmid, or even between different strains, including strains of different species and genus, depending on the characteristics of each genetic element (van Hoek *et al*., 2011). Currently, thanks to all the advances that have been made of sequences available in databases such as GenBank, it is known that mobile elements are the most abundant genetic elements in all domains of living beings. They are considered to be an essential force in the modification of genes and genomes throughout evolution, probably as a result of the development of a symbiotic relationship with its host (Koonin, 2016). In addition, either by themselves or by the fact that they sometimes spread between strains from different species by conjugative plasmids and/or phages, and/or by transformation, they can be transferred from one bacterium to another, constituting a shared source of genetic material. As a result of the usual lateral transfer, the linkage of their evolution to that of their host organisms or other genes in the organism is prevented. The mobile genetic elements (MGE) bring about a series of reorganizations in the genomes: inversions of genetic segments, deletions, cointegrations between two replicons, duplications, etc (Darmon and Leach, 2014; Shapiro, 1995). It should be noted that these elements are not replicons, that is, they do not possess the autonomous replication capacity characteristic of the chromosome and the plasmids, but they are integral parts of the genomes of many bacteria, conferring genetic plasticity on which evolutionary pathways can proceed. This movable characteristic of the MGE makes the bacterial genomes, dynamic and malleable on an evolutionary scale (Koonin, 2016).

MGE related to AR incidence in *Lactobacillus*

MGE are DNA sequences that usually have clear ends that can move within the genome by excision and insertion processes that are independent of homologous recombination.

There are several types of MGE that move in bacterial genomes, such as (i) insertion sequences or IS elements; (ii) composite transposons; (iii) simple transposons; (iv) conjugative transposons or genomic islands; and (v) integron/gene cassettes system.

The simplest of the MGE are insertion sequences (IS), which consist of the gene required for element mobility (transposase) and the inverted repeat (IR) at the end of the element (van Hoek *et al.*, 2011). To be able to mobilize, most MGE have short terminal IR and employ the transposases that identify, cut and paste the ends of these elements in a novel target site selected also by this enzyme. During this process, the transposase usually duplicates the target sequence in which they integrate, creating a short direct-repeat sequence called a target site duplication. Some MGE invade particular target sites, whereas other exhibit modest target specificity. In turn, each MGE must regulate their mobility to avoid unnecessary mutagenesis that would be disadvantageous for the cell.

When the IS contains accessory genes not involved in element translocation they are known as transposons. The simplest transposon contains an accessory gene, often a gene encoding AR, together with the transposase (van Hoek *et al.*, 2011). Since this chapter deals mainly with acquired resistance, attention is focused on MGE associated with AR genes, and involved in the dispersion of antimicrobial determinants between different bacteria.

Composite transposons

Composite transposons MGE have two copies of the same IS, which may or may not be exact replicas. They usually have border genes that encode some adaptive function, usually a mechanism of antibiotic or heavy metal resistance, and as result of that, they have bigger size than an IS (Bellanger *et al.*, 2014). Each of these IS in turn has short inverted repeats at both ends of the transposon. Some examples of composite transposon are the Tn*10* and Tn*5405* transposons, conferring resistance to tetracycline and aminoglycosides, respectively (Derbise *et al.*, 1996; Hillen and Berens, 1994). An example of composite transposons is the transposon Tn*10* which only one of the IS*10* elements encodes a functional transposase (Hillen and Berens, 1994). However, these transposons were not found in the genus *Lactobacillus.*

Simple transposons

Single transposons have long repeated sequences at the ends (35–40 base pairs) without repetition of the insertion sequences. A typical example of this type of transposon is the Tn3 transposon, which has a gene that codes for a transposase (*tnp*), a site called *res*, a site involved in the resolution of the transposition intermediate called cointegrated, and a site-specific recombinase called resolvase *tnp*R (Nicolas *et al.*, 2015). An example of this type of transposon is the Tn*1546*, a Tn3-related transposon conferring glycopeptide resistance by synthesis of depsipeptide peptidoglycan precursors in the lactic acid bacterium *Enterococcus faecium* BM4147 (Arthur *et al.*, 1993). However, to date, this class of transposon has not been described in *Lactobacillus.*

Conjugative transposons

Conjugative transposons have the property of promoting rearrangements within the cell but also they can spread from one bacterium to another by conjugation (Guédon *et al.*, 2017). They do not duplicate their target sequence upon integration, and they are inserted anywhere or at a specific site on the recipient chromosome or plasmid. Many of the so-called 'genomic islands' or 'alien islands' belong to this type of transposons, and their size can range from 10 to 500 kb, which are characterized by being present in certain strains of one species but absent in others, and they usually confer remarkable adaptive advantages that allow them to occupy diverse ecological niches. In turn, the genomic islands are classified into several types according to the adaptive advantage they confer to the strains that harbour them. The first to be discovered were the so-called 'islands of pathogenicity', which have the ability to confer remarkable pathogenic capacities on animals, for example, because they carry genes for toxins, adhesins, or siderophores, among other factors. Later, antibiotic resistance islands were identified, which can have up to 40 genes of resistance to several families of antibiotics.

In short, these conjugative transposons allow prokaryotic genomes to be considered 'modular' and 'malleable', and because they can be mobilized between strains and different species by themselves or by the effect of plasmids and phages, they provide rapid mechanisms of change evolutionary. In the case of the acquisition of genomic islands, it is observed that the phenotype of the bacterium changes dramatically, for example it goes from being non-pathogenic to pathogenic, from susceptible to antibiotics to resistant, so this phenomenon is called 'evolution by quantum leaps'.

The conjugative transposon Tn*916* belongs to this type of transposon (Fig. 10.2), which is common among the human oral commensal bacteria, which act as a reservoir for resistance genes (Roberts and Mullany, 2011). Tn*916* is, together with the related conjugative transposon Tn*1545*, the paradigm of a large family of related conjugative transposons known as the Tn*916*/Tn*1545* family, which are found in several species of the genus *Lactobacillus* and in diverse range of bacteria (Devirgiliis *et al.*, 2009; Mendonça *et al.*, 2016). Thus, it was also discovered that the *tet*(M) gene was spreading among isolates of *L. paracasei* from starter cultures (Devirgiliis *et al.*, 2009) as well as in industrial strains of *L. vini* (Mendonça *et al.*, 2016). With the huge increase in bacterial genomic sequence data available, due to the widespread use of next generation sequencing, more putative conjugative transposons belonging to the Tn*916*/Tn*1545* family are being reported (Santoro *et al.*, 2014). Most Tn*916*/Tn*1545*-like elements encode tetracycline resistance gene (Ammor *et al.*, 2008c; Devirgiliis *et al.*, 2009; Mendonça *et al.*, 2016). Plasmid pCTN1046, harbours by the porcine isolate *L. salivarius* JCM1046, has a single copy of an integrated conjugative transposon (Tn*6224*) that appears to be functionally intact and includes the tetracycline resistance gene *tet*(M) (Raftis *et al.*, 2014).

In addition, some Tn*916*-like elements possess macrolide resistance genes *erm*B (e.g. Tn*1545*, Tn*6002* and Tn*6003*) and *mef*(A) (e.g. Tn*2009* and Tn*2017*), as well as kanamycin (*aphA-3*) (Tn*1545*) and mercury (*mer*) (Tn*6009*) resistance genes (Radulović *et al.*, 2012). In *L. crispatus* CHCC3692, a 3165-bp chromosomally integrated transposon, designated Tn*3692*, contains an *erm*B gene conferring resistance to erythromycin (Strøman *et al.*, 2003).

Integron/gene cassettes system

The gene cassettes are the MGE of the integron/gene cassette system. The gene cassette does not move *per se*, but is completely dependent on the integrase of the integron, which is in charge of identifying, splitting and inserting each gene cassette into a new site. In other words, it is a genetic engineering system of *in vivo* cloning and expressing genes, ancient in bacterial genomes (Labbate *et al.*, 2009). The basic structure of an integron is composed of three elements: a gene that codes for integrase (*intI*), a recombination site (*attI*) and a promoter that will allow the expression of the gene cassette (s) that are incorporated in the variable region of integrons. The gene cassettes are composed of an open reading frame that can code for an

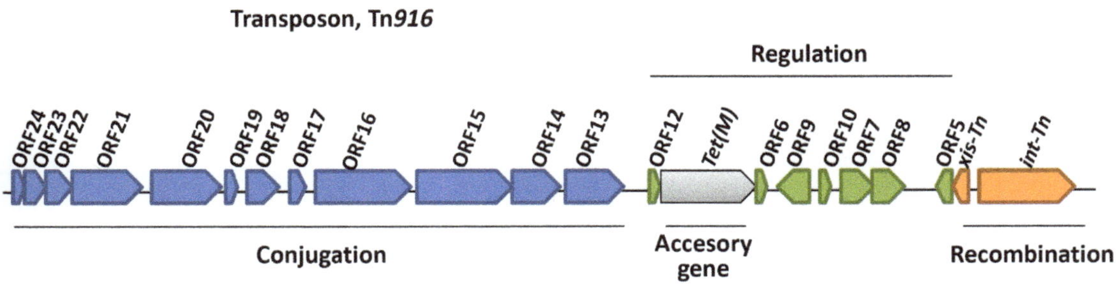

Figure 10.2 Tetracycline resistance determinant associated with mobile elements in *Lactobacillus*. Conjugative transposon Tn*916*. The arrows represent the individual ORF pointing in the probable direction of transcription. The grey arrow indicate *tet*(M) encoding tetracycline resistance. Blue arrows indicate the predicted proteins for conjugation, green arrows predicted proteins for transcriptional regulation and orange arrows insertion and excision (recombination). Adapted from Roberts and Mullany (2011).

antibiotic resistance gene, or a pathogenicity factor (Labbate *et al.*, 2009). Downstream of this gene, there is a DNA sequence that ranges from 58 to 141 base pairs, called the *attC* site which is recognized by the integrase to mediate site-specific recombination with the so-called *attI* site. It should be noted that if the gene cassette is not inserted in the variable zone of the integron, it is not expressed as it cannot be regulated. To date, more than 130 antibiotic resistance gene cassettes have been described, covering almost all antibiotic families available in the clinic, such as β-lactams, aminoglycosides, trimethoprim, fluoroquinolones, chloramphenicol, rifampicin, lincosamines and macrolides. In clinical samples, the integrons, mostly class 1 integrons, are associated with other genetic platforms, such as transposons, insertion sequences, genomic islands and/or conjugative plasmids, which allows them to be disseminated to other bacterial species by placing them on the scenario of the HGT, being in this way, one of the biggest contributors to the spread of antibiotic resistance.

Class 1 integrons are involved in the antibiotic multiresistance phenotype mainly in Gram-negative bacteria. They have also been described with low frequency in Gram-positive bacteria such as *Staphylococcus aureus, Enterococcus faecalis* and *E. faecium* (Xu *et al.*, 2010, 2011). However, to our knowledge, class 1 integrons have not been detected in other Gram-positive bacteria such as LAB up to date.

Acquired AR genes in *Lactobacillus*

For several decades, studies on the emergence and spread of antibiotic resistance focused mainly on bacteria of clinical importance. However, in recent years commensal and environmental bacteria appear as a reservoir of the determinants of resistance to antibiotics found in bacteria of clinical origin. In particular, the food chain was proposed as the main route for the introduction of resistant bacteria associated with animals and the environment (Economou and Gousia, 2015). In animals, antibiotics are used as therapeutics to treat sick animals, as prophylactics to prevent infection and/or as growth promoters to convert feed into more body mass (Economou and Gousia, 2015; Pomba *et al.*, 2017). Moreover, the use of antimicrobials is not restricted to animal husbandry but also occurs in horticulture (Teale, 2002). Carlet (2012) considers the gut as the epicentre of antibiotic resistance. In the human gastrointestinal tract, beneficial LAB can participate in the genetic exchange between resident bacteria and those that transiently colonize the gastrointestinal tract (Ammor *et al.*, 2008b; Arioli *et al.*, 2014; González-Zorn and Escudero, 2012; Witte, 2000; Zhou *et al.*, 2012).

The mechanisms involved in the transfer of resistance of *Lactobacillus* and several Gram-positive bacteria could be evolutionarily similar to those identified in Gram-negative bacteria (van Reenen and Dicks, 2011). For this reason, the selection of microorganisms to be use as food additives should be aim at those lacking the genetic determinants of antibiotics resistance (EFSA, 2012). Therefore, it is important to make a distinction between intrinsic and acquired resistance. As we mentioned above, intrinsic or natural resistance is inherent to the bacterial species and involves different mechanisms; in certain species, the mechanisms of resistance and the potential transferability are known so the clinical impact is low. Acquired antibiotic resistance occurs by mutation of preexisting genes or by HGT (van Hoek *et al.*, 2011). The acquisition of exogenous genes by HGT, mediated by MGE (Ammor *et al.*, 2007, Normark and Normark, 2002; van Reenen and Dicks, 2011), has the greatest clinical impact because the resistance genes are located in genetic platforms that allow their mobilization between bacteria of the same or different species, favouring their dissemination (Devirgiliis *et al.*, 2009; Feld *et al.*, 2009).

In general, the studies found in the literature on LAB antibiotic resistance describe the sensitivity of different species to a panel of antibiotics; however, both the methodology used and the class of antimicrobial evaluated are variables and the genes involved in AR are not always identified (Ammor *et al.*, 2008b; Klare *et al.*, 2007; Liu *et al.*, 2009; Muñoz-Atienza *et al.*, 2013). Only a few reports describe the MGE associated with the acquired resistance genes and the frequency of their dissemination to other species (Ammor *et al.*, 2008c; Feld *et al.*, 2009; Mater *et al.*, 2008; Nawaz *et al.*, 2011).

Below is a summary of the acquired antibiotic resistance genes to tetracyclines, erythromycin and aminoglycosides described for the *Lactobacillus* genus of food origin. Resistance to these three types of antibiotics has a great impact on human health due to its clinical use.

Tetracyclines

Tetracyclines (TET) are broad-spectrum antibiotics that have a bacteriostatic effect due to the reversible association of aminoacyl-tRNA with the bacterial ribosome subunit 30S during the initial phase of protein synthesis (Chopra and Roberts, 2001). This class of antibiotic has been widely used in the prophylaxis and therapy of human and animals infections caused by Gram-positive and Gram-negative bacteria (Chopra and Roberts, 2001). However, after its introduction in animal feed as growth promoters (Wegener, 2003), tetracycline-resistance increased considerably during the past few years; being the most frequent resistance associated with food-related bacteria (Devirgiliis *et al.*, 2011; Thaker *et al.*, 2010). Several TET resistance genes were identified in *Lactobacillus*, conferring resistance through different mechanisms: ribosomal protein protection [*tet*(M), *tet*(O), *tet*(S), *tet*(W) genes], and efflux [*tet*(K), *tet*(L) genes]. Table 10.1 shows the resistance genetic determinants found in several lactobacilli from food origin.

The *tet*(M) gene prevails in the genus *Lactobacillus* and it is widely distributed in different species from different sources such as *L. sakei, L. curvatus, L. paracasei, L. salivarius, L. plantarum, L. brevis, L. animalis, L. fermentum* (Ammor *et al.*, 2008b; Comunian *et al.*, 2010; Devirgiliis *et al.*, 2009; Nawaz *et al.*, 2011; Pan *et al.*, 2011) (Table 10.1). The remaining *tet*(R) determinants were describe less frequently (Guo *et al.*, 2017; Huys *et al.*, 2008; Nawaz *et al.*, 2011). The presence of more than one *tet*-resistance determinant encoding for the same or different mechanisms in the same bacterium was also documented (Ammor *et al.*, 2008a; Huys *et al.*, 2008; Nawaz *et al.*, 2011; Zonenschain *et al.*, 2009). Many of these genes are associated with mobile elements highly transmissible i.e. plasmids [*tet*(L) in *L. sakei* from dairy source described by Ammor *et al.*

Table 10.1 Acquired resistance genes to antibiotics and mobile genetic elements identified in *Lactobacillus* species from different food sources

Source	Species	Genotype	Location	References
Fermented meat	*L. curvatus*	*erm*B, *tet*(M)	nd[a]	Zonenschain *et al.* (2009)
		tet(M)	Chromosome	Gevers *et al.* (2003)
		*erm*B, *tet*(M), *tet*(W)	nd	Zonenschain *et al.* (2009)
		*erm*B, *msr*A/B	nd	Toomey *et al.* (2010)
	L. paracasei	*erm*B, *tet*(M)	nd	Zonenschain *et al.* (2009); Comunian *et al.* (2010)
		*erm*B, *msr*A/B	nd	Toomey *et al.* (2010)
	L. reuteri	*erm*B, *tet*(W), *tet*(L)	nd	Thumu and Halami (2012)
		*erm*B, *msr*A/B	nd	Toomey *et al.* (2010)
		*erm*B, *tet*(M)	nd	Zonenschain *et al.* (2009
	L. sakei	*erm*B, tet(M), *tet*(W)	nd	Zonenschain *et al.* (2009)
		tet(M)	Plasmid	Gevers *et al.* (2003)
	L. plantarum	*erm*B, *erm*C, *tet*(M), *tet*(S), *tet*(W)	nd	Zonenschain *et al.* (2009)
		tet(M)	Plasmid	Gevers *et al.* (2003)
		*erm*B, *tet*(M)	Plasmid	Gevers *et al.* (2003)
		*erm*B, *tet*(M)	Plasmid	Pan *et al.* (2011)
		tet(M)	Plasmid	Pan *et al.* (2011)
	L. rhamnosus	*erm*B, *tet*(W)	nd	Zonenschain *et al.* (2009)
	L. salivarius	*erm*B, *tet*(M), *tet*(W), *tet*(L), *tet*(O)	nd	Thumu and Halami (2012)
	L. brevis	*erm*B, *tet*M	nd	Zonenschain *et al.* (2009)
	L. alimentarius	*tet*(M)	Plasmid	Gevers *et al.* (2003)

Table 10.1 Continued

Source	Species	Genotype	Location	References
Dairy products	*L. salivarius*	*tet*(M)	nd	Nawaz *et al.* (2011)
	L. vaginalis	*erm*B	nd	Nawaz *et al.* (2011)
	L. fermentum	*erm*B, *msr*C	nd	Thumu and Halami (2012)
		*msr*C	nd	Thumu and Halami (2012)
		tet(K), *tet*(L)	nd	Thumu and Halami (2012)
	L. plantarum	*erm*B	nd	Thumu and Halami (2012)
		tet(W), *tet*(L)	nd	Thumu and Halami (2012)
		tet(M)	nd	Zago *et al.* (2011)
	L. paracasei	*erm*B, *tet*(W)	nd	Comunian *et al.* (2010)
	L. helveticus	*erm*B, *tet*(W), *aph*(3′)-III	nd	Guo *et al.* (2017)
Milk	*L. paracasei*	*tet*(M)	Tn*916*	Devirgiliis *et al.* (2009)
		tet(M)	nd	Comunian *et al.* (2010)
		tet(M), *tet*(W)	nd	Comunian *et al.* (2010)
	L. reuteri	*tet*(W)	nd	Egervärn *et al.* (2009)
	L. paracasei	*erm*B, *tet*(M)	nd	Huys *et al.* (2008)
		*erm*B, *tet*(W)	nd	Huys *et al.* (2008)-Comunian *et al.* (2010)
		tet(M)	nd	Huys *et al.* (2008)-Comunian *et al.* (2010)
Cheese	*L. sakei*	*tet*(M), *tet*(L)	transposon, Plasmid	Ammor *et al.* (2008b)
	L. curvatus	*erm*B, *tet*(M)	nd	van Hoeck *et al.* (2008)
	L. plantarum	*erm*B	plasmid	Feld *et al.* (2009)
		tet(M)	nd	Zago *et al.* (2011)
		*van*X	nd	Liu *et al.* (2009)
	L. casei	*aad*A, *aad*E, *aph*(3′)-III	nd	Ouoba *et al.* (2008)
	L. fermentum	*erm*B	Plasmid	Gfeller *et al.* (2003)
	L. plantarum	*erm*B, *tet*(M)	nd	Nawaz *et al.* (2011)
Yoghurt	*L. acidophilus*	*erm*B	nd	Nawaz *et al.* (2011)
	L. fermentum	*erm*B	nd	Nawaz *et al.* (2011)
	L. brevis	*tet*(M), *tet*(S)	nd	Nawaz *et al.* (2011)
	L. kefiri	*tet*(S)	nd	Nawaz *et al.* (2011)
	L. delbrueckii subsp. bulgaricus	*tet*(S)	nd	Nawaz *et al.* (2011)
		*erm*B	nd	Nawaz *et al.* (2011)
		tet(M), *ant*(6)	nd	Zhou *et al.* (2012)
		aph(3′)-III	nd	Zhou *et al.* (2012)
	L. plantarum	*tet*(M)	nd	Nawaz *et al.* (2011)
		tet(S)	nd	Nawaz *et al.* (2011)
		aph(3′)-III	Plasmid	Pan *et al.* (2011)
		*erm*B, *tet*(M)	Plasmid	Pan *et al.* (2011)
		*mef*A, *tet*(M)	Chromosome	Pan *et al.* (2011)

Table 10.1 Continued

Source	Species	Genotype	Location	References
Fermented vegetables	*L. salivarirus*	*erm*B, *tet*(M)	nd	Nawaz *et al.* (2011)
	L. animalis	*erm*B	nd	Nawaz *et al.* (2011)
		tet(M)	nd	Nawaz *et al.* (2011)
	L. vaginalis	*erm*B	nd	Nawaz *et al.* (2011)
	L. brevis	*tet*(M), *tet*S	nd	Nawaz *et al.* (2011)
		*erm*B	Plasmid	Pan *et al.* (2011)
	L. fermentum	*tet*(M)	Chromosome	Pan *et al.* (2011)
	L. namurensis	*tet*(M), *aph*(3′)-III	Chromosome, plasmid	Pan *et al.* (2011)
	L. vini	*tet*(M), *int*	Tn*916*	Mendonça *et al.* (2016)

n.d., not determined.

(2008c) or transposons like *Tn*916-like (Ammor *et al.*, 2008c; Devirgiliis *et al.*, 2009; Mendonça *et al.*, 2016)]. The later conjugative transposon has been found in a wide range of hosts, both Gram-positive and Gram-negative, including clinical isolates of *Staphylococcus aureus, Streptococcus agalactiae, Streptococcus pneumoniae, Enterococcus* sp. (Mingoia *et al.*, 2013; Roberts and Mullany, 2011). In addition, it has been documented the mobilization of *tet*(M)-transposon from *L. plantarum, L. salivarius* and *L. brevis* to *E. faecalis* clinical strain (Nawaz *et al.*, 2011).

Erythromycin

Erythromycin belongs to the macrolide family of antibiotic, which acts by inhibiting protein synthesis at the 50S subunit level (Abriouel *et al.*, 2015). Erythromycin is active against a wide range of clinical pathogens and useful for the treatment of several bacterial infections. In addition, this antibiotic was intensively used as growth promoter in animal feed. As a consequence, a high frequency of resistance in lactobacilli was also reported (Comunian *et al.*, 2010; Nawaz *et al.*, 2011; Thumu and Halami, 2012; Toomey *et al.*, 2010; Zonenschain *et al.*, 2009) (Table 10.1). Among the main mechanisms of erythromycin resistance we can mention: (a) methylases (*erm*A, *erm*B, *erm*C), (b) efflux pumps (*msr*A/B, *msr*C, *mef*A/E, *mef*A, *mef*E, *mph*A), and (c) esterases (*ere*A, *ere*B) (Abriouel *et al.*, 2015).

The *erm*B gene has been detected among *Lactobacillus* strains isolated from different origins such as fermented meat (Comunian *et al.*, 2010; Thumu and Halami, 2012; Zonenschain *et al.*, 2009), fermented milk products (Guo *et al.*, 2017), cheese (Feld *et al.*, 2009; Huys *et al.*, 2008; van Hoek *et al.*, 2008), yoghurt (Nawaz *et al.*, 2011), and fermented vegetables (Nawaz *et al.*, 2011; Pan *et al.*, 2011). Moreover, *erm*B and *tet*M can co-exist in the same strain; they are widely distributed among *Lactobacillus* (Table 10.1). For example, it was found in *L. paracasei* (Huys *et al.*, 2008), *L. fermentum, L. salivarius, L. animalis* (Nawaz *et al.*, 2011) and *L. brevis* (Zonenschain *et al.*, 2009).

The *erm*B determinant is often associated with different plasmids. For instance, pLME300, a 19.3 kb plasmid was reported in several erythromycin-resistant strains of *L. fermentum* (Ammor *et al.*, 2008b; Gfeller *et al.*, 2003). pLME300 contains at least four main regions, region I bears the ORF1, encoding a theta-replicating like protein followed by a tandem repeat of a 22 bp sequence; meanwhile region II with ORF3 encodes a methylase and ORF4 showing high identity to proteins related to restriction-modification systems, a region III harbouring multiple antibiotic resistance genes and a region IV carries ORFs involved in plasmid mobilization. *erm*B was also described in pLFE1, a 4.03 kb plasmid from the food-isolate *L.*

plantarum M345 (Feld *et al.*, 2009) Moreover, this plasmid encodes an open reading frame following a *oriT* site encoding for a putative truncated mobilization (Mob) protein (Feld *et al.*, 2009). Furthermore, *erm*B is also located in pLEM3, a 5.7 kb plasmid from *L. fermentum* LEM89 (Fons *et al.*, 1997). Besides its plasmid localization, *erm*B gene may be also chromosomally encoded as it was reported in the erythromycin-resistant *L. salivarius* BFE7441 strain (Hummel *et al.*, 2007).

The transfer of erythromycin resistance between bacterial species has been previously evaluated. Thus, *erm*B gene from *L. fermentum* was successfully transferred to *E. faecalis* strain (Nawaz *et al.*, 2011). The plasmid pLFE1 from *L. plantarum* M345 was able to be transferred to a broad host-range such as *L. rhamnosus, Lactococcus lactis, Listeria innocua, Enterococcus faecalis,* and *Listeria monocytogenes* (Feld *et al.*, 2009).

Aminoglycosides

Aminoglucosides (AMG) are a broad group of antibiotics of natural or semisynthetic origin, which act on the 30S subunit, inhibiting irreversibly the protein synthesis (van Hoek *et al.*, 2011). AMG such as gentamicin, neomycin, streptomycin and amikacin are broad-spectrum antibiotics used to treat serious human infections. Particularly, gentamicin and amikacin are the most active antibiotics against multiresistant bacteria such as carbapenem-resistant Gram-negative bacilli, which produce infections with limited therapeutic options (Livermore *et al.*, 2011). Currently, a high frequency of carbapenem-resistant Gram-negative bacilli infections is registered at global level, so the use of AMG in combination with other antimicrobials, such as tigecycline or polymyxin B, is the therapeutic alternative for these types of infections (Morrill *et al.*, 2015). They are also used in veterinary medicine for prophylaxis and treatment of diseases in livestock. The use of AMG is strictly regulated in countries such as the United States, England and the European Union, in order to avoid the appearance of resistant bacterial strains present in the intestinal microbiota (Abriouel *et al.*, 2015). However, in the last decades there has been an increase of the AMG-resistance (AMG-R) in LAB as a result of intensive use in farm animals (Devirgiliis *et al.*, 2013; Jaimee and Halami, 2016; Liu *et al.*, 2009).

Several mechanisms of AMG-R have been described (Ramirez and Tolmasky, 2010). The inactivation of the drugs by aminoglycoside-modifying enzymes is the main mechanism of resistance described in Gram-positive bacteria, particularly in LAB (Ouoba *et al.*, 2008; Zhou *et al.*, 2012).These proteins are classified into three major classes according to the type of modification: AAC [acetyltransferases, AAC(1), AAC(2′), AAC(3), and AAC(6′)], ANT [nucleotidyltransferases or adenyltransferases, ANT(2″), ANT(3″), ANT(4′), ANT(6), and ANT(9)] and APH [phosphotransferases, APH(2″), APH(3′), APH(3″), APH(4), APH(6), APH(7″), and APH(9)]. Furthermore, there also exists a bifunctional enzyme, AAC(6′)–APH(2″), that can acetylate and phosphorylate its substrates sequentially (van Hoek *et al.*, 2011).

George and Halami (2017) recently reported a bifunctional AAC(6′)Ie-aph(2″)Ia enzyme in *L. plantarum* MCC 3011. Moreover, a constant exposure to gentamicin have induced the chaperon [*gro*EL] and this bifunctional gene up to 9-fold. The sub inhibitory doses of gentamicin may facilitate, chaperon action, biofilm formation and resistance gene induction together making the bacteria fight against other antibiotics.

The *aph*(3′)-III gene that confer resistance to kanamycin was detected in *L. casei, L. delbrueckii* subsp. *bulgaricus* and *L. helveticus* from dairy products (Guo *et al.*, 2017; Ouoba *et al.*, 2008; Zhou *et al.*, 2012), as well as in *L. plantarum* from fermented vegetables (Pan *et al.*, 2011). *aad*A and *ant*(6) genes that confer streptomycin resistance have been described in *L. casei* and *L. delbrueckii* subsp. *bulgaricus* strains (Ouoba *et al.*, 2008; Zhou *et al.*, 2012).

The irrational use of AMG in animal husbandry has elevated the emergence of AMG resistance in food-grade LAB (Devirgiliis *et al.*, 2011). Since currently AMG are widely used as the last resort to severe infections in humans, it is very important the surveillance of intensive use of AMG in animals to stop spreading the resistance genes.

Bioinformatic approach to detect antibiotic resistance genes in *Lactobacillus* genomes

The complete sequencing of a bacterial genome (whole-genome sequencing strategy) has recently emerged as a cost-effective and convenient approach to test genes related to safety. This achievement has tremendous impact especially in detecting and characterizing microbial pathogens of public health concern (Didelot *et al.*, 2012) Among the 124,338 bacterial genomes available at NCBI database (09/11/17), only 1.2% belong to *Lactobacillus* genus in comparison with other members of the LAB group. Comparative genomics studies of this group of microorganisms have provided new perspectives regarding their natural evolution and laboratory scale, and their interaction with the environment. These studies have shown the remarkable diversity among the members of the group at different taxonomic levels, result of the interaction between the genome and the environment; with extensive gene loss from a common ancestor, the acquisition of key functions by HGT and the diversification of certain biological activities through gene duplication (Douillard and de Vos, 2014). However, only a few genomic analysis of antibiotic resistance genes (ARG) in *Lactobacillus* are reported so far in the literature (Kazimierczak and Scott, 2007).

One approach often used for mining of ARG in bacterial genomes is the use of a bioinformatics tool called ARG-ANNOT (Gupta *et al.*, 2014). ARG-ANNOT uses a local BLAST program in Bio-Edit software that allows the user to analyse sequences without a Web interface. One of the biases of this database is the low number of ARG sequences from *Lactobacillus* deposited at GenBank. However, in spite of this issue, this bioinformatic tool detects specific genes based on homology and helps in predicting putative new AR genes with low sequence similarities (Gupta *et al.*, 2014).

As a first approach to study the impact of antibiotics in *Lactobacillus*, we analysed ARG in 27 complete genomes of several reference *Lactobacillus* genomes as shown in Table 10.2. We also included the draft genome of *L. delbrueckii* subsp. *lactis* CRL581, a proteolytic strain extensively study by our group; isolated from a homemade Argentinian hard cheese (Hebert *et al.*, 2013, Pescuma *et al.*, 2013). A BLAST in Bioedit against the ARG-ANNOT database was carried out (Hall, 2013). We used moderately stringent conditions (E-Value ≤ 1e-50), the BLOSUM 62 substitution matrix and allowed for gapped BLAST hits. Only those results with 40% sequence identity were taken into account. The ARG included in the database correspond to different antibiotic families (aminoglycosides, beta-lactams, fosfomycin, fluoroquinolones, glycopeptides, macrolide-lincosamide-streptogramin, phenicols, rifampicin, sulfonamides, tetracyclines and trimethoprim).

Among all genomes analysed, *L. amylovorus* 30SC (NC_015214.1) carry the highest number of ARG (22) in its genome. Genes *ermB* conferring resistance to erythromycin and *tet*(W) part of the tetracycline (*tet*) gene resistance family showed the highest score (98.78 and 96.87%, respectively) of identity among all genes detected (Table 10.2). Furthermore, this strain was the only one carrying *ermB* gene which was located in a plasmid named pRKC30SC1 of 7.19 kb (NC_015213.1). Moreover, one ORF detected in the same plasmid encodes a mobilization protein suggesting the potential transferability of this ARG to other strains. In addition, *tet*-32, *tet* -36, *tet* -44, *tet*(M), *tet*(O), *tet*(Q), *tet*(S), *tet*(T) and *erm*33, *erm*A, *erm*C, *erm*G, *erm*T, *erm*Y were also detected with 40 to 70% percentage of identity. *L. mucosae* LM1 (NZ_CP011013.1) also bear a set of tetracycline resistance genes in its genome (*tet*-32, *tet* -36, *tet* -44, *tet*(M), *tet*(O), *tet*(Q), *tet*(S), *tet*(T) with *tet*(W) showing 96.71% of identity. Sequence analysis of the regions flanking *tet*(W) gene in both strains revealed ORF associated with mobile genetic elements, encoding mobilization proteins and relaxase and recombinase enzymes (Fig. 10.3). *tet*(W) is one of the most abundant tetracycline resistance genes found in bacteria from the mammalian gut; it has also been described in *Lactobacillus* previously (Gueimonde *et al.*, 2013). The majority of tetracycline resistance genes are located on mobilizable or conjugative elements, which may partially explain their wide distribution among bacterial species (Kazimierczak *et al.*, 2006, 2009; Kazimierczak *et al.*, 2008).

Another interesting feature related to resistance to macrolides, lincosamides and streptogramins (MLS) is the presence of a *lsa* variant (A) gene in 60% of the evaluated strains. *lsa* variant (A) gene is generally located in the chromosome, encoding a putative ABC protein that confers resistance to lincosamides (lincomycin and clindamycin) and streptogramins A (dalfopristin, pristinamycin II, and virginiamycin M) (Singh and Murray, 2005). The same gene variant was detected in *E. faecalis*, other LAB member, where Lsa phenotype is also an

Table 10.2 Antibiotic resistance genes in *Lactobacillus* genomes detected by the bioinformatic tool ARG-ANNOT

Strain	GenBank accession number	Antibiotic resistance genes																										
		Bla[a]			GLY[a]				AG[a]	MLS[a]										Tet[a]								
		PBP1a	PBP1b	AmpC1	VanR-C	VanR-F	VanR-L	VanR-M	APH-Stph	Erm33	ErmA	ErmB	ErmC	ErmG	ErmT	ErmY	LsaA	LsaB	LsaC	Tet-32	Tet-36	Tet-44	TetM	TetO	TetQ	TetS	TetT	TetW
L. acidophilus NCFM	NC_006814.3	yellow				yellow		yellow	yellow								yellow											
L. amylovorus 30SC	NC_015214.1	yellow				yellow		yellow		yellow	yellow	red	yellow	yellow	yellow	yellow	yellow	yellow	yellow	orange	yellow	orange	orange	orange	yellow	orange	yellow	red
L. brevis ATCC 367	NC_008497.1	yellow				yellow		yellow																				
L. buchneri CD034	NC_018610.1	yellow				yellow	yellow	yellow									yellow	yellow	yellow									
L. casei BL23	FM177140.1	yellow	yellow			yellow		yellow									yellow	yellow	yellow									
L. coryniformis DSM 20001	NZ_AZCN01000001.1	yellow	yellow			yellow		yellow	yellow								yellow	yellow	yellow									
L. crispatus ST1	FN692037.1	yellow				yellow											yellow	yellow	yellow									
L. curvatus MRS6	CP022474.1	yellow				yellow		yellow	yellow								yellow	yellow	yellow									
L. delbrueckii subsp. *bulgaricus* ATCC 11842	NC_008054.1	yellow				yellow																						
L. delbrueckii subsp. *lactis* CRL 581	KE145374.1	yellow				yellow																						
L. fermentum IFO 3956	NC_010610.1	yellow				yellow		yellow																				
L. gasseri ATCC 33323	NC_008530.1	yellow				yellow		yellow									yellow	yellow	yellow									
L. helveticus CAUH18	CP012381.1	yellow		yellow		yellow											yellow											
L. iners DSM 13335	GG700812.1																											
L. jensenii SNUV360	CP018809.1	yellow				yellow											yellow	yellow	yellow									
L. johnsonii NCC 533	AE017198.1	yellow				yellow		yellow									yellow	yellow	yellow									
L. kefiranofaciens ZW3	CP002764.1	yellow				yellow		yellow									yellow											
L. kunkeei MP2	CP012920.1	yellow				yellow		yellow																				
L. mucosae LM1	NZ_CP011013.1	yellow						yellow												orange	yellow	orange	orange	orange	yellow	orange	yellow	red
L. paracasei ATCC 334	NC_008526.1	yellow				yellow		yellow									yellow	yellow	yellow									
L. pentosus SLC13	CP022130.1	yellow				yellow		yellow	yellow																			
L. plantarum WCFS1	NC_004567.2	yellow				yellow		yellow	yellow																			
L. reuteri DSM 20016	CP000705.1	yellow				yellow		yellow									yellow											
L. rhamnosus GG	FM179322.1	yellow				yellow		yellow									yellow	yellow	yellow									
L. ruminis ATCC 27782	CP003032.1	yellow				yellow		yellow																				
L. sakei FAM18311	NZ_CP020459.1	yellow				yellow		yellow	yellow								yellow	yellow	yellow									
L. salivarius UCC118	CP000233.1		yellow		yellow	yellow	yellow	yellow	yellow								yellow	yellow	yellow									
L. sanfranciscensis TMW	NC_015978.1	yellow																										

[a]Bla: β-Lactam, AG, aminoglycoside; Gly, glycopeptide; MLS, macrolide–lincosamide–streptogramin B; Tet, tetracycline.
Colours depicted in the table indicate percentage of gene identity detected in bacterial genomes by ARG-ANNOT using parameters mentioned in the text. Red >90%; orange 60–90%; and yellow 40–60% identity.

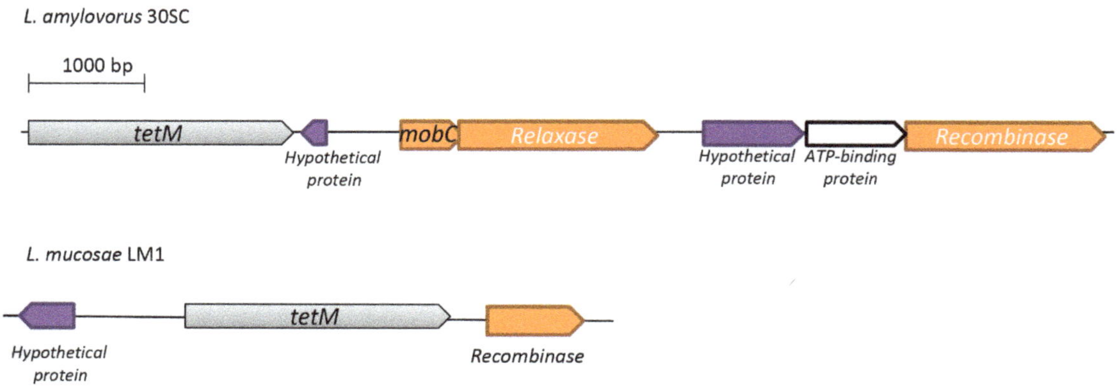

Figure 10.3 Schematic representation of *tet*(M) mobilization platforms in *L. amylovorus* 30SC and *L. mucosae* LM1. The arrows represent the individual ORF pointing in the probable direction of transcription. The grey arrow indicate *tet*(M) encoding tetracycline resistance and orange arrows indicate the predicted proteins for mobilization.

intrinsic feature (Singh and Murray, 2005). Besides, other variants of *lsa*(A) gene, i.e. variants B, C, E, have been described. These gene variants have been found to be carried by MGE, implying putative mobility in *Streptococcus* (Douarre *et al.*, 2015). Variants B and C are also detected in thirteen *Lactobacillus* strains here analysed, whereas C variant always accompanies variant B and all strains carrying B and C also bear variant A (Table 10.2). *lsa*(A) variant was the only gene detected by ARG-NOOT in twelve *L. rhamnosus* complete genome sequences retrieved from the NCBI database (11/09/2017) using the same parameters as described above. The analysed *L. rhamnosus* strains were the following: GG (NC_013198), 1505 (ATBI00000000.1), ATCC 8530 (NC_017491), DSM 14870 (NZ_CP006804), LOCK900 (NC_021723), LOCK908 (NC_021725), ASCC290 (NZ_CP014645), BPL5 (NZ_LT220504), LR5 (NZ_CP017063), WQ2 (CP020016), LRB (NZ_CP016823), Lc705 (NC_013199) and Pen (NZ_CP020464). BLAST analyses demonstrated that *lsa*(A) gene is widely distributed among *L. rhamnosus* and it seems to be intrinsic to the species; however, no phenotypically resistant strains were described. In addition, *L. rhamnosus lsa*(A) gene was phylogenetically distant from mobile *lsa* gene (B,C,E) (Fig. 10.4). *L. rhamnosus* is widely used in the manufacture of cheese and other dairy products. This organism has also been shown to stimulate the immune system and have antibacterial activity against intestinal pathogens, indicating that it may be useful as a probiotic (Salva *et al.*, 2010, Taranto *et al.*, 2013). Moreover, *L. rhamnosus* CRL1505 has been included into official Nutritional Programs in Argentina (Villena *et al.*, 2012) while *L. rhamnosus* GG is widely used as a commercial probiotic (Mantegazza *et al.*, 2017).

Intrinsic resistance to vancomycin has been reported for different species of *Lactobacillus* (Gueimonde *et al.*, 2013). The vancomycin resistance in lactobacilli is chromosomally encoded and not inducible or transferable (Gueimonde *et al.*, 2013). The biochemical mechanism of vancomycin resistance in *Leuconostoc mesenteroides* VR1 and in *L. rhamnosus* ATCC 7469 (previous named *L. casei*) has been shown to be based on D-Ala–D-lactate modification, as in enterococci suggesting that the *van* genes may have originated from an intrinsically resistant bacterium (Handwerger *et al.*, 1994). Using our bioinformatic approach, we were able to detect *van*R as the only gene from *van* cluster based in our search parameters (40 to 60% identity at the amino acid level). In agreement with our results, previous studies reported no hybridization of enterococcal *van*A, *van*B and *van*C genes in *Lactobacillus* species (Klein *et al.*, 2000, Patel, 2000)

Regarding β -lactam antibiotics, ARG-NOOT detected a penicillin-binding protein (PBP) *1a* gene more frequent than *pbp1b* among analysed strains (Table 10.2). Our results are in agreement with the notion that the predominant mechanism of resistance to β-lactams in Gram-positive organisms is mostly achieved by modifications of their target site. PBP are enzymes responsible for peptidoglycan synthesis. Each bacterium harbours an array of PBP, which catalyse the polymerization

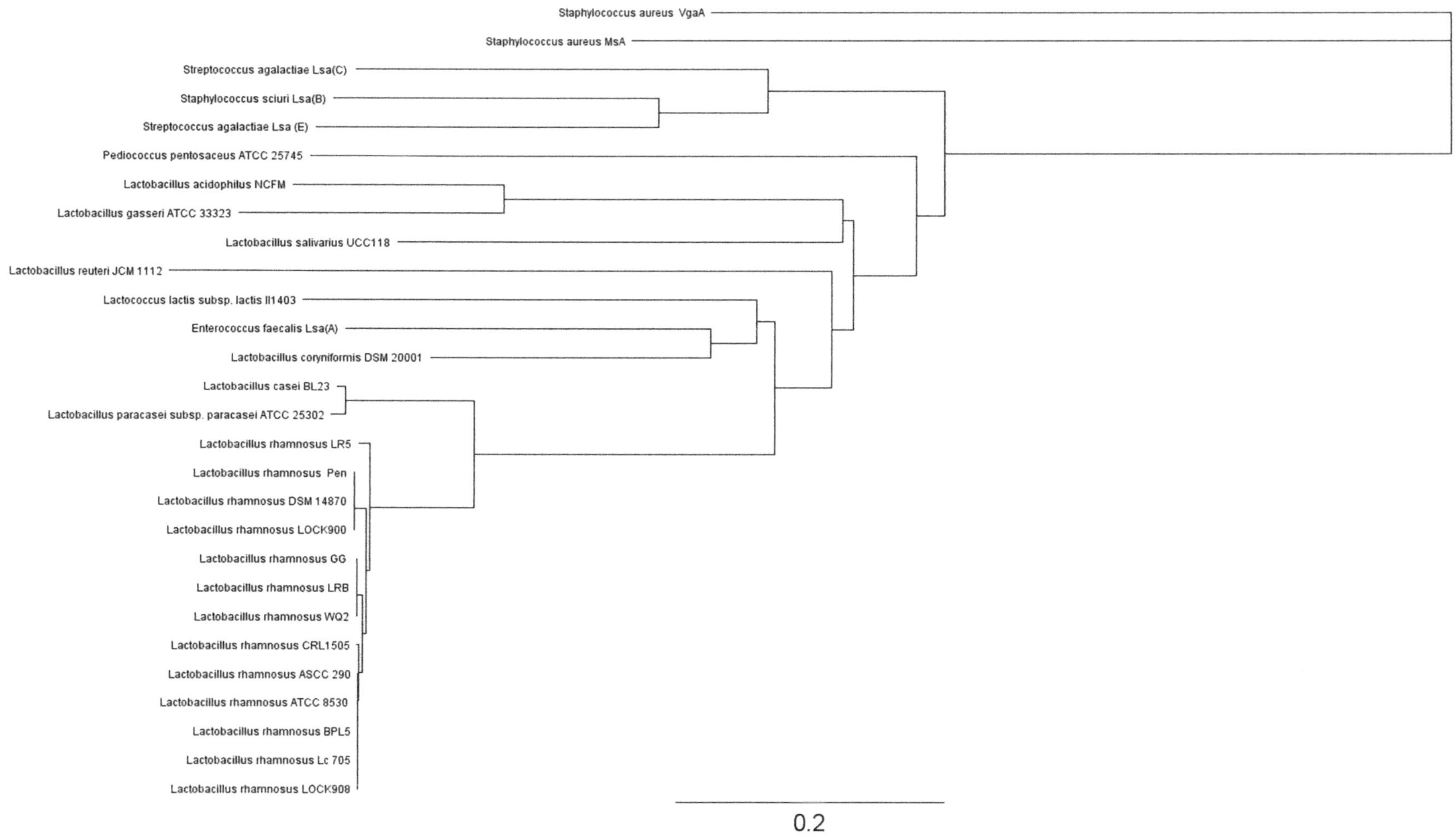

Figure 10.4 Lsa homologous protein sequences were identified by using BLASTP (cutoff e-value $<10^{-60}$) in 28 *Lactobacillus* strains. All sequence alignments were generated using the Geneious Alignment tool with default settings (Geneious Pro 4.8.5). Phylogenetic trees were constructed using the Geneious Pro PhyML plugin50 with the WAG amino acid substitution model and 1000 boot-strapping.

of polysaccharide chains and the cross-linking of pentapeptide side chains. β-lactam antibiotics inhibit the PBP activity; in turn, peptide glycan integrity is compromised causing bacterial death. The PBP system of *Escherichia coli* is currently the most extensively studied. *E. coli* harbours five high molecular weight (HMW) PBPs (PBP1a, -1b, -1c, -2, and -3) and eight or low molecular weight (LMW) PBP (PBP4, -5, -6, -7, and -8 and DacD, AmpC, and AmpH) (Ghosh *et al.*, 2008). Recently, ten PBP were identified in *L. paracasei* strains, seven of which had apparent molecular sizes similar to those of *E. coli* (Hattori *et al.*, 2017). Moreover, Piuri *et al.* (2005) demonstrated three essential PBP in *L. casei* ATCC 393. However, these studies demonstrated the relationship of PBP protein with key factors functioning in the maintenance peptidoglycan under stress conditions.

Conclusions

As we have shown here, mining bacterial genomes using bioinformatic tools represent a valuable approach for the detection of antibiotic resistance in commensal bacteria such as *Lactobacillus. In silico* studies need to be confirmed by phenotypic analysis with minimal inhibitory concentration determination following strict guidelines of antibiotic resistance authorities such as European Food Safety Authority (EFSA), Clinical and Laboratory Standards Institute (CLSI). In general, reports of multidrug resistant bacteria indicate that they carry the same resistance genes as environmental or commensal bacteria, supporting the notion of a 'potential reservoir' for ARG in these bacteria. It is therefore very important to elucidate how antibiotic resistance genes are maintained and spread through commensal bacterial communities.

Acknowledgements

Consejo Nacional de Investigaciones Científicas y Técnicas (PIP 272 – PIP 883) and FONCYT (PICT2011 N°0175, PICT2014 N°3428) from Argentina.

References

Abriouel, H., Casado Muñoz, M.D.C., Lavilla Lerma, L., Pérez Montoro, B., Bockelmann, W., Pichner, R., Kabisch, J., Cho, G.S., Franz, C.M.A.P., Gálvez, A., *et al.* (2015). New insights in antibiotic resistance of *Lactobacillus* species from fermented foods. Food Res. Int. 78, 465–481.

Aminov, R.I. (2010). A brief history of the antibiotic era: Lessons learned and challenges for the future. Front. Microbiol. *1*, 134. https://doi.org/10.3389/fmicb.2010.00134

Ammor, M.S., Flórez, A.B., Alvarez-Martín, P., Margolles, A., and Mayo, B. (2008a). Analysis of tetracycline resistance *tet*(W) genes and their flanking sequences in intestinal *Bifidobacterium* species. J. Antimicrob. Chemother. *62*, 688–693. https://doi.org/10.1093/jac/dkn280

Ammor, M.S., Flórez, A.B., and Mayo, B. (2007). Antibiotic resistance in non-enterococcal lactic acid bacteria and bifidobacteria. Food Microbiol. *24*, 559–570.

Ammor, M.S., Flórez, A.B., van Hoek, A.H., de Los Reyes-Gavilán, C.G., Aarts, H.J., Margolles, A., and Mayo, B. (2008b). Molecular characterization of intrinsic and acquired antibiotic resistance in lactic acid bacteria and bifidobacteria. J. Mol. Microbiol. Biotechnol. *14*, 6–15.

Ammor, M.S., Gueimonde, M., Danielsen, M., Zagorec, M., van Hoek, A.H., de Los Reyes-Gavilán, C.G., Mayo, B., and Margolles, A. (2008c). Two different tetracycline resistance mechanisms, plasmid-carried *tet*(L) and chromosomally located transposon-associated *tet*(M), coexist in *Lactobacillus sakei* Rits 9. Appl. Environ. Microbiol. *74*, 1394–1401. https://doi.org/10.1128/AEM.01463-07

Arioli, S., Guglielmetti, S., Amalfitano, S., Viti, C., Marchi, E., Decorosi, F., Giovannetti, L., and Mora, D. (2014). Characterization of *tetA*-like gene encoding for a major facilitator superfamily efflux pump in *Streptococcus thermophilus*. FEMS Microbiol. Lett. *355*, 61–70. https://doi.org/10.1111/1574-6968.12449

Arthur, M., Molinas, C., Depardieu, F., and Courvalin, P. (1993). Characterization of Tn1546, a Tn3-related transposon conferring glycopeptide resistance by synthesis of depsipeptide peptidoglycan precursors in *Enterococcus faecium* BM4147. J. Bacteriol. *175*, 117–127.

Bellanger, X., Payot, S., Leblond-Bourget, N., and Guédon, G. (2014). Conjugative and mobilizable genomic islands in bacteria: Evolution and diversity. FEMS Microbiol. Rev. *38*, 720–760. https://doi.org/10.1111/1574-6976.12058

Boucher, Y., Douady, C.J., Papke, R.T., Walsh, D.A., Boudreau, M.E., Nesbø, C.L., Case, R.J., and Doolittle, W.F. (2003). Lateral gene transfer and the origins of prokaryotic groups. Annu. Rev. Genet. *37*, 283–328. https://doi.org/10.1146/annurev.genet.37.050503.084247

Carlet, J. (2012). The gut is the epicentre of antibiotic resistance. Antimicrob. Resist. Infect. Control *1*, 39. https://doi.org/10.1186/2047-2994-1-39.

CDC. (2013). Antibiotic Resistance Threats in the United States (CDC). https://www.cdc.gov/drugresistance/threat-report-2013/index.html

Comunian, R., Daga, E., Dupré, I., Paba, A., Devirgiliis, C., Piccioni, V., Perozzi, G., Zonenschain, D., Rebecchi, A., Morelli, L., *et al.* (2010). Susceptibility to tetracycline and erythromycin of *Lactobacillus paracasei* strains isolated from traditional Italian fermented foods. Int. J. Food Microbiol. *138*, 151–156. https://doi.org/10.1016/j.ijfoodmicro.2009.11.018

Chatterjee, S., Mondal, A., Mitra, S., and Basu, S. (2017). *Acinetobacter baumannii* transfers the blaNDM-1 gene via outer membrane vesicles. J. Antimicrob. Chemother. 72, 2201–2207. https://doi.org/10.1093/jac/dkx131

Chopra, I., and Roberts, M. (2001). Tetracycline antibiotics: Mode of action, applications, molecular biology, and epidemiology of bacterial resistance. Microbiol. Mol. Biol. Rev. 65, 232–260. https://doi.org/10.1128/MMBR.65.2.232-260.2001

Danielsen, M., and Wind, A. (2003). Susceptibility of *Lactobacillus* spp. to antimicrobial agents. Int. J. Food Microbiol. 82, 1–11.

Darmon, E., and Leach, D.R. (2014). Bacterial genome instability. Microbiol. Mol. Biol. Rev. 78, 1–39. https://doi.org/10.1128/MMBR.00035-13

Davies, J., and Davies, D. (2010). Origins and evolution of antibiotic resistance. Microbiol. Mol. Biol. Rev. 74, 417–433. https://doi.org/10.1128/MMBR.00016-10

Derbise, A., Dyke, K.G., and el Solh, N. (1996). Characterization of a *Staphylococcus aureus* transposon, Tn5405, located within Tn5404 and carrying the aminoglycoside resistance genes, *aphA*-3 and *aadE*. Plasmid 35, 174–188.

Devirgiliis, C., Barile, S., and Perozzi, G. (2011). Antibiotic resistance determinants in the interplay between food and gut microbiota. Genes Nutr. 6, 275–284. https://doi.org/10.1007/s12263-011-0226-x

Devirgiliis, C., Coppola, D., Barile, S., Colonna, B., and Perozzi, G. (2009). Characterization of the Tn916 conjugative transposon in a food-borne strain of *Lactobacillus paracasei*. Appl. Environ. Microbiol. 75, 3866–3871. https://doi.org/10.1128/AEM.00589-09

Devirgiliis, C., Zinno, P., and Perozzi, G. (2013). Update on antibiotic resistance in foodborne *Lactobacillus* and *Lactococcus* species. Front. Microbiol. 4, 301. https://doi.org/10.3389/fmicb.2013.00301

Didelot, A., Le Corre, D., Luscan, A., Cazes, A., Pallier, K., Emile, J.F., Laurent-Puig, P., and Blons, H. (2012). Competitive allele specific TaqMan PCR for KRAS, BRAF and EGFR mutation detection in clinical formalin fixed paraffin embedded samples. Exp. Mol. Pathol. 92, 275–280. https://doi.org/10.1016/j.yexmp.2012.03.001

Douarre, P.E., Sauvage, E., Poyart, C., and Glaser, P. (2015). Host specificity in the diversity and transfer of lsa resistance genes in group B *Streptococcus*. J. Antimicrob. Chemother. 70, 3205–3213. https://doi.org/10.1093/jac/dkv277

Douillard, F.P., and de Vos, W.M. (2014). Functional genomics of lactic acid bacteria: From food to health. Microb. Cell Fact. 13 (Suppl. 1), S8. https://doi.org/10.1186/1475-2859-13-S1-S8

Economou, V., and Gousia, P. (2015). Agriculture and food animals as a source of antimicrobial-resistant bacteria. Infect. Drug Resist. 8, 49–61. https://doi.org/10.2147/IDR.S55778

EFSA. (2012). Guidance on the assessment of bacterial susceptibility to antimicrobials of human and veterinary importance. EFSA Journal 10, 2740. https://doi.org/10.2903/j.efsa.2012.2740

Egervärn, M., Roos, S., and Lindmark, H. (2009). Identification and characterization of antibiotic resistance genes in *Lactobacillus reuteri* and *Lactobacillus plantarum*. J. Appl. Microbiol. 107, 1658–1668. https://doi.org/10.1111/j.1365-2672.2009.04352.x

Feld, L., Bielak, E., Hammer, K., and Wilcks, A. (2009). Characterization of a small erythromycin resistance plasmid pLFE1 from the food-isolate *Lactobacillus plantarum* M345. Plasmid 61, 159–170.

Fons, M., Hégé, T., Ladiré, M., Raibaud, P., Ducluzeau, R., and Maguin, E. (1997). Isolation and characterization of a plasmid from *Lactobacillus fermentum* conferring erythromycin resistance. Plasmid 37, 199–203.

George, J., and Halami, P.M. (2017). Sub-inhibitory concentrations of gentamicin triggers the expression of *aac*(6')Ie-*aph*(2")Ia, chaperons and biofilm related genes in *Lactobacillus plantarum* MCC 3011. Res. Microbiol. 168, 722–731.

Gevers, D., Danielsen, M., Huys, G., and Swings, J. (2003). Molecular characterization of *tet*(M) genes in *Lactobacillus* isolates from different types of fermented dry sausage. Appl. Environ. Microbiol. 69, 1270–1275.

Gfeller, K.Y., Roth, M., Meile, L., and Teuber, M. (2003). Sequence and genetic organization of the 19.3-kb erythromycin- and dalfopristin-resistance plasmid pLME300 from *Lactobacillus fermentum* ROT1. Plasmid 50, 190–201.

Ghosh, A.S., Chowdhury, C., and Nelson, D.E. (2008). Physiological functions of D-alanine carboxypeptidases in *Escherichia coli*. Trends Microbiol. 16, 309–317. https://doi.org/10.1016/j.tim.2008.04.006

González-Zorn, B., and Escudero, J.A. (2012). Ecology of antimicrobial resistance: Humans, animals, food and environment. Int. Microbiol. 15, 101–109. https://doi.org/10.2436/20.1501.01.163

Guédon, G., Libante, V., Coluzzi, C., Payot, S., and Leblond-Bourget, N. (2017). The obscure world of integrative and mobilizable elements, highly widespread elements that pirate bacterial conjugative systems. Genes 8, E337.

Gueimonde, M., Sánchez, B., G de Los Reyes-Gavilán, C., and Margolles, A. (2013). Antibiotic resistance in probiotic bacteria. Front. Microbiol. 4, 202. https://doi.org/10.3389/fmicb.2013.00202

Guo, H., Pan, L., Li, L., Lu, J., Kwok, L., Menghe, B., Zhang, H., and Zhang, W. (2017). Characterization of antibiotic resistance genes from *Lactobacillus* isolated from traditional dairy products. J. Food Sci. 82, 724–730. https://doi.org/10.1111/1750-3841.13645

Gupta, S.K., Padmanabhan, B.R., Diene, S.M., Lopez-Rojas, R., Kempf, M., Landraud, L., and Rolain, J.M. (2014). ARG-ANNOT, a new bioinformatic tool to discover antibiotic resistance genes in bacterial genomes. Antimicrob. Agents Chemother. 58, 212–220. https://doi.org/10.1128/AAC.01310-13

Hall, T. (2013). BioEdit: Biological Sequence Alignment Editor for Win95/98/NT/2K/XP. http://www.mbio.ncsu.edu/BioEdit/bioedit.html

Handwerger, S., Pucci, M.J., Volk, K.J., Liu, J., and Lee, M.S. (1994). Vancomycin-resistant *Leuconostoc* mesenteroides and *Lactobacillus casei* synthesize cytoplasmic peptidoglycan precursors that terminate in lactate. J. Bacteriol. 176, 260–264.

Hattori, M., Torres, G.A., Tanaka, N., Okada, S., Endo, A., and Nakagawa, J. (2017). Detection and analysis of *Lactobacillus paracasei* penicillin-binding proteins revealed the presence of cholate-sensitive penicillin-binding protein 3 and an elongated cell shape in a cholate-sensitive strain. Biosci. Microbiota Food Health *36*, 65–72. https://doi.org/10.12938/bmfh.16-019

Hebert, E.M., Raya, R.R., Brown, L., Font de Valdez, G., Savoy de Giori, G., and Taranto, M.P. (2013). Genome sequence of the cheese-starter strain *Lactobacillus delbrueckii* subsp. *lactis* CRL 581. Genome Announc. *1*, e00602–13. https://doi.org/10.1128/genomeA.00602-13

Hillen, W., and Berens, C. (1994). Mechanisms underlying expression of Tn10 encoded tetracycline resistance. Annu. Rev. Microbiol. *48*, 345–369. https://doi.org/10.1146/annurev.mi.48.100194.002021

Hummel, A.S., Hertel, C., Holzapfel, W.H., and Franz, C.M. (2007). Antibiotic resistances of starter and probiotic strains of lactic acid bacteria. Appl. Environ. Microbiol. *73*, 730–739.

Huys, G., D'Haene, K., Danielsen, M., Mättö, J., Egervärn, M., and Vandamme, P. (2008). Phenotypic and molecular assessment of antimicrobial resistance in *Lactobacillus paracasei* strains of food origin. J. Food Prot. *71*, 339–344.

Jaimee, G., and Halami, P.M. (2016). Emerging resistance to aminoglycosides in lactic acid bacteria of food origin: An impending menace. Appl. Microbiol. Biotechnol. *100*, 1137–1151. https://doi.org/10.1007/s00253-015-7184-y

Kazimierczak, K.A., Flint, H.J., and Scott, K.P. (2006). Comparative analysis of sequences flanking *tet*(W) resistance genes in multiple species of gut bacteria. Antimicrob. Agents Chemother. *50*, 2632–2639.

Kazimierczak, K.A., Rincon, M.T., Patterson, A.J., Martin, J.C., Young, P., Flint, H.J., and Scott, K.P. (2008). A new tetracycline efflux gene, *tet*(40), is located in tandem with *tet*(O/32/O) in a human gut firmicute bacterium and in metagenomic library clones. Antimicrob. Agents Chemother. *52*, 4001–4009. https://doi.org/10.1128/AAC.00308-08

Kazimierczak, K.A., and Scott, K.P. (2007). Antibiotics and resistance genes: influencing the microbial ecosystem in the gut. Adv. Appl. Microbiol. *62*, 269–292.

Kazimierczak, K.A., Scott, K.P., Kelly, D., and Aminov, R.I. (2009). Tetracycline resistome of the organic pig gut. Appl. Environ. Microbiol. *75*, 1717–1722. https://doi.org/10.1128/AEM.02206-08

Klare, I., Konstabel, C., Werner, G., Huys, G., Vankerckhoven, V., Kahlmeter, G., Hildebrandt, B., Muller-Bertling, S., Witte, W., and Goossens, H. (2007). Antimicrobial susceptibilities of *Lactobacillus*, *Pediococcus* and *Lactococcus* human isolates and cultures intended for probiotic or nutritional use. J. Antimicrob. Chemother. *59*, 900–912. https://doi.org/10.1093/jac/dkm035.

Klein, G., Hallmann, C., Casas, I.A., Abad, J., Louwers, J., and Reuter, G. (2000). Exclusion of *vanA*, *vanB* and *vanC* type glycopeptide resistance in strains of *Lactobacillus reuteri* and *Lactobacillus rhamnosus* used as probiotics by polymerase chain reaction and hybridization methods. J. Appl. Microbiol. *89*, 815–824.

Klimentová, J., and Stulík, J. (2015). Methods of isolation and purification of outer membrane vesicles from gram-negative bacteria. Microbiol. Res. *170*, 1–9. https://doi.org/10.1016/j.micres.2014.09.006

Koonin, E.V. (2016). Viruses and mobile elements as drivers of evolutionary transitions. Philos. Trans. R. Soc. B. Biol. Sci. *371*, 20150442. https://doi.org/10.1098/rstb.2015.0442

Kwon, D.H., Dore, M.P., Kim, J.J., Kato, M., Lee, M., Wu, J.Y., and Graham, D.Y. (2003). High-level beta-lactam resistance associated with acquired multidrug resistance in *Helicobacter pylori*. Antimicrob. Agents Chemother. *47*, 2169–2178.

Labbate, M., Case, R.J., and Stokes, H.W. (2009). The integron/gene cassette system: an active player in bacterial adaptation. Methods Mol. Biol. *532*, 103–125. https://doi.org/10.1007/978-1-60327-853-9_6

Liu, C., Zhang, Z.Y., Dong, K., Yuan, J.P., and Guo, X.K. (2009). Antibiotic resistance of probiotic strains of lactic acid bacteria isolated from marketed foods and drugs. Biomed. Environ. Sci. 22, 401–412. https://doi.org/10.1016/S0895-3988(10)60018-9

Livermore, D.M., Mushtaq, S., Warner, M., Zhang, J.C., Maharjan, S., Doumith, M., and Woodford, N. (2011). Activity of aminoglycosides, including ACHN-490, against carbapenem-resistant *Enterobacteriaceae* isolates. J. Antimicrob. Chemother. *66*, 48–53. https://doi.org/10.1093/jac/dkq408

Lorenz, M.G., and Wackernagel, W. (1994). Bacterial gene transfer by natural genetic transformation in the environment. Microbiol. Rev. *58*, 563–602.

Mantegazza, C., Molinari, P., D'Auria, E., Sonnino, M., Morelli, L., and Zuccotti, G.V. (2018). Probiotics and antibiotic-associated diarrhea in children: A review and new evidence on *Lactobacillus rhamnosus* GG during and after antibiotic treatment. Pharmacol. Res. *128*, 63–72.

Mater, D.D., Langella, P., Corthier, G., and Flores, M.J. (2008). A probiotic *Lactobacillus* strain can acquire vancomycin resistance during digestive transit in mice. J. Mol. Microbiol. Biotechnol. *14*, 123–127.

Mendonça, A.A., de Lucena, B.T., de Morais, M.M., and de Morais, M.A. (2016). First identification of Tn916-like element in industrial strains of *Lactobacillus vini* that spread the *tet*-M resistance gene. FEMS Microbiol. Lett. *363*, fnv240. https://doi.org/10.1093/femsle/fnv240

Mingoia, M., Morici, E., Tili, E., Giovanetti, E., Montanari, M.P., and Varaldo, P.E. (2013). Characterization of Tn5801.Sag, a variant of *Staphylococcus aureus* Tn916 family transposon Tn5801 that is widespread in clinical isolates of *Streptococcus agalactiae*. Antimicrob. Agents Chemother. *57*, 4570–4574. https://doi.org/10.1128/AAC.00521-13

Morrill, H.J., Pogue, J.M., Kaye, K.S., and LaPlante, K.L. (2015). Treatment options for carbapenem-resistant *Enterobacteriaceae* infections. Open Forum Infect. Dis. 2, ofv050. https://doi.org/10.1093/ofid/ofv050

Muñoz-Atienza, E., Gómez-Sala, B., Araújo, C., Campanero, C., del Campo, R., Hernández, P.E., Herranz, C., and Cintas, L.M. (2013). Antimicrobial activity, antibiotic susceptibility and virulence factors of lactic acid bacteria of aquatic origin intended for use as probiotics in aquaculture. BMC Microbiol. *13*, 15. https://doi.org/10.1186/1471-2180-13-15

Nawaz, M., Wang, J., Zhou, A., Ma, C., Wu, X., Moore, J.E., Millar, B.C., and Xu, J. (2011). Characterization and transfer of antibiotic resistance in lactic acid bacteria from fermented food products. Curr.

Microbiol. *62*, 1081–1089. https://doi.org/10.1007/s00284-010-9856-2

Nicolas, E., Lambin, M., Dandoy, D., Galloy, C., Nguyen, N., Oger, C.A., and Hallet, B. (2015). The Tn3-family of replicative transposons. Microbiol. Spectr. 3,. https://doi.org/10.1128/microbiolspec.MDNA3-0060-2014

Normark, B.H., and Normark, S. (2002). Evolution and spread of antibiotic resistance. J. Intern. Med. *252*, 91–106.

Ouoba, L.I., Lei, V., and Jensen, L.B. (2008). Resistance of potential probiotic lactic acid bacteria and bifidobacteria of African and European origin to antimicrobials: Determination and transferability of the resistance genes to other bacteria. Int. J. Food Microbiol. *121*, 217–224.

Pan, L., Hu, X., and Wang, X. (2011). Assessment of antibiotic resistance of lactic acid bacteria in Chinese fermented foods. Food Control *22*, 1316–1321. https://doi.org/10.1016/j.foodcont.2011.02.006

Patel, R. (2000). Enterococcal-type glycopeptide resistance genes in non-enterococcal organisms. FEMS Microbiol. Lett. *185*, 1–7.

Pescuma, M., Espeche Turbay, M.B., Mozzi, F., Font de Valdez, G., Savoy de Giori, G., and Hebert, E.M. (2013). Diversity in proteinase specificity of thermophilic Lactobacilli as revealed by hydrolysis of dairy and vegetable proteins. Appl. Microbiol. Biotechnol. *97*, 7831–7844. https://doi.org/10.1007/s00253-013-5037-0

Piuri, M., Sanchez-Rivas, C., and Ruzal, S.M. (2005). Cell wall modifications during osmotic stress in *Lactobacillus casei*. J. Appl. Microbiol. *98*, 84–95.

Pomba, C., Rantala, M., Greko, C., Baptiste, K.E., Catry, B., van Duijkeren, E., Mateus, A., Moreno, M.A., Pyörälä, S., Ružauskas, M., *et al.* (2017). Public health risk of antimicrobial resistance transfer from companion animals. J. Antimicrob. Chemother. 72, 957–968. https://doi.org/10.1093/jac/dkw481

Radulović, Z., Petrović, T., and Bulajić, S. (2012). Antibiotic susceptibility of probiotic bacteria, antibiotic resistant bacteria. In A Continuous Challenge in the New Millennium, Pana, M., ed. (Intech), pp. 550–576. https://cdn.intechopen.com/pdfs-wm/34710.pdf

Raftis, E.J., Forde, B.M., Claesson, M.J., and O'Toole, P.W. (2014). Unusual genome complexity in *Lactobacillus salivarius* JCM1046. BMC Genomics *15*, 771. https://doi.org/10.1186/1471-2164-15-771

Ramirez, M.S., and Tolmasky, M.E. (2010). Aminoglycoside modifying enzymes. Drug Resist. Updat. *13*, 151–171. https://doi.org/10.1016/j.drup.2010.08.003

Roberts, A.P., and Mullany, P. (2011). Tn916-like genetic elements: A diverse group of modular mobile elements conferring antibiotic resistance. FEMS Microbiol. Rev. *35*, 856–871. https://doi.org/10.1111/j.1574-6976.2011.00283.x

Salva, S., Villena, J., and Alvarez, S. (2010). Immunomodulatory activity of *Lactobacillus rhamnosus* strains isolated from goat milk: Impact on intestinal and respiratory infections. Int. J. Food Microbiol. *141*, 82–89. https://doi.org/10.1016/j.ijfoodmicro.2010.03.013

Santoro, F., Vianna, M.E., and Roberts, A.P. (2014). Variation on a theme; an overview of the Tn916/Tn1545 family of mobile genetic elements in the oral and nasopharyngeal *Streptococci*. Front. Microbiol. *5*, 535. https://doi.org/10.3389/fmicb.2014.00535

Shapiro, J.A. (1995). The discovery and significance of mobile genetic elements. In Mobile Genetic Elements, Sherratt, D., ed. (IRL Press, UK). http://shapiro.bsd.uchicago.edu/Shapiro.1995.DiscSignifMobElem.pdf

Singh, K.V., and Murray, B.E. (2005). Differences in the *Enterococcus faecalis lsa* locus that influence susceptibility to quinupristin-dalfopristin and clindamycin. Antimicrob. Agents Chemother. *49*, 32–39.

Spellberg, B., Guidos, R., Gilbert, D., Bradley, J., Boucher, H.W., Scheld, W.M., Bartlett, J.G., and Edwards, J. (2008). The epidemic of antibiotic-resistant infections: A call to action for the medical community from the Infectious Diseases Society of America. Clin. Infect. Dis. *46*, 155–164. https://doi.org/10.1086/524891

Strøman, P., Müller, C.C., and Sørensen, K.I. (2003). Heat shock treatment increases the frequency of loss of an erythromycin resistance-encoding transposable element from the chromosome of *Lactobacillus crispatus* CHCC3692. Appl. Environ. Microbiol. *69*, 7173–7180.

Taranto, M.P., Villena, J., Salva, S., Alvarez, S., Savoy de Giori, G., Font de Valdez, G., and Hebert, E.M. (2013). Draft genome sequence of *Lactobacillus rhamnosus* CRL1505, an immunobiotic strain used in social food programs in Argentina. Genome Announc. *1*, e00627–13. https://doi.org/10.1128/genomeA.00627-13

Teale, C.J. (2002). Antimicrobial resistance and the food chain. J. Appl. Microbiol. *92*, 85S–9S.

Thaker, M., Spanogiannopoulos, P., and Wright, G.D. (2010). The tetracycline resistome. Cell. Mol. Life Sci. *67*, 419–431. https://doi.org/10.1007/s00018-009-0172-6

Thomas, C.M., and Nielsen, K.M. (2005). Mechanisms of, and barriers to, horizontal gene transfer between bacteria. Nat. Rev. Microbiol. *3*, 711–721.

Thumu, S.C., and Halami, P.M. (2012). Presence of erythromycin and tetracycline resistance genes in lactic acid bacteria from fermented foods of Indian origin. Antonie Van Leeuwenhoek *102*, 541–551. https://doi.org/10.1007/s10482-012-9749-4

Toomey, N., Bolton, D., and Fanning, S. (2010). Characterisation and transferability of antibiotic resistance genes from lactic acid bacteria isolated from Irish pork and beef abattoirs. Res. Microbiol. *161*, 127–135. https://doi.org/10.1016/j.resmic.2009.12.010

Trzciński, K., Thompson, C.M., and Lipsitch, M. (2004). Single-step capsular transformation and acquisition of penicillin resistance in *Streptococcus pneumoniae*. J. Bacteriol. *186*, 3447–3452. https://doi.org/10.1128/JB.186.11.3447-3452.2004

Tynkkynen, S., Singh, K.V., and Varmanen, P. (1998). Vancomycin resistance factor of *Lactobacillus rhamnosus* GG in relation to enterococcal vancomycin resistance (*van*) genes. Int. J. Food Microbiol. *41*, 195–204.

Van Boeckel, T.P., Brower, C., Gilbert, M., Grenfell, B.T., Levin, S.A., Robinson, T.P., Teillant, A., and Laxminarayan, R. (2015). Global trends in antimicrobial use in food animals. Proc. Natl. Acad. Sci. U.S.A. *112*, 5649–5654. https://doi.org/10.1073/pnas.1503141112

Van Boeckel, T.P., Glennon, E.E., Chen, D., Gilbert, M., Robinson, T.P., Grenfell, B.T., Levin, S.A., Bonhoeffer, S., and Laxminarayan, R. (2017). Reducing antimicrobial use in food animals. Science 357, 1350–1352. https://doi.org/10.1126/science.aao1495

van Hoek, A.H., Mayrhofer, S., Domig, K.J., Flórez, A.B., Ammor, M.S., Mayo, B., and Aarts, H.J. (2008). Mosaic tetracycline resistance genes and their flanking regions in *Bifidobacterium thermophilum* and *Lactobacillus johnsonii*. Antimicrob. Agents Chemother. *52*, 248–252.

van Hoek, A.H., Mevius, D., Guerra, B., Mullany, P., Roberts, A.P., and Aarts, H.J. (2011). Acquired antibiotic resistance genes: An overview. Front. Microbiol. *2*, 203. https://doi.org/10.3389/fmicb.2011.00203

van Reenen, C.A., and Dicks, L.M. (2011). Horizontal gene transfer amongst probiotic lactic acid bacteria and other intestinal microbiota: What are the possibilities? A review. Arch. Microbiol. *193*, 157–168. https://doi.org/10.1007/s00203-010-0668-3

Villena, J., Chiba, E., Tomosada, Y., Salva, S., Marranzino, G., Kitazawa, H., and Alvarez, S. (2012). Orally administered *Lactobacillus rhamnosus* modulates the respiratory immune response triggered by the viral pathogen-associated molecular pattern poly(I:C). BMC Immunol. *13*, 53.

Wegener, H.C. (2003). Antibiotics in animal feed and their role in resistance development. Curr. Opin. Microbiol. *6*, 439–445.

WHO. (2017). Global priority list of antibiotic-resistant bacteria to guide research, discovery, and development of new antibiotics. News release by the WHO. http://www.who.int/medicines/publications/global-priority-list-antibiotic-resistant-bacteria/en/

Witte, W. (2000). Selective pressure by antibiotic use in livestock. Int. J. Antimicrob. Agents 16 (Suppl. 1), S19–24.

Xu, Z., Li, L., Shi, L., and Shirtliff, M.E. (2011). Class 1 integron in *Staphylococci*. Mol. Biol. Rep. *38*, 5261–5279. https://doi.org/10.1007/s11033-011-0676-7

Xu, Z., Li, L., Shirtliff, M.E., Peters, B.M., Peng, Y., Alam, M.J., Yamasaki, S., and Shi, L. (2010). First report of class 2 integron in clinical *Enterococcus faecalis* and class 1 integron in *Enterococcus faecium* in South China. Diagn. Microbiol. Infect. Dis. *68*, 315–317. https://doi.org/10.1016/j.diagmicrobio.2010.05.014

Zago, M., Fornasari, M.E., Carminati, D., Burns, P., Suàrez, V., Vinderola, G., Reinheimer, J., and Giraffa, G. (2011). Characterization and probiotic potential of *Lactobacillus plantarum* strains isolated from cheeses. Food Microbiol. *28*, 1033–1040. https://doi.org/10.1016/j.fm.2011.02.009

Zhou, N., Zhang, J.X., Fan, M.T., Wang, J., Guo, G., and Wei, X.Y. (2012). Antibiotic resistance of lactic acid bacteria isolated from Chinese yoghurts. J. Dairy Sci. *95*, 4775–4783.

Zonenschain, D., Rebecchi, A., and Morelli, L. (2009). Erythromycin- and tetracycline-resistant lactobacilli in Italian fermented dry sausages. J. Appl. Microbiol. *107*, 1559–1568. https://doi.org/10.1111/j.1365-2672.2009.04338.x

Index

A

B

E

F

G

N

O

P

Q

R

S

T

U

V

W

X

Y

Z

www.ingramcontent.com/pod-product-compliance
Ingram Content Group UK Ltd.
Pitfield, Milton Keynes, MK11 3LW, UK
UKHW061959041026
15591UKWH00016B/1087

9 781910 190890